SCRIPTA SERIES IN MATHEMATICS

Tikhonov and Arsenin • Solutions of Ill-Posed Problems, 1977

Rozanov • Innovation Processes, 1977

Pogorelov • The Minkowski Multidimensional Problem, 1978

Kolchin, Sevast'yanov, and Chistyakov • Random Allocations, 1978

Boltianskiĭ • Hilbert's Third Problem, 1978

Besov, Il'in, and Nikol'skiĭ • Integral Representations of Functions and Imbedding Theorems, Volume I, 1978

INTEGRAL REPRESENTATIONS OF FUNCTIONS AND IMBEDDING THEOREMS

Volume I

INTEGRAL REPRESENTATIONS OF FUNCTIONS AND IMBEDDING THEOREMS

Volume I

Oleg V. Besov
Valentin P. Il'in
Sergey M. Nikol'skiĭ
Steklov Institute of Mathematical Sciences,
Moscow

Edited by
Mitchell H. Taibleson
Washington University

1978

V. H. WINSTON & SONS
Washington, D.C.

A HALSTED PRESS BOOK

JOHN WILEY & SONS
New York Toronto London Sydney

Copyright © 1978, by V. H. Winston & Sons, a Division of Scripta Technica, Inc.

All rights reserved. Printed in the United States of America. No part of this publication may be reproduced, stored in a retrieval system, or transmitted, in any form or by any means, electronic, mechanical, photocopying, recording, or otherwise, without prior written permission of the publisher.

V. H. Winston & Sons, a Division of Scripta Technica, Inc., Publishers
1511 K Street, N.W., Washington, D.C. 20005

Distributed solely by Halsted Press, a Division of John Wiley & Sons, Inc.

Library of Congress Cataloging in Publication Data

Besov, Oleg Vladimirovich.
  Integral representations of functions and imbedding theorems.

  (Scripta series in mathematics)
  Translation of Integral'nye predstavleniia funktsiĭ i teoremy vlozheniia.
  Bibliography: p.
  Includes index.
  1. Functions of several complex variables. 2. Embedding theorems. I. Il'in, Valentin Petrovich, 1921– joint author. II. Nikol'skiĭ, Sergeĭ Mikhaĭlovich, joint author. III. Title. IV. Series.
  QA331.B4313   515'.94   78-13337
  ISBN 0-470-26540-X

Composition by **Isabelle Sneeringer**, Scripta Technica, Inc.

CONTENTS

TRANSLATION EDITOR'S PREFACE TO VOLUME I

A major trend in contemporary mathematics, for a period exceeding 50 years, has been the study of spaces of functions that satisfy difference conditions (such as the Hölder continuity) and functions that satisfy differentiability conditions, *plus* the imbedding relations among and between these various spaces. From the beginning of this study, the Russian school has been a central contributor and, in recent years, their acknowledged leader has been Sergey M. Nikol'skiĭ. In his 1969 book Nikol'skiĭ summarized the contributions of his school, using approximation by entire functions of exponential type as his main tool. In this book, Nikol'skiĭ and his colleagues Valentin P. Il'in and Oleg V. Besov bring us up to date. Integral representations using kernels that are adapted to the "shape" of the domain of the function constitute the main tool used in *Integral Representations of Functions and Imbedding Theorems.*

The Russian text was written in a somewhat informal style and we have attempted to preserve the liveliness of the original. As the translation editor, I should like to add a personal note. A substantial part of my early mathematical work was built on the studies described in this book. In particular, the pioneering work of what most properly are known as Besov spaces was fundamental to my studies. I hope my efforts on this edition will repay, in part, my debt to Oleg V. Besov and his co-workers.

There are two volumes in the English-language edition of *Integral Representations of Functions and Imbedding Theorems.* The first three chapters appear in Volume I and the last three in Volume II. Chapter 1 is concerned with various integral inequalities, and in particular with a version of the Calderón–Zygmund theory in §4. Chapter 2 introduces the major idea of the book, integral representations. In Chapter 3 the authors introduce anisotropic Sobolev spaces on domains that satisfy a horn-condition (a generalization of a cone-condition) and study imbedding theorems among these spaces.

Mitchell H. Taibleson
Chairman, Dept. of Mathematics
Washington University, Missouri

INTRODUCTION

The theory of imbedding of spaces of differentiable functions of several real variables developed as a new trend in mathematics in the 1930's as a result of the works of S. L. Sobolev. In 1950, he organized these results in the form of a monograph [2]. This theory studies important connections and relationships of differentiability properties of functions in different metrics. Apart from its independent interest from the point of view of the theory of functions, it has numerous effective applications in the theory of partial differential equations. Sobolev included such applications in his monograph. He studied isotropic spaces $W_p^{(l)}(G)$ of functions $f(x)$ defined on a region G contained in E^n with norm

$$\sum_{|\alpha| \leqslant l} \| D^\alpha f \|_{p,\, G},$$

where l is a natural number, $p \geqslant 1$, and $\| f \|_{p,G} = \left\{ \int_G | f(x) |^p \, dx \right\}^{1/p}$.

Sobolev obtained the first imbedding theorems for regions in n-dimensional spaces, specifically, the theorems on q-summability of the derivatives $D^\beta f$ over a region or over manifolds of lower dimension contained in a region.

In recent years, imbedding theory has been intensively developed in various directions by the efforts of many mathematicians and has acquired new interesting and important applications.

S. M. Nikol'skiĭ developed a theory of imbedding of spaces

$$H_p^l(E^n), \quad l=(l_1, \ldots, l_n), \quad 1 \leqslant p \leqslant \infty,$$

the functions in which are characterized, first, by differential indices of both integral and nonintegral orders (Hölder conditions) and, second, by the fact that, in general, they have different properties with respect to the different variables. He first obtained a generalization of those imbedding theorems that deal with a restriction to manifolds of lower dimension. The method that he used was based on approximation of functions by trigonometric polynomials or entire functions of the exponential type (see [1], [9]).

The first definitive results in the problem of traces of functions in Sobolev spaces were obtained for $p = 2$ by Aronszajn [1] and independently by V. M. Babič and L. N. Slobodeckiĭ [1] and by Freud and Kralik [1]. Slobodeckiĭ [2] developed for $p = 2$ a complete theory of anisotropic Sobolev spaces $W_2^l(E^n)$, where $l=(l_1, \ldots, l_n)$, with both integral and fractional indices of differentiation of functions. Gagliardo [1] characterized, for $1 \leqslant p < \infty$, the traces of functions in the Sobolev space $W_p^{(1)}(E^n)$ on an $(n-1)$-dimensional cross-section of E^n.

O. V. Besov [3] used approximation methods to develop a theory of the spaces $B_{p,\theta}^l(E^n)$, which are interesting in that they, like H_p^l-spaces, form a closed system with respect to the imbedding theorems and also have a close connection with Sobolev (and Slobodeckiĭ) spaces since, for a suitable choice of the parameters, they coincide with $W_2^l(E^n)$ and also with the spaces of traces on E^m (for $m < n$) of functions in $W_p^l(E^n)$, where $1 < p < \infty$.

Sobolev proved his imbedding theorems by means of integral representations of functions in terms of their derivatives. This method of integral representations was then further developed in the works of V. P. Il'in and, in particular, they were carried over to cases of representation in terms of differences. One of the important advantages of the method of integral representations is that the representation of a function at a given point x is constructed from the values of that function at points of a bounded cone (or horn) with vertex at the point x. This made it possible to study function spaces of functions defined on an open set of rather general form (a star-shaped region with respect to a ball, an open set with a cone condition or with an l-horn condition, etc.).

In the development of various aspects of the theory of imbedding of function spaces presented in the present book, other mathematicians made their contributions: P. I. Lizorkin, S. V. Uspenskiĭ, K. K. Golovkin, V. A. Solonnikov, V. I. Burenkov, and others. References to their contributions will be made in the course of the exposition.

The monograph of S. M. Nikol'skiĭ [9] appeared in 1969. Among other questions, this monograph illuminated a certain aspect of the theory of imbedding of spaces of differentiable functions. It was devoted primarily to the study of functions and function spaces defined on the entire n-dimensional Euclidean space. In it, an instrument of study is the apparatus of approximation of functions by means of entire functions of the exponential type.

The present book and Nikol'skiĭ's monograph can be regarded as two parts of a single work, presenting the results of the development of the basic trends in imbedding theory over a period of many years.*

These two books differ both in approach and in subject matter. Here, the basic apparatus is integral representation of functions and its subject is functions defined on regions in a Euclidean space. We shall treat anisotropic Sobolev spaces and imbedding theorems for them, various families of spaces of functions characterized by difference relations, the behavior of differentiable functions of particular classes at infinity, estimates of mixed derivatives in terms of differential operators, the dependence of imbedding theorems on the structure of a region, a generalization of the Zygmund-Calderón theorem on estimates of singular integrals together with applications

*Familiarity with Nikol'skiĭ's monograph is not a prerequisite for the present book.

of it, traces of functions on manifolds, questions of compactness of sets of functions, and classes of the Morrey and Campanato types.

We do not discuss the theory of imbedding of weighted function spaces. We mention only that the ideas and methods expounded have direct and broad applications in that theory. The book is designed for readers familiar with the Lebesgue integral.

The numbering of the formulas begins afresh in each section (indicated by §). If reference is made to a formula in the section in which the formula occurs, it will be indicated by the formula number in parentheses, for example (39). If the formula is in another section, the formula number in parentheses is preceded by the number of the section in which the formula occurs, for example, 2(17). The numbers of the theorems, lemmas, etc., coincide with the numbers of the subsections in which they appear. They are referred to by the subsection number.

In conclusion, the authors consider it their pleasant duty to express their deep gratitude to Viktor Ivanovič Burenkov and Petr Ivanovič Lizorkin, who read the book in manuscript and made a number of valuable comments. Many of these were followed and they made possible an improvement of the book.

Chapter 1

INTEGRAL INEQUALITIES

An important part of the investigation tools used in the present book is constituted by estimates of integral operators of different kinds. These estimates rest primarily on classical integral inequalities such as those of Hölder and Minkowski, the generalized Minkowski inequality, Hardy's inequality, the Hardy-Littlewood inequality for fractional integrals, the Mihlin–Calderón–Zygmund inequality for singular integrals, and various generalizations of them.

In the present chapter, we shall present the basic integral inequalities to be used later on and we shall also present the necessary information on L_p spaces of real functions.

§1. L_p spaces

1.1. In this section, we shall state certain properties of spaces $L_p(G)$ of real functions $f(x)$ defined on a measurable not necessarily bounded subset G of E^n, where E^n denotes the n-dimensional Euclidean space of points $x = (x_1, \ldots, x_n)$. Throughout, measurability should be understood in the sense of Lebesgue.

Let p denote a real number such that $1 \leqslant p < \infty$. We denote by $L_p(G)$ the space of functions $f(x)$ that are measurable on G and for which the function $|f(x)|^p$ is Lebesgue-integrable on G.

The number

$$\|f\|_{L_p(G)} = \|f\|_{p,G} = \left(\int_G |f(x)|^p \, dx\right)^{1/p}$$

is called the norm of the element $f \in L_p(G)$.

Consider also the space $L_\infty(G)$, that is, the space of functions that are measurable and essentially bounded. The norm for this space is

$$\|f\|_{L_\infty(G)} = \|f\|_{\infty,G} = \operatorname*{ess\,sup}_{x \in G} |f(x)|.$$

The notation L_∞ is justified by the fact that, for bounded G,

$$\|f\|_{\infty,G} = \lim_{p \to \infty} \|f\|_{p,G}$$

(see, for example, Nikol'skiĭ [9], 1.1).

An important subspace of the space $L_\infty(G)$ is the space $C(G)$ of functions $f(x)$ that are uniformly continuous on G with norm

$$\|f\|_{C(G)} = \sup_{x \in G} |f(x)|.$$

Thus, the spaces $L_p(G)$ are defined for all real p such that $1 \leqslant p \leqslant \infty$.

Let $\boldsymbol{p} = (p_1, \ldots, p_n)$ denote a vector with components satisfying the inequalities $1 \leqslant p_i \leqslant \infty$ for $i = 1, \ldots, n$.

We denote by $L_{\boldsymbol{p}}(E^n)$ the space of functions $f(x)$ defined on E^n that are measurable and for which the norm

$$\|f\|_{\boldsymbol{p},E^n} = \|f\|_{(p_1,\ldots,p_n),E^n} = \| \ldots \|\|f\|_{p_1,x_1}\|_{p_2,x_2} \ldots \|_{p_n,x_n} =$$

$$\left\{\int_{E^1}\left[\cdots\left\{\int_{E^1}\left(\int_{E^1} |f(x)|^{p_1} dx_1\right)^{p_2/p_1} dx_2\right\}^{p_3/p_2} \cdots\right]^{p_n/p_{n-1}} dx_n\right\}^{1/p_n} \tag{1}$$

is finite. We note that the order in which the norms are taken with respect to the different variables is significant; in general,*

$$\left[\int_{E^1} dx_2\left(\int_{E^1} |f(x_1, x_2)|^{p_1} dx_1\right)^{p_2/p_1}\right]^{1/p_2}$$
$$\neq\left[\int_{E^1} dx_1\left(\int_{E^1} |f(x_1, x_2)|^{p_2} dx_2\right)^{p_1/p_2}\right]^{1/p_1}.$$

Let G denote an arbitrary measurable subset of E^n and let f denote a measurable function defined on G. Then, we set

$$\|f\|_{\boldsymbol{p},\,G}=\|\tilde{f}\|_{\boldsymbol{p},\,E^n}, \tag{2}$$

where $\tilde{f}(x)=f(x)$ for $x\in G$ and $\tilde{f}(x)=0$ for $x\in E^n\setminus G$. If $\|f\|_{\boldsymbol{p},\,G}$ is finite, we write $f\in L_{\boldsymbol{p}}(G)$.

For simplicity, when $G=E^n$, we shall often write $\|f\|_{\boldsymbol{p}}$ instead of $\|f\|_{\boldsymbol{p},\,E^n}$.

We note that, if $\boldsymbol{p}=(p,\ldots,p)$, then

$$\|f\|_{\boldsymbol{p},\,G}=\|f\|_{p,\,G}.$$

In what follows, we shall write $\boldsymbol{p}\geqslant\boldsymbol{q}$ and $\boldsymbol{p}>\boldsymbol{q}$, where $\boldsymbol{p}=(p_1,\ldots,p_n)$ and $\boldsymbol{q}=(q_1,\ldots,q_n)$ to mean respectively that $p_i\geqslant q_i$ and $p_i>q_i$ for $i=1,\ldots,n$. In particular, the notation $\mathbf{1}\leqslant\boldsymbol{p}\leqslant\boldsymbol{\infty}$ (where $(\mathbf{1}=\overbrace{(1,\ldots,1)}^{n}$ and $\boldsymbol{\infty}=\overbrace{(\infty,\ldots,\infty)}^{n})$ means that $1\leqslant p_i\leqslant\infty$ for $i=1,\ldots,n$.

Let us list a number of facts involving properties of spaces $L_{\boldsymbol{p}}(G)$ that we shall need.**

First of all, the space $L_{\boldsymbol{p}}(G)$, where $\mathbf{1}\leqslant\boldsymbol{p}\leqslant\boldsymbol{\infty}$, is a Banach space of functions with the norm defined above. In the present case, this means that the following properties hold:

*If $p_1\leqslant p_2$, then $\|f\|_{(p_1,\,p_2),\,E^2}\leqslant\|f\|_{(p_2,\,p_1),\,E^2}$ (see subsection 2.11 below).

**The article by Benedek and Panzone [1] is devoted to properties of L_p spaces.

1) $\| f \|_{p, G} = 0$ is equivalent to $f(x) = 0$ for almost all $x \in G$.
2) $\| cf \|_{p, G} = | c | \| f \|_{p, G}$.
3) $\| f_1 + f_2 \|_{p, G} \leqslant \| f_1 \|_{p, G} + \| f_2 \|_{p, G}$.
4) The space $L_p(G)$ is complete; that is, the inclusion and limit relations

$$f_k \in L_p(G) \quad (k = 1, 2, \ldots), \quad \| f_k - f_l \|_{p, G} \to 0 \text{ as } k, \quad l \to \infty, \tag{3}$$

imply the existence of a function $f \in L_p(G)$ such that $\| f_k - f \|_{p, G} \to 0$ as $k \to \infty$.

Properties 1) and 2) are obvious. Inequality 3) is known as Minkowski's inequality (it will be proven in subsection 2.7). Let us prove the completeness of the space $L_p(G)$.

Obviously, we can confine ourselves to the case $G = E^n$. Let $\{f_k\}_1^\infty$ denote a sequence of functions in $L_p(G)$ for which (3) holds.

To prove completeness, we shall use the corollary of proposition 2.6. According to the formula in that corollary, for any measurable function φ and any p such that $1 \leqslant p \leqslant \infty$, we have

$$\| \varphi \|_p = \sup_{\| g \|_{p'} = 1} \int_{E^n} | \varphi(x) g(x) | \, dx, \tag{4}$$

where $p' = (p_1', \ldots, p_n')$ and $\frac{1}{p_i} + \frac{1}{p_i'} = 1$ for $i = 1, \ldots, n$.

Let $\{I_m\}_1^\infty$ denote a family of bounded measurable sets such that $\bigcup^\infty I_m = E^n$ and let χ_{I_m} denote the characteristic function of the set I_m.

Then, on the basis of (4), we have

$$\| f_k - f_l \|_p = \sup_{\| g \|_{p'} = 1} \int_{E^n} | f_k - f_l | \, | g | \, dx \geqslant \| \chi_{I_m} \|_{p'}^{-1} \int_{I_m} | f_k - f_l | \, dx.$$

By virtue of (3) and the completeness of the space L_1, this inequality implies the existence of a function f defined on E^n such that $\|(f - f_k)\chi_{I_m}\|_1 \to 0$ for every m as $k \to \infty$. By means of a diagonal process, let us extract from the sequence $\{f_k\}_1^\infty$ a subsequence $\{f_{k_i}\}_{i=1}^\infty$ that converges to f almost everywhere in E^n. If we apply Fatou's lemma to the sequence $\{|f_{k_i} - f_{k_j}||g|\}$ (for $k_j = k_i, k_{i+1}, \ldots$), where $\|g\|_{p'} = 1$, and then use equation (4), we obtain

$$\int_{E^n} |f_{k_i} - f||g|\,dx \leqslant \sup_{k_s \geqslant k_i} \int_{E^n} |f_{k_i} - f_{k_j}||g|\,dx$$

$$\leqslant \sup_{k_j \geqslant k_i} \|f_{k_i} - f_{k_j}\|_p .$$

Since this inequality holds for arbitrary $g \in L_{p'}$ such that $\|g\|_{p'} = 1$, it follows that $(f - f_{k_i}) \in L_p$. This in turn implies (by Minkowski's inequality) that $f = (f - f_{k_i}) + f_{k_i} \in L_p$. It then follows from the last inequality on the basis of (3) that $\|f_k - f\|_p \to 0$ as $k \to \infty$. This proves property 4).

In what follows, we shall identify two equivalent functions (that is, two functions that coincide almost everywhere on G and hence have the same norm in the sense of $L_p(G)$), treating them as a single element of the space $L_p(G)$.

REMARK. Let f denote a member of $L_{\boldsymbol{p}}$, so that $\|f\|_{\boldsymbol{p}} < \infty$, and suppose that $\boldsymbol{p} = (\overline{\boldsymbol{p}}, \overline{\overline{\boldsymbol{p}}})$, where $\overline{\boldsymbol{p}} = (p_1, \ldots, p_j)$ and $\overline{\overline{\boldsymbol{p}}} = (p_{j+1}, \ldots, p_n)$ for $1 \leqslant j \leqslant n - 1$. Then, the norm $\|f(\cdot, \overline{\overline{x}})\|_{\overline{\boldsymbol{p}}}$, as a function of the point $\overline{\overline{x}} = (x_{j+1}, \ldots, x_n)$, is measurable and almost everywhere finite. It is easy to show this successively for $j = 1, 2, \ldots, n-1$ by using Fubini's theorem. For p_j representing infinite components of the vector $\boldsymbol{p}$, we also use the relation (see, for example, Nikol'skiĭ [9], 1.1)

$$\lim_{q \to \infty} \left(\int_a^b |\varphi(t)|^q\,dt \right)^{1/q} = \operatorname{ess\,sup} |\varphi| \quad (b - a < \infty) \tag{5}$$

and the fact that the pointwise limit of measurable functions is a measurable function.

In what follows, we shall have occasion to use the completeness of the space $\widetilde{L}_p$, where $p = (p_1, \ldots, p_n)$, with a norm differing from the norm of L_p defined in (1) only in that instead of taking the essential supremum (ess sup) we take the least upper bound (sup) with respect to some or all of the variables x_i corresponding to any infinite components $p_i = \infty$.

We shall consider the space $\widetilde{L}_p$ to contain exactly those measurable functions $f(x)$ for which the functions $\| f(\cdot, \overline{\overline{x}}) \|_{\widetilde{L}_{\overline{p}}}$, for $\overline{p} = (p_1, \ldots, p_j)$, as functions of $\overline{\overline{x}} = (x_{j+1}, \ldots, x_n)$, for all $j = 1, 2, \ldots, n-1$, are measurable.

Thus, the space $\widetilde{L}_p$ is *a priori* narrower than the corresponding space L_p. However, it is easy to show that it is possible to modify any function $f \in L_p$ on a set of n-dimensional measure zero in such a way that the resulting function $\tilde{f}$ belongs to the space $\widetilde{L}_p$. To do this, we need only compare the two norms L_p and $\widetilde{L}_p$, which differ only in that, for some i in $[1, n]$, we take ess sup with respect to x_i in the first case and sup in the second. For definiteness, let us assume that $1 < i < n$ (the cases $i = 1$ and $i = n$ are simpler),

$$x = (x_1, \ldots, x_{i-1}, x_i, x_{i+1}, \ldots, x_n) = (x', x_i, x''),$$
$$p = (p_1, \ldots, p_{i-1}, p_i, p_{i+1}, \ldots, p_n) = (p', p_i, p'').$$

Suppose that $\| f \|_{L_p} < \infty$. Consequently, for almost all points x'',

$$\operatorname*{ess\,sup}_{x_i} \| f(\cdot, x_i, x'') \|_{L_{p'}} = M(x'') < \infty,$$

where the norm $L_{p'}$ coincides with the corresponding "internal part" of the norms of L_p and $\widetilde{L}_p$. For every fixed point x'' such that $M(x'') < \infty$, the inequality $\| f(\cdot, x_i, x'') \|_{p'} > M(x'')$ is possible only on a set of points x_i of one-dimensional measure zero.

Let us replace the function $f(x)$ with the function $\tilde{f}(x)$, setting

$$\tilde{f}(x) = \begin{cases} 0 & \text{if} \quad \| f(\cdot, x_i, x'') \|_{p'} > M(x''), \\ f(x) & \text{if} \quad \| f(\cdot, x_i, x'') \|_{p'} \leqslant M(x''). \end{cases}$$

The functions $f(x)$ and $\tilde{f}(x)$ differ only on a set of n-dimensional measure zero. To see this, note that the functions $\varphi(x_i, x'') = \| f(\cdot, x_i, x'') \|_{p'}$ and $M(x'')$ are measurable as functions of the pairs (x_i, x''). Consequently, by virtue of the properties of measurable functions, the set of pairs

$$\{(x_i, x''): \varphi(x_i, x'') - M(x'') > 0\}$$

is measurable, and hence obviously so is the set of points $x \in E^n$ on which the functions $f(x)$ and $\tilde{f}(x)$ differ. Since this last set intersects almost every straight line parallel to the x_i-axis on a set of points of one-dimensional measure zero, we conclude on the basis of Fubini's theorem that mes $\{x: f(x) \neq \tilde{f}(x)\} = 0$.

It follows from the construction of $\tilde{f}$ that

$$\sup_{x_i} \| \tilde{f}(\cdot, x_i, x'') \|_{L_{p'}} = \operatorname{ess\,sup}_{x_i} \| f(\cdot, x_i, x'') \|_{L_{p'}}$$

for all x''. Consequently, $\| \tilde{f} \|_{\tilde{L}_p} = \| f \|_{L_p}$, as we needed to show.

The completeness of the space $\tilde{L}_p$ follows in an obvious way from the completeness of the space L_p with the aid of the relationship just established between the elements of these spaces.

1.2. We present without proof the following assertion, which generalizes a familiar theorem of Lebesgue:

If $1 \leqslant p < \infty$, $f_k \in L_p(G)$, $|f_k| \leqslant g \in L_p(G)$ *(for* $k = 1, \ldots$*) and* $f_k \to f$ *almost everywhere on* G *as* $k \to \infty$, *then* $f \in L_p(G)$ *and* $\lim_{k\to\infty} \| f_k - f \|_{p, G} = 0$.

It follows from this, for example, that, if $f \in L_p(G)$ (where $1 \leqslant p < \infty$), if $\{G_k\}_1^\infty$ is an increasing sequence of bounded measurable sets such that $G_k \subset G$ (for $k = 1, \ldots$) and $\bigcup_1^\infty G_k = G$, and if χ_{G_k} is the characteristic function of the set G_k, then

$$\lim_{k\to\infty} \| f\chi_{G_k} - f \|_{p, G} = 0. \tag{6}$$

If some components of the vector p are infinite, this last relation may not hold. However, for arbitrary p such that $1 \leqslant p \leqslant \infty$,

$$\lim_{k\to\infty} \| f\chi_{G_k} \|_{p, G} = \| f \|_{p, G} \tag{7}$$

whether $\| f \|_{p, G}$ is finite or not.

1.3. Suppose that f and f_k belong to $L_p(G)$ for $k = 1, 2, \ldots$ and that

$$\lim_{k\to\infty} \| f_k - f \|_{p,\,G} = 0 \qquad (1 \leqslant p \leqslant \infty).$$

Then, there exists a subsequence $\{k_i\}$ of natural numbers such that

$$\lim_{i\to\infty} f_{k_i}(x) = f(x) \tag{8}$$

for almost all $x \in G$ (see proof of completeness of $L_p(G)$).

It follows that if, in addition to (8), we have

$$\lim_{k\to\infty} \| f_k - f^* \|_{q,\,G} = 0$$

for some function $f^*(x)$ and some q such that $1 \leqslant q \leqslant \infty$, then the functions $f(x)$ and $f^*(x)$ are equivalent on G.

This fact can be generalized. Let $G = G' \times G''$ denote the Cartesian product of a measurable set $G' \subset E^m$ (where $1 \leqslant m < n$) and a measurable set $G'' \subset E^{n-m}$ such that every point $x = (x_1, \ldots, x_n)$ belonging to G can be represented in the form $x = (x', x'')$, where $x' = (x_1, \ldots, x_m) \in G'$ and $x'' = (x_{m+1}, \ldots, x_n) \in G''$. Then, we may assume that $f(x) = f(x', x'')$.

Let $\boldsymbol{q} = (q_1, \ldots, q_m)$ denote an m-tuple such that $1 \leqslant \boldsymbol{q} \leqslant \infty$. We set

$$\| f(\cdot, x'') \|_{\boldsymbol{q},\,G'} = \| \tilde{f}(\cdot, x'') \|_{\boldsymbol{q},\,E^m} = \| \ldots \| \tilde{f}(\cdot, x'') \|_{q_1,\,x_1} \ldots \|_{q_m,\,x_m},$$

where $\tilde{f}(x', x'') = f(x', x'')$ for $x' \in G'$ and $\tilde{f}(x', x'') = 0$ for $x' \in E^m \setminus G'$.

1.4. Theorem.* *Suppose that* $G = G' \times G''$. *Let* $\boldsymbol{p} = (p_1, \ldots, p_n)$ *denote an* n*-tuple such that* $1 \leqslant \boldsymbol{p} \leqslant \infty$. *Let* f *and* f_k, *for* $k = 1, 2, \ldots,$ *denote functions belonging to* $L_{\boldsymbol{p}}(G)$. *Suppose that*

*This theorem was proven for the case $\boldsymbol{p} = (p, \ldots, p)$ and $\boldsymbol{q} = (q, \ldots, q)$ in Nikol'skiĭ's book [9] (see §1.3). The proof for arbitrary $\boldsymbol{p}$ and $\boldsymbol{q}$ is analogous.

$$\lim_{k\to\infty} \| f_k - f \|_{p,\, G} = 0.$$

Suppose also that there exists an m-tuple $\boldsymbol{q} = (q_1, \ldots, q_m)$ *such that* $1 \leqslant \boldsymbol{q} \leqslant \infty$ *and a function* $f^*(x)$ *such that*

$$\lim_{k\to\infty} \| f_k(\cdot\,, x'') - f^*(\cdot\,, x'') \|_{q,\, G'} = 0$$

for almost all $x'' \in G''$. *Then,* $f = f^*$*; that is, the functions* $f(x)$ *and* $f^*(x)$ *are equivalent on G.*

1.5. Suppose that $f(x)$ is defined on a measurable set G. Let us extend it to E^n by assigning it the value 0 on $E^n \setminus G$, keeping our previous notation for it.

A function $f \in L_p(G)$ is said to be continuous (in the wide sense) in $L_p(G)$ if, for every $\varepsilon > 0$, there exists a $\delta > 0$ such that

$$\| f(\cdot + y) - f \|_{p,\, E^n} < \varepsilon,$$

whenever $| y | = \left(\sum_{i=1}^{n} y_i^2 \right)^{1/2} < \delta.$

Theorem.* *Every function* $f \in L_p(G)$, *for* $1 \leqslant p < \infty$, *is continuous in the wide sense in* $L_p(G)$.

We note that this assertion becomes false if any components of the vector $\boldsymbol{p}$ are infinite.

1.6. The *support* of a function is defined as the closure of the set of all points at which it assumes a value other than zero. We shall denote the support of a function φ by supp φ. Let G denote an open subset of E^n. If the support of a function φ defined on a subset (proper or otherwise) of E^n is contained in G, we shall say that φ is *concentrated* in G.

We introduce the following notation: We denote by $C^\infty(G)$ the set of infinitely differentiable functions on G and we denote by $C_0^\infty(G)$ the set of functions of bounded support belonging to

*For $\boldsymbol{p} = (p, \ldots, p)$, this theorem is proven, for example, in Sobolev's book [2]. The proof is essentially the same for arbitrary $\boldsymbol{p}$.

$C^\infty(G)$. We denote by $\overset{\circ}{L}_\infty(G)$ the set of measurable essentially bounded functions whose support is bounded and contained in G.

A subset S of the space $L_p(G)$ is said to be *dense* in $L_p(G)$ if, for every $f \in L_p(G)$ and every $\varepsilon > 0$, there exists an element $\varphi_\varepsilon \in S$ such that

$$\| \varphi_\varepsilon - f \|_{p,\,G} < \varepsilon.$$

Theorem. *The set $C_0^\infty(G)$ is dense in $L_p(G)$ for every p such that $1 \leqslant p < \infty$.*

Thus, for every function $f \in L_p(G)$, where $1 \leqslant p < \infty$, there exists a sequence of functions $\varphi_k \in C_0^\infty(G)$ such that

$$\lim_{k\to\infty} \| \varphi_k - f \|_{p,\,G} = 0.$$

We shall find it convenient to wait and prove this theorem in §5 although we shall use it before then.

1.7. The method of integral representations of functions is largely based on a theorem in the theory of differentiation of multiple integrals due to Jessen, Marcinkiewicz, and Zygmund (Theorem 6 of [1]).

Let G denote an open subset of E^n. We shall say that a function f defined on G belongs to the class $L_p^{\mathrm{loc}}(G)$ if the restriction of f to any compact subset F of G belongs to $L_p(F)$. If $p = 1 = (1, \ldots, 1)$, we shall write simply $L^{\mathrm{loc}}(G)$. The theorem just mentioned is the following:

Theorem. *Let $\psi_i(v)$, for $i = 1, \ldots, n$, denote a positive nondecreasing function defined for $v > 0$ such that $\lim_{v\to 0} \psi_i(v) = 0$. Let $|I_v|$ denote the volume of the rectangular parallelepiped*

$$I_v = \{x : a_i \leqslant x_i \leqslant b_i,\ \ b_i - a_i = \psi_i(v)\ \ (i = 1, \ldots, n)\}.$$

If $f \in L^{\mathrm{loc}}(G)$ and if a point x belongs to the parallelepiped I_v for every $v > 0$, then

$$\lim_{v\to 0}\frac{1}{|I_v|}\int_{I_v}|f(y)-f(x)|\,dy=0 \tag{9}$$

and, a fortiori,

$$\lim_{v\to 0}\frac{1}{|I_v|}\int_{I_v}f(y)\,dy=f(x) \tag{10}$$

*for almost every point $x\in G$.**

One can easily see that (9) and (10) remain valid when x does not belong to the parallelepipeds I_v but there exist parallepepipeds I'_v and a positive number N such that $|I'_v|/|I_v|<N$ for every $v>0$, where I'_v is the smallest parallelepiped with edges parallel to the coordinate axes that is centered at the point x and contains I_v.

REMARK. Let us define

$$I=\{x:\ a_i\leqslant x_i\leqslant b_i \quad (i=1,\ \dots,\ n)\}.$$

We shall say that $I\to x$ if x belongs to I and $b_i - a_i$ approaches 0 for all i. It turns out that (9) and (10) do not hold in general for a function $f\in L^{\mathrm{loc}}(G)$ when we replace I_v with I and let $I\to x$. However, we do have the following: *If $f\in L_p^{\mathrm{loc}}(G)$ for some $p>1$, then, for almost every $x\in G$, we have*

$$\lim_{I\to x}\frac{1}{|I|}\int_{I}|f(y)-f(x)|\,dy=0 \tag{9'}$$

and

*In the work of the three authors mentioned above, it was assumed that $f\in L(\Delta)$, where Δ is a closed cube. The validity of the assertions for $f\in L^{\mathrm{loc}}(G)$ follows from the fact that G is the union of a denumerable set of cubes and the closure of each cube is contained in G. With regard to this theorem and strengthenings of it, see also the article by Riviére [1].

$$\lim_{I \to x} \frac{1}{|I|} \int_I f(y)\, dy = f(x). \qquad (10')$$

If I is a cube of the form $b_i - a_i = v$ for $i = 1, \ldots, n$, then (9′) and (10′) hold also for $f \in L^{\text{loc}}(G)$. This follows immediately from the theorem given above.*

1.8. In what follows, we shall use the familiar Banach-Steinhaus theorem on the convergence of a sequence of linear operators. We shall state here only a particular consequence of that theorem that we shall actually have occasion to use.

Suppose that $\{K_i(x, y)\}_1^\infty$ is a sequence of measurable functions on $E_x^n \times E_y^n$, where E_x^n and E_y^n are n-dimensional Euclidean spaces of the points $x = (x_1, \ldots, x_n)$ and $y = (y_1, \ldots, y_n)$ respectively. Let f denote a member of $L_p(E^n)$. We define

$$F_i(f)(x) = \int_{E^n} f(y) K_i(x, y)\, dy \qquad (i = 1, \ldots). \qquad (11)$$

Theorem. *Let* $\mathbf{p} = (p_1, \ldots, p_n)$ *and* $\mathbf{q} = (q_1, \ldots, q_n)$ *denote n-tuples such that* $\mathbf{1} \leqslant \mathbf{p} < \infty$ *and* $\mathbf{p} \leqslant \mathbf{q} \leqslant \infty$. *Suppose that the sequence of integral operators* (11) *treated as operators from* $L_{\mathbf{p}}(E^n)$ *into* $L_{\mathbf{q}}(E^n)$ *satisfies the following conditions:*

1) *For every function* $f \in L_{\mathbf{p}}(E^n)$, *we have***

$$\| F_i(f) \|_{\mathbf{q}} \leqslant M \| f \|_{\mathbf{p}} \qquad (i = 1, \ldots),$$

where M is a constant independent of f and i;

2) *for every function* $\varphi \in C_0^\infty(E^n)$,

*These results can be found in the article of the three authors mentioned above and also in the book [1] by Zygmund and in part in the book [1] by Dunford and Schwartz.

**Here, we write $\| \ \|_{\mathbf{p}}$ instead of $\| \ \|_{\mathbf{p}, E^n}$.

$$\lim_{i,j\to\infty} \| F_i(\varphi) - F_j(\varphi) \|_q = 0.$$

Then, for every function $f \in L_p(E^n)$ *there exists a function* $F(f) \in L_q(E^n)$ *such that*

$$\lim_{i\to\infty} \| F_i(f) - F(f) \|_q = 0, \qquad \| F(f) \|_q \leqslant M \| f \|_p.$$

For proof, see the book [1] by Kantorovic and Akilov.

§2. The basic integral inequalities

2.1. Let us adopt certain conventions regarding notation to be used later.

For a given number p in $[1, \infty]$, we denote by p' the number conjugate to p, that is, the number such that $1/p + 1/p' = 1$. In particular, $p' = \infty$ if $p = 1$ and $p' = 1$ if $p = \infty$.

For an n-tuple $\boldsymbol{p} = (p_1, \ldots, p_n)$ such that $p_i \in [1,\infty]$ for $i = 1, \ldots, n$, the symbol $\boldsymbol{p}'$ will denote $(p_1', \ldots, p_n')$ and $1/\boldsymbol{p}$ will denote $(1/p_1, \ldots, 1/p_n)$. As usual, we define the sum of vectors $\boldsymbol{p}$ and $\boldsymbol{q} = (q_1, \ldots, q_n)$ as the vector $\boldsymbol{p} + \boldsymbol{q} = (p_1 + q_1, \ldots, p_n + q_n)$.

We recall (see 1.1) that the inequality $\boldsymbol{p} \geqslant \boldsymbol{q}$ (resp. $\boldsymbol{p} > \boldsymbol{q}$) means that $p_i \geqslant q_i$ (resp. $p_i > q_i$) for $i = 1, \ldots, n$. As in subsection 1.1, we shall write

$$\mathbf{1} = (\overbrace{1, \ldots, 1}^{n}), \quad \boldsymbol{\infty} = (\overbrace{\infty, \ldots, \infty}^{n}).$$

All the inequalities given below are formulated for functions defined on the entire space E^n except when the contrary is obvious from the nature of the situation. For simplicity, we shall write $\|f\|_p$ instead of $\| f \|_{p, E^n}$ and we shall sometimes write $\int f(x)\, dx$ instead of $\int\limits_{E^n} f(x)\, dx$.

2.2. Hölder's inequality. *Suppose that* $1 \leqslant p \leqslant \infty$, $f_1 \in L_p(E^n)$, *and* $f_2 \in L_{p'}(E^n)$. *Then, the function* $f_1(x) f_2(x)$ *is integrable over* E^n *and*

$$\int_{E^n} | f_1(x) f_2(x) | \, dx \leqslant \| f_1 \|_p \| f_2 \|_{p'}. \tag{1}$$

This is Hölder's inequality.

PROOF. Inequality (1) is obvious for $p = 1$ and $p = \infty$. Suppose that $1 < p < \infty$. Consider the function $\varphi(t) = t^p/p + t^{-p'}/p'$. Since its derivative is positive for $t > 1$ and negative for $0 < t < 1$, it assumes its minimum value $\varphi(1) = 1$ at $t = 1$. Let us set $t = a^{1/p'} b^{-1/p}$, where a and b are arbitrary positive numbers. Then, the inequality $\varphi(t) \geqslant 1$ (which holds for $t > 0$) implies that

$$ab \leqslant \frac{a^p}{p} + \frac{b^{p'}}{p'}$$

with equality holding if and only if $a^p = b^{p'}$ (at $t = 1$).

We assume in what follows that neither of the norms in the right-hand member of (1) is equal to zero (otherwise, inequality (1) is trivial). In the last inequality, we set

$$a = \frac{| f_1(x) |}{\| f_1 \|_p}, \qquad b = \frac{| f_2(x) |}{\| f_2 \|_{p'}}.$$

We then obtain

$$\frac{| f_1(x) f_2(x) |}{\| f_1 \|_p \| f_2 \|_{p'}} \leqslant \frac{1}{p} \frac{| f_1(x) |^p}{\| f_1 \|_p^p} + \frac{1}{p'} \frac{| f_2(x) |^{p'}}{\| f_2 \|_{p'}^{p'}}. \tag{2}$$

Since the right-hand member is integrable, so is the left-hand member and the integral of the left-hand member does not exceed the integral of the right-hand member; that is, it does not exceed

$$\frac{1}{p} + \frac{1}{p'} = 1.$$

If we integrate both sides of (2) and multiply the result by $\|f_1\|_p \|f_2\|_{p'}$, we get inequality (1).

We note that equality is possible in (1) only when there exists a constant C such that $|f_1(x)|^p = C|f_2(x)|^{p'}$ for almost all $x \in E^n$.

The following more general inequality is easily proven by induction:

2.3. Hölder's inequality for several functions. *Suppose that* $1 \leqslant p_i \leqslant \infty$ *for* $i = 1, \ldots, m$, *that* $\frac{1}{p_1} + \ldots + \frac{1}{p_m} = 1$, *and that* $f_i \in L_{p_i}(E^n)$ *for* $i = 1, \ldots, m$. *Then, the product* $f_1(x) \ldots f_m(x)$ *is integrable over* E^n *and*

$$\int_{E^n} |f_1(x) \ldots f_m(x)| \, dx \leqslant \|f_1\|_{p_1} \cdots \|f_m\|_{p_m}. \tag{3}$$

2.4. By successive application of inequality (1) with respect to each variable separately, we obtain Hölder's inequality for vector-valued $\boldsymbol{p} = (p_1, \ldots, p_n)$:

$$\int_{E^n} |f_1(x) f_2(x)| \, dx \leqslant \|f_1\|_{\boldsymbol{p}} \|f_2\|_{\boldsymbol{p}'}, \tag{4}$$

where $\mathbf{1} \leqslant \boldsymbol{p} \leqslant \boldsymbol{\infty}$ and $\frac{1}{\boldsymbol{p}} + \frac{1}{\boldsymbol{p}'} = \mathbf{1}$.

2.5. If in inequality (1) we assume $f_1(x)$ and $f_2(x)$ to be finite-valued functions, we can replace the integrals with sums. We then have *Hölder's inequality for sums*:

$$\sum_{i=1}^{N} |a_i b_i| \leqslant \left(\sum_{i=1}^{N} |a_i|^p \right)^{1/p} \left(\sum_{i=1}^{N} |b_i|^{p'} \right)^{1/p'} \qquad (1 \leqslant p \leqslant \infty), \tag{5}$$

where N is a natural number—although inequality (5) remains valid for $N = \infty$.

2.6. The following proposition is a converse to Hölder's inequality:

If $1 \leqslant p \leqslant \infty$ *and* $f(x)$ *is a measurable function on* E^n *such that the inequality*

$$\int_{E^n} f(x)\, g(x)\, dx \leqslant M \| g \|_{p'}, \tag{6}$$

holds for every function g *belonging to the class** $\overset{\circ}{L}_{\infty}(E^n)$, *then* $f \in L_p(E^n)$ *and* $\| f \|_p \leqslant M$.

PROOF. For definiteness, let us assume that s components of the vector $\boldsymbol{p}$, where $0 \leqslant s \leqslant n$, are infinite and that the remaining ones are finite numbers.

Let us suppose now that our assertion is false, that is, that, for some $\varepsilon > 0$,

$$(1 + \varepsilon)^{-s} \| f \|_p > M,$$

where $\|f\|_p$ denotes the right-hand member of 1(1) without our assuming that it is finite. Then, on the basis of 1(7), there exist a parallelepiped $I = I_1 \times \ldots \times I_n$, where $I_i = \{x_i\colon a_i < x_i < b_i\}$ for $i = 1, \ldots, n$, and a natural number k such that, for the function

$$\varphi(x) = \begin{cases} |f(x)| & \text{if} \quad |f(x)| < k \quad \text{and} \quad x \in I, \\ k & \text{if} \quad |f(x)| \geqslant k \quad \text{and} \quad x \in I, \\ 0 & \text{if} \quad x \in E^n \setminus I, \end{cases}$$

which belongs to $L_p(E^n)$, we have

$$(1 + \varepsilon)^{-s} \| \varphi \|_p > M.$$

Let ψ_i, for $i = 1, \ldots, n$, denote n functions defined as follows: for $p_i < \infty$,

*For definition of the class $\overset{\circ}{L}_{\infty}(E^n)$, see subsection 1.6.

$$\psi_i(x_i, \ldots, x_n) = \left(\| \varphi \|_{(p_1, \ldots, p_{i-1}),\, I_1 \times \ldots \times I_{i-1}} \| \varphi \|^{-1}_{(p_1, \ldots, p_i),\, I_1 \times \ldots \times I_i}\right)^{p_i - 1}$$

(for $i = 1$, we take $\| \varphi \|_{(p_1, \ldots, p_{i-1})} = | \varphi |$). Here, we take $\psi_i = 0$ if $\| \varphi \|_{(p_1, \ldots, p_i)} = 0$.

For $p_i = \infty$, we take

$$\psi_i(x_i, \ldots, x_n) = \begin{cases} \dfrac{\chi_{F_{i\varepsilon}}(x_i, \ldots, x_n)}{m_i F_{i\varepsilon}} & \text{if} \quad m_i F_{i\varepsilon} \neq 0, \\ 0 & \text{if} \quad m_i F_{i\varepsilon} = 0, \end{cases}$$

where

$$F_{i\varepsilon} = \{(x_i, \ldots, x_n) : \| \varphi \|_{(p_1, \ldots, p_{i-1}),\, I_1 \times \ldots \times I_{i-1}} > (1 + \varepsilon)^{-1} \| \varphi \|_{(p_1, \ldots, p_i),\, I_1 \times \ldots \times I_i}\},$$

$m_i F_{i\varepsilon}$ is the measure of the one-dimensional cross-section of the set $F_{i\varepsilon}$ with respect to the variable x_i for fixed $x_{i+1}, \ldots, x_n$, and $\chi_{F_{i\varepsilon}}$ is the characteristic function of the set $F_{i\varepsilon}$.

Let us now set

$$g(x_1, \ldots, x_n) = (\operatorname{sgn} f) \prod_{i=1}^{n} \psi_i.$$

One can easily show that

$$g \in \mathring{L}_\infty(E^n),$$

$$\| g \|_{p'} = 1,$$

and

$$\int_{E^n} | \varphi g | \, dx \geqslant (1 + \varepsilon)^{-s} \| \varphi \|_p .$$

Then,

$$\int fg\,dx = \int |fg|\,dx \geqslant \int |\varphi g|\,dx$$
$$\geqslant (1+\varepsilon)^{-s} \| \varphi \|_p \| \| g \|_{p'} > M \| g \|_{p'},$$

which contradicts inequality (6). This proves assertion 2.6.

We note that, since $\mathring{L}_\infty(E^n) \subset L_{p'}(E^n)$ for arbitrary p' in $1 \leqslant p' \leqslant \infty$, this assertion holds *a fortiori* if g is any function in $L_{p'}(E^n)$.

Corollary. *If f is a measurable function on E^n, then*

$$\| f \|_p = \sup_{\| g \|_{p'}=1} \int_{E^n} | f(x)\, g(x) |\, dx = \sup_{\| g \|_{p'}=1} \int_{E^n} f(x)\, g(x)\, dx.$$

PROOF. On the one hand, we have from Hölder's inequality

$$\sup_{\| g \|_{p'}=1} \int | f(x)\, g(x) |\, dx \leqslant \| f \|_p.$$

On the other hand, on the basis of proposition 2.6, we have

$$\| f \|_p \leqslant \sup_{\| g \|_{p'}=1} \int f(x)\, g(x)\, dx$$

since we may take for the M of inequality (6) the right-hand member of this last inequality. If we combine the last two inequalities, we get the result asserted.

2.7. Minkowski's inequality. *Suppose that $1 \leqslant p \leqslant \infty$ and that $f_i \in L_p(E^n)$ for $i = 1, \ldots, m$. Then, $\{f_1 + \ldots + f_m\} \in L_p(E^n)$ and*

$$\left\| \sum_{i=1}^{m} f_i \right\|_p \leqslant \sum_{i=1}^{m} \| f_i \|_p. \tag{7}$$

(This is Minkowski's inequality.)

PROOF. Inequality (7) is obvious for $p = 1$ and $p = \infty$. Suppose that $1 < p < \infty$. We define

$$S(x) = \sum_{i=1}^{m} f_i(x).$$

Applying Hölder's inequality for sums (5), we obtain

$$|S| \leqslant m^{1/p'} \left(\sum_{i=1}^{m} |f_i|^p \right)^{1/p},$$

from which it follows that $S \in L_p(E^n)$ and, since $p'(p-1) = p$, we have $|S|^{p-1} \in L_{p'}(E^n)$.

If we integrate both sides of the inequality

$$|S|^p \leqslant \sum_{i=1}^{m} |f_i| |S|^{p-1}$$

and then apply Hölder's inequality to each term in the right-hand member, we get

$$\|S\|_p^p \leqslant \|S\|_p^{p/p'} \sum_{i=1}^{m} \|f_i\|_p \qquad \text{or} \qquad \|S\|_p \leqslant \sum_{i=1}^{m} \|f_i\|_p,$$

which proves (7).

2.8. By successively applying inequality (7) to each variable, we obtain the following inequality for *vector-valued* $\mathbf{p}$:

If $1 \leqslant \mathbf{p} \leqslant \infty$ *and* $f_i \in L_{\mathbf{p}}(E^n)$ *for* $i = 1, \ldots, m$, *then*

$$\left\| \sum_{i=1}^{m} f_i \right\|_{\mathbf{p}} \leqslant \sum_{i=1}^{m} \|f_i\|_{\mathbf{p}}. \tag{8}$$

2.9. We mention also *Minkowski's inequality for sums,* which is a particular case of inequality (7):

$$\left\{ \sum_{i=1}^{N} \left| \sum_{j=1}^{m} a_{ij} \right|^p \right\}^{1/p} \leqslant \sum_{j=1}^{m} \left(\sum_{i=1}^{N} | a_{ij} |^p \right)^{1/p} \quad (1 \leqslant p \leqslant \infty), \tag{9}$$

where m and N are natural numbers.

In what follows, we shall often need to use an inequality known as Minkowski's generalized inequality, which differs from (7) in that summation is replaced on both sides of the inequality with integration.

We denote by E_x the n-dimensional space of points x, we denote by E_y the m-dimensional space of points y, and we denote by $E_x \times E_y$ their Cartesian product.

2.10. Minkowski's generalized inequality. *If* $1 \leqslant p \leqslant \infty$ *and* $\varphi(x, y)$ *is a measurable function defined on* $E_x \times E_y$, *then*

$$\left\| \int_{E_y} \varphi(\cdot, y)\, dy \right\|_{p, E_x} \leqslant \int_{E_y} \| \varphi(\cdot, y) \|_{p, E_x}\, dy. \tag{10}$$

REMARK. In the formulation of proposition 2.10, it is not assumed *a priori* that the right-hand member of inequality (10) is finite. Therefore, we need to understand (10) as saying that, if the right-hand member is finite, so is the left-hand member and inequality (10) holds.

Almost all the inequalities given in this book will be formulated in an analogous form.

PROOF. For $p = 1$, inequality (10) is a consequence of Fubini's theorem and, for $p = \infty$, it is obvious. Suppose that $1 < p < \infty$. We set

$$S(x) = \int_{E_y} | \varphi(x, y) |\, dy.$$

Let $g(x)$ denote an arbitrary function in $L_{p'}(E_x)$. Then,

$$\int_{E_x} S(x)\, g(x)\, dx \leqslant \int_{E_x} |g(x)| \left(\int_{E_y} |\varphi(x, y)|\, dy \right) dx$$

$$= \int_{E_y} dy \int_{E_x} |g(x)|\, |\varphi(x, y)|\, dx \leqslant \| g \|_{p'} \int_{E_y} \| \varphi(\cdot, y) \|_p\, dy.$$

We note that equality of the iterated integrals follows from Fubini's theorem for a nonnegative function for which one of these integrals exists. The inequality obtained and the converse to Hölder's inequality imply that $S \in L_p(E_x)$ and

$$\| S \|_p \leqslant \int_{E_y} \| \varphi(\cdot, y) \|_p\, dy.$$

This proves inequality (10).

2.11. If in (10) we replace $\varphi(x, y)$ with $|\varphi(x, y)|^{\mu}$ and set $p = \frac{\nu}{\mu}$, we obtain the following more general assertion:

If $0 < \mu \leqslant \nu \leqslant \infty$ *and* $\varphi(x, y)$ *is a measurable function on* $E_x \times E_y$, *then*

$$\left[\int_{E_x} \left(\int_{E_y} |\varphi(x, y)|^{\mu}\, dy \right)^{\nu/\mu} dx \right]^{1/\nu}$$

$$\leqslant \left[\int_{E_y} \left(\int_{E_x} |\varphi(x, y)|^{\nu}\, dx \right)^{\mu/\nu} dy \right]^{1/\mu}. \tag{11}$$

If $1 \leqslant \mu \leqslant \nu \leqslant \infty$, *we can write* (11) *in the form*

$$\| \varphi \|_{(\mu, \nu), E_y \times E_x} \leqslant \| \varphi \|_{(\nu, \mu), E_x \times E_y}. \tag{11'}$$

2.12. We shall have frequent occasion to use the following obvious generalization of inequality (10) to the case of vector-valued $\boldsymbol{p} = (p_1, \ldots, p_n)$, where $\mathbf{1} \leqslant \boldsymbol{p} \leqslant \boldsymbol{\infty}$:

$$\left\| \int_{E_y} \varphi(\cdot, y)\, dy \right\|_{p, E_x} \leqslant \int_{E_y} \| \varphi(\cdot, y) \|_{p, E_x}\, dy. \tag{12}$$

For $m = n$, we also have

$$\left\| \int_{E_y} \varphi(\cdot, y)\, dy \right\|_{p, E_x} \leqslant \left\| \int_{E^1_{y_n}} dy_n \ldots \right.$$
$$\left. \ldots \left\| \int_{E^1_{y_2}} dy_2 \left\| \int_{E^1_{y_1}} \varphi(\cdot, y)\, dy_1 \right\|_{p_1, E^1_{x_1}} \right\|_{p_2, E^1_{x_2}} \ldots \right\|_{p_n, E^1_{x_n}}. \tag{13}$$

2.13. Young's inequality. *Let p, q, and r denote real numbers such that*

$$1 \leqslant p \leqslant q \leqslant \infty, \qquad 1 - \frac{1}{p} + \frac{1}{q} = \frac{1}{r}. \tag{14}$$

Let $f(x)$ and $K(x)$ denote functions defined on E^1 such that $f \in L_p(E^1)$ and $K \in L_r(E^1)$. Define

$$\mathcal{Y}(x) = \int_{E^1} f(y)\, K(y - x)\, dy.$$

Then,

$$\| \mathcal{Y} \|_q \leqslant \| K \|_r \| f \|_p. \tag{15}$$

Proof. We note first of all that, if $q = \infty$, inequality (15) is a consequence of Hölder's inequality since in this case $r = p'$. Therefore, we shall assume in what follows that $q < \infty$.

It follows from (14) that any of the following three combinations of relationships between p, q, and r are possible:

1) $1 < p < q,\ r < q$; 2) $1 = p < q,\ r = q$; 3) $p = q,\ r = 1$.

Let us assume first that case 1) holds. We write $|fK|$ in the form

$$|fK| = (|f|^p |K|^r)^{\frac{1}{q}} |K|^{1-\frac{r}{q}} |f|^{1-\frac{p}{q}}. \tag{16}$$

To estimate the integral inside the norm bars in the left-hand member of (15), we use Hölder's inequality for three functions with

$$p_1 = q, \quad p_2 = p' = \frac{r}{1 - \frac{r}{q}},$$

$$p_3 = \frac{p}{1 - \frac{p}{q}} \quad \left(\frac{1}{p_1} + \frac{1}{p_2} + \frac{1}{p_3} = 1\right).$$

We then obtain

$$|\mathcal{I}(x)| \leqslant \left(\int |f(y)|^p |K(y-x)|^r \, dy\right)^{\frac{1}{q}} \|K\|_r^{1-\frac{r}{q}} \|f\|_p^{1-\frac{p}{q}}.$$

Therefore,

$$\|\mathcal{I}\|_q \leqslant \|K\|_r^{1-\frac{r}{q}} \|f\|_p^{1-\frac{p}{q}} \left(\int dx \int |f(y)|^p |K(y-x)|^r \, dy\right)^{\frac{1}{q}}$$

$$= \|K\|_r^{1-\frac{r}{q}} \|f\|_p^{1-\frac{p}{q}} \left(\int |f(y)|^p \, dy \int |K(x)|^r \, dx\right)^{\frac{1}{q}} = \|K\|_r \|f\|_p,$$

which proves inequality (15).

The proof is analogous for cases 2) and 3). We note only that the right-hand member of (16) in these cases is the product of two factors and that the integral $\mathcal{I}(x)$ is estimated with the aid of Hölder's inequality for two functions.

2.14. With the aid of inequality (13) and n-fold application of inequality (15), we obtain the following generalization of inequality (15) to the multidimensional case:

Suppose that $\boldsymbol{p} = (p_1, \ldots, p_n)$, $\boldsymbol{q} = (q_1, \ldots, q_n)$, $\boldsymbol{r} = (r_1, \ldots, r_n)$,

$$1 \leqslant p \leqslant q \leqslant \infty, \quad 1 - \frac{1}{p} + \frac{1}{q} = \frac{1}{r}, \tag{17}$$

$$\mathcal{J}(x) = \int_{E^n} f(y) K(y - x)\, dy.$$

Then,

$$\| \mathcal{J} \|_q \leqslant \| K \|_r \| f \|_p. \tag{18}$$

In particular, for $q = p$,

$$\| \mathcal{J} \|_p \leqslant \| K \|_1 \| f \|_p. \tag{19}$$

2.15. A generalization of Hardy's inequality. In this subsection, we shall prove an important inequality of Hardy for functions of a single variable defined on the semi-axis $E^1_+ = (0, \infty)$. As usual, we define

$$\| f \|_{p, E^1_+} = \left(\int_0^\infty | f(x) |^p \, dx \right)^{1/p}. \tag{20}$$

Suppose that $1 \leqslant p \leqslant q \leqslant \infty$, $\alpha \neq 0$, $\gamma > 0$, $f \in L_p(E^1_+)$. *If*

$$F_{\alpha, \gamma}(x) = \int_0^{x^\gamma} f(y)\, y^{-\frac{1}{p'} + \alpha} dy \quad \text{for} \quad \alpha > 0 \tag{21}$$

and

$$F_{\alpha, \gamma}(x) = \int_{x^\gamma}^\infty f(y)\, y^{-\frac{1}{p'} + \alpha} dy \quad \text{for} \quad \alpha < 0, \tag{22}$$

then

$$\left\| x^{-\frac{1}{q}-\alpha\gamma} F_{\alpha,\gamma} \right\|_{q,E_+^1} \leqslant \gamma^{-\frac{1}{q}} \left(\frac{\mu}{|\alpha|}\right)^{\mu} \| f \|_{p,E_+^1}, \tag{23}$$

where $\mu = 1 - \frac{1}{p} + \frac{1}{q}$. (If $p = 1$ and $q = \infty$, we define $\left(\frac{\mu}{|\alpha|}\right)^{\mu} = 1$.)

PROOF. Let us suppose first that $1 < p \leqslant q < \infty$. Let δ denote a number such that $0 < \delta < |\alpha|$. By Hölder's inequality, we have

$$|F_{\alpha,\gamma}(x)| \leqslant \left(\int_0^{x^\gamma} |f(y)|^p y^{\delta p} \, dy \right)^{1/p} \left(\int_0^{x^\gamma} y^{-1+(\alpha-\delta)p'} \, dy \right)^{1/p'}$$

$$\leqslant C_1 x^{(\alpha-\delta)\gamma} \left(\int_0^{x^\gamma} |f(y)|^p y^{\delta p} \, dy \right)^{1/p} \quad \text{if} \quad \alpha > 0, \tag{24}$$

$$|F_{\alpha,\gamma}(x)| \leqslant C_2 x^{(\alpha+\delta)\gamma} \left(\int_{x^\gamma}^{\infty} |f(y)|^p y^{-\delta p} \, dy \right)^{1/p} \quad \text{if} \quad \alpha < 0, \tag{25}$$

where $C_1 = [(\alpha - \delta) p']^{-1/p'}$ and $C_2 = [-(\alpha + \delta) p']^{-1/p'}$. Then, if we apply Minkowski's generalized inequality (11), we obtain

$$\left\| x^{-\frac{1}{q}-\alpha\gamma} F_{\alpha,\gamma} \right\|_{q,E_+^1} \leqslant C_1 \left[\int_0^{\infty} \left(\int_0^{x^\gamma} |f(y)|^p y^{\delta p} x^{-\left(\frac{1}{q}+\delta\gamma\right)p} \, dy \right)^{\frac{q}{p}} dx \right]^{1/q}$$

$$\leqslant C_1 \left[\int_0^{\infty} |f(y)|^p y^{\delta p} \left(\int_{y^{1/\gamma}}^{\infty} x^{-1-\delta\gamma q} \, dx \right)^{\frac{p}{q}} dy \right]^{1/p}$$

$$\leqslant C_1 C_3 \| f \|_{p,E_+^1} \quad (\alpha > 0),$$

$$\left\| x^{-\frac{1}{q}-\alpha\gamma} F_{\alpha,\gamma} \right\|_{q,E_+^1} \leqslant C_2 C_3 \| f \|_{p,E_+^1} \quad (\alpha < 0),$$

where $C_3 = (\delta\gamma q)^{-1/q}$.

If we set $\delta = \frac{|\alpha| p'}{p' + q}$ in the expressions for C_1, C_2, and C_3, we easily obtain inequality (23).

If $1 = p \leqslant q < \infty$ (so that $1/p' = 0$), inequality (23) is immediately obtained by applying Minkowski's generalized inequality. If $1 \leqslant p \leqslant q = \infty$, then inequality (23) follows from (24) and (25) with $\delta = 0$.

Let us put proposition 2.15 in a somewhat different form. To do this, we shall introduce the following notation, which we shall use again later:

$$[a]_1 = \min\{a, 1\}.$$

Then, proposition 2.15 can be rephrased as follows:

2.15′. *Suppose that* $1 \leqslant p \leqslant q \leqslant \infty$, $\alpha > 0$, $\gamma > 0$, $\mu = 1 - \frac{1}{p} + \frac{1}{q}$, $f \in L_p(E^1_+)$, *and*

$$h(y, x) = y^{-\frac{1}{p'} - \alpha} x^{-\frac{1}{q} + \alpha\gamma} \left[\frac{y}{x^\gamma}\right]_1^{2\alpha}.$$

Then,

$$\left\| \int_0^\infty f(y)\, h(y, \cdot)\, dy \right\|_{q, E^1_+} \leqslant \gamma^{-1/q} \left(\frac{\mu}{\alpha}\right)^\mu \| f \|_{p, E^1_+}. \tag{23'}$$

2.16. The following inequality can be obtained from (23) by renaming the parameters but, because of its simplicity, we shall give a direct proof of it.

If $1 \leqslant p \leqslant \infty$, $\beta \neq 1/p$,

$$F(x) = \int_0^x f(y)\, dy \quad \text{for} \quad \beta > \frac{1}{p},$$

and

$$F(x) = \int_x^\infty f(y)\,dy \quad \text{for} \quad \beta < \frac{1}{p},$$

then,

$$\| x^{-\beta} F \|_{p, E_+^1} \leqslant \frac{1}{\left| \beta - \frac{1}{p} \right|} \| x^{-\beta+1} f \|_{p, E_+^1}. \tag{26}$$

PROOF. Suppose that $\beta > 1/p$. Making the change of variable $y = xt$ in the integral and applying Minkowski's generalized inequality (10), we obtain

$$\left\| x^{-\beta} \int_0^x f(y)\,dy \right\|_p = \left\| \int_0^1 f(xt)\, x^{-\beta+1}\,dt \right\|_p \leqslant \int_0^1 \| x^{-\beta+1} f(xt) \|_p\,dt$$
$$= \int_0^1 t^{\beta-1-\frac{1}{p}} \| x^{-\beta+1} f \|_p\,dt = \frac{1}{\beta - \frac{1}{p}} \| x^{-\beta+1} f \|_p.$$

Analogously, if $\beta < 1/p$, we have

$$\left\| x^{-\beta} \int_x^\infty f(y)\,dy \right\|_p = \left\| \int_1^\infty f(xt)\, x^{-\beta+1}\,dt \right\|_p \leqslant \int_1^\infty \| x^{-\beta+1} f(xt) \|_p\,dt$$
$$= \int_1^\infty t^{\beta-1-\frac{1}{p}} \| x^{-\beta+1} f \|_p\,dt = \frac{1}{\frac{1}{p} - \beta} \| x^{-\beta+1} f \|_p.$$

2.17. To prove the Hardy-Littlewood inequality for fractional integrals and certain analogues of it, we shall need a lemma on rearrangements of functions in decreasing order.* (This is a special

*Translation editor's note. Decreasing is to be understood in the nonstrict sense, i.e., nonincreasing.

case of the well-known theorem of F. Riesz* on the rearrangement of three functions.)

Let $f(x)$ denote a nonnegative measurable function of a single variable x defined on the entire real axis. Let $f^*(x)$ denote a function symmetrically decreasing with respect to the coordinate origin and equimeasurable with $f(x)$, that is, a function such that $f^*(-x) = f^*(x)$ for all $x \in E^1$ and mes $\{x: f^*(x) \geqslant y\}$ = mes $\{x: f(x) \geqslant y\}$ for all y.

Lemma. *Let $f(x)$, $g(x)$, and $h(x)$ denote nonnegative functions of a single variable defined on E^1 such that $f \in L_p(E^1)$ for some $p \geqslant 1$ $g \in L_q(E^1)$ for some $q \geqslant 1$, and $h \in L^{\text{loc}}(E^1)$. Suppose also that $h(x)$ decreases symmetrically about the point $x = 0$. Let $f^*(x)$ and $g^*(x)$ denote the corresponding equimeasurable symmetrically decreasing functions. Then,*

$$\mathcal{I} = \int_{E^1} \int_{E^1} f(y)\, g(x)\, h(y-x)\, dy\, dx \leqslant \int_{E^1} \int_{E^1} f^*(y)\, g^*(x)\, h(y-x)\, dy\, dx = \mathcal{I}^*. \quad (27)$$

PROOF. Let us first prove inequality (27) for the case in which the functions $f(x)$ and $g(x)$ assume only the values 0 and 1. Here, because of the assumption of summability of the functions, we may suppose that the sets on which they assume the value 1 are of finite measure. These sets can be represented as the union of a finite system of disjoint half-closed half-open intervals and sets of arbitrarily small measure. Since f and g do not exceed unity, the sets of arbitrarily small measure make a small correction to the integrals $\mathcal{I}$ and $\mathcal{I}^*$. Therefore, we may assume that the sets on which $f = 1$ and $g = 1$ are formed by finite systems of half-closed half-open intervals.

Furthermore, by the exact same reasoning, we may assume that the end-points of these intervals are rational numbers. By suitable changes of variable, we can make them integers.

Thus, it will be sufficient to prove inequality (27) for the

*See the book by Hardy, Littlewood, and Pólya [1], Chapter X.

functions

$$f(x)=\sum_{i=1}^{i_0}\chi_{m_in_i}(x) \qquad (m_i<n_i<m_{i+1}) \tag{28}$$

and

$$g(x)=\sum_{j=1}^{j_0}\chi_{k_jl_j}(x) \qquad (k_j<l_j<k_{j+1}), \tag{29}$$

where $\chi_{m_in_i}(x)$ and $\chi_{k_jl_j}(x)$ are the characteristic functions of the intervals $[m_i, n_i)$ and $[k_j, l_j)$ with integer-valued end-points.

We note that

$$\int_{E^1}\int_{E^1} f(y)\, g(x)\, h(y-x)\, dy\, dx = \int_{E^1}\left(\sum_{i=1}^{i_0}\chi_{m_in_i}(y)\right)\left(\sum_{j=1}^{j_0}\mathrm{X}_{k_jl_j}(y)\right) dy, \tag{30}$$

where the function

$$\mathrm{X}_{k_jl_j}(y)=\int_{E^1}\chi_{k_jl_j}(x)\, h(y-x)\, dx$$

decreases symmetrically with respect to the mid-point of $[k_j, \ldots l_j)$. It follows that, if $i_0=j_0=1$, that is, if $f(y)=\chi_{mn}(y)$ and $g(x)=\chi_{kl}(x)$, then inequality (27) holds since the midpoints of the supports of the functions $\chi^*_{mn}(x)$ and $\chi^*_{kl}(x)$ coincide.

Let us suppose now that $i_0+j_0>2$ in (28) and (29). If the midpoints of the supports in the last terms in (28) and (29) coincide, we displace their graphs one unit to the left. More precisely, we replace the functions $\chi_{mi_0n_{i_0}}(x)$ and $\chi_{k_{j_0}l_{j_0}}(x)$ with the functions $\chi_{m_{i_0}-1,\, n_{i_0}-1}(x)$ and $\chi_{k_{j_0}-1,\, l_{j_0}-1}(x)$ respectively. On the other hand, if the midpoints of the supports in these terms do not coincide, we displace one unit to the left the graph of that function the mid-point of the support of which lies to the right. Obviously, such a transformation of the functions f and g does not decrease the integral (30) and i_0 and j_0 in (28) and (29) can only decrease. After a finite number of transformations, we arrive at the case $i_0 = j_0 = 1$, for which we have already proven inequality (27).

Thus, inequality (27) is proven for functions f and g that assume only the values 0 and 1.

Let us suppose now that f and g assume only a finite number of values. Suppose, for example, that f assumes the values $\alpha_0, \alpha_1, \ldots, \alpha_n$, where $0 = \alpha_0 < \alpha_1 < \ldots < \alpha_n$. Then, $f(x)$ can be represented in the form

$$f(x) = \sum_{i=1}^{n} c_i f_i(x), \quad c_i = \alpha_i - \alpha_{i-1} > 0 \qquad (i = 1, \ldots, n),$$

where f_i assumes only the values 0 and 1 and

$$\operatorname{supp} f_i \supset \operatorname{supp} f_{i+1} \quad (i = 1, \ldots, n-1).$$

One can easily see that

$$f^*(x) = \sum_{i=1}^{n} c_i f_i^*(x).$$

If we represent the function g in this way, inequality (27) for the functions f and g that we are now considering follows from a linear combination of analogous inequalities containing the f_i and the g_i.

The shift from the case that we have been considering to the general one is done by approximating the functions f and g with functions that assume only a finite number of values. For example, let us approximate the function f with functions f_n defined by

$$\begin{aligned} f_n &= \frac{k}{n} \qquad \left(\frac{k}{n} \leqslant f \leqslant \frac{k+1}{n}, \quad k = 0, 1, \ldots, n^2 - 1\right), \\ f_n &= n \qquad (f \geqslant n). \end{aligned}$$

Then, $f_n \leqslant f$ and $f_n^* \leqslant f^*$. Analogously, we approximate g with functions g_n. Since inequality (27) is proven for the f_n and the g_n, we have

$$\mathcal{I}_n = \int\limits_{E^1} \int\limits_{E^1} f_n(y)\, g_n(x)\, h(y-x)\, dy\, dx \leqslant \mathcal{I}_n^* \leqslant \mathcal{I}^*,$$

so that $\mathcal{I} = \lim\limits_{n\to\infty} \mathcal{I}_n \leqslant \mathcal{I}^*$. This completes the proof of the lemma.

We shall later use also the following inequality of Čebyšev:

2.18. Čebyšev's inequality. *If $f(x)$ is a nondecreasing and $g(x)$ a nonincreasing function, both summable and defined on an interval $[a, b]$, then*

$$\int\limits_a^b f(x)\, g(x)\, dx \leqslant \frac{1}{b-a} \int\limits_a^b f(x)\, dx \int\limits_a^b g(x)\, dx. \tag{31}$$

PROOF.* Let us define

$$\varphi(x) = f(x) - \frac{1}{b-a} \int\limits_a^b f(t)\, dt.$$

The monotonicity of $f(x)$ implies the existence of a point c in $[a, b]$ such that

$$\varphi(x) \leqslant 0 \quad \text{if} \quad a \leqslant x < c, \qquad \varphi(x) \geqslant 0 \quad \text{if} \quad c < x \leqslant b.$$

Therefore,

$$\int\limits_a^b g(x) \left[f(x) - \frac{1}{b-a} \int\limits_a^b f(t)\, dt \right] dx$$

$$= \int\limits_a^c g(x)\, \varphi(x)\, dx + \int\limits_c^b g(x)\, \varphi(x)\, dx \leqslant g(c) \int\limits_a^b \varphi(x)\, dx = 0,$$

from which inequality (31) follows.

*This proof is due to Sapogov [1].

2.19. The Hardy-Littlewood inequality. *Suppose that* $1 < p < q < \infty$, $\mu = 1 - \frac{1}{p} + \frac{1}{q}$, $f \in L_p(E^1)$, *and*

$$\mathcal{I}(x) = \int_{E^1} f(y) \, | y - x |^{-\mu} \, dy.$$

Then,

$$\| \mathcal{I} \|_q \leqslant K(p, q) \| f \|_p. \tag{32}$$

Proof. By virtue of the result of subsection 2.6, that is, the converse of Hölder's inequality, it will be sufficient to show that the inequality

$$I = \int_{E^1} \int_{E^1} g(x) f(y) \, | y - x |^{-\mu} \, dy \, dx \leqslant K(p, q) \| f \|_p \| g \|_{q'}$$

holds for any function g in $L_{q'}(E^1)$, where $\left(\frac{1}{q} + \frac{1}{q'} = 1 \right)$. Obviously, we may assume that the functions f and g are nonnegative and, on the basis of Lemma 2.17, that they decrease symmetrically about the point $x = 0$.

Thus, it will be sufficient to prove inequality (32) for a nonnegative symmetrically decreasing function $f(x)$. For such a function,

$$f(x) \leqslant | 2x |^{-\frac{1}{p}} \left(\int_{-|x|}^{|x|} f^p(y) \, dy \right)^{\frac{1}{p}} \leqslant | 2x |^{-\frac{1}{p}} \| f \|_p,$$

$$\mathcal{I}(x) \leqslant \int_{E^1} f^{\frac{p}{q}}(y) \, | 2y |^{-\frac{1}{p}\left(1 - \frac{p}{q}\right)} \| f \|_p^{1 - \frac{p}{q}} \, | y - x |^{-\mu} \, dy$$

$$= 2^{\mu - 1} \| f \|_p^{1 - \frac{p}{q}} \int_{E^1} f^{\frac{p}{q}}(y) \, | y - x |^{-\mu} \, | y |^{\mu - 1} \, dy$$

$$= 2^{\mu - 1} \| f \|_p^{1 - \frac{p}{q}} \int_{E^1} f^{\frac{p}{q}}(tx) \, | t - 1 |^{-\mu} \, | t |^{\mu - 1} \, dt.$$

With the aid of Minkowski's generalized inequality (10), we now have

$$\|\mathcal{I}\|_q \leqslant$$
$$\leqslant 2^{\mu-1}\|f\|_p^{1-\frac{p}{q}}\left\{\int_{E^1}\left[\int_{E^1} f^{\frac{p}{q}}(tx)|t-1|^{-\mu}|t|^{\mu-1}dt\right]^q dx\right\}^{\frac{1}{q}}$$
$$\leqslant 2^{\mu-1}\|f\|_p^{1-\frac{p}{q}}\int_{E^1}|t-1|^{-\mu}|t|^{\mu-1}\left(\int_{E^1} f^p(tx)\,dx\right)^{\frac{1}{q}}dt$$
$$\leqslant 2^{\mu-1}\int_{E^1}|t-1|^{-\mu}|t|^{-\frac{1}{p}}dt\,\|f\|_p.$$

Thus, we have proven inequality (32) with constant

$$K(p,d)=2^{\frac{1}{q}-\frac{1}{p}}\int_{E^1}|t-1|^{\frac{1}{p}-\frac{1}{q}-1}|t|^{-\frac{1}{p}}dt. \tag{33}$$

2.20. This subsection is devoted to a generalization of inequality (32) to the multidimensional case. We shall find the following simple inequality useful: *If* a, λ, *and* μ *are positive numbers and* $1 \leqslant r \leqslant \infty$, *then there exists a number* C *independent of* a *such that*

$$\left\|\left(a+|t|^{\frac{1}{\lambda}}\right)^{-\mu-\frac{\lambda}{r}}\right\|_{r,E^1} \leqslant Ca^{-\mu}. \tag{34}$$

For $r=\infty$, the inequality is obvious. If $r<\infty$, when we make the substitution $t=a^{\lambda}\tau$, we get

$$\left\|\left(a+|t|^{\frac{1}{\lambda}}\right)^{-\mu-\frac{\lambda}{r}}\right\|_r = a^{-\mu}\left\|\left(1+|\tau|^{\frac{1}{\lambda}}\right)^{-\mu-\frac{\lambda}{r}}\right\|_r = Ca^{-\mu}.$$

Theorem. *Let* $\boldsymbol{p}=(p_1,\ldots,p_n)$, *and* $\boldsymbol{q}=(q_1,\ldots,q_n)$ *denote vectors such that* $\mathbf{1}\leqslant\boldsymbol{p}\leqslant\boldsymbol{q}\leqslant\boldsymbol{\infty}$ *and*

$$1<p_n<q_n<\infty. \tag{35}$$

Let λ_i denote positive numbers for $i = 1, \ldots, n$. Define the number

$$\mu = \sum_{i=1}^{n} \lambda_i \left(1 - \frac{1}{p_i} + \frac{1}{q_i}\right)$$

and the functions

$$\rho(y) = \sum_{i=1}^{n} | y_i |^{1/\lambda_i}$$

and

$$\mathcal{J}(x) = \int_{E^n} f(y) [\rho(y - x)]^{-\mu} \, dy,$$

where $f \in L_p(E^n)$. Then, there exists a constant C independent of f such that

$$\| \mathcal{J} \|_q \leqslant C \| f \|_p . \tag{36}$$

PROOF. For any vector $s = (s_1, \ldots, s_n)$, let us define $s^{(n)} = (s_1, \ldots, s_{n-1})$ and let us identify s and $(s^{(n)}, s_n)$.

Using successively Minkowski's generalized inequality (12) and Young's inequality (18), we obtain

$$\begin{aligned} \| \mathcal{J} \|_q &= \| \mathcal{J} \|_{(q^{(n)}, q_n)} \\ &\leqslant \Big\| \int_{E^1} dy_n \Big\| \int_{E^{n-1}} f(y^{(n)}, y_n) \, \rho^{-\mu}(y^{(n)} - x^{(n)}, y_n - x_n) \, dy^{(n)} \Big\|_{q^{(n)}} \Big\|_{q_n} \\ &\leqslant \Big\| \int_{E^1} \| f(\cdot, y_n) \|_{p^{(n)}} \| \rho^{-\mu}(\cdot, y_n - x_n) \|_{r^{(n)}} \, dy_n \Big\|_{q_n}, \end{aligned}$$

where $r = (r_1, \ldots, r_n)$ with $\dfrac{1}{r_i} = 1 - \dfrac{1}{p_i} + \dfrac{1}{q_i}$ for $i = 1, \ldots, n$.

Let us estimate the norm of the kernel $\rho^{-\mu}$ by applying inequality (34) $n-1$ times. We have

$$\| \rho^{-\mu}(\cdot, y_n) \|_{r(n)} = \| \ldots \| \rho^{-\mu} \|_{r_1} \ldots \|_{r_{n-1}}.$$

By virtue of the definition of the numbers r_i, we have

$$\mu = \sum_1^n \frac{\lambda_i}{r_i}.$$

By again applying inequality (34), we get

$$\| \rho^{-\mu} \|_{r_1} = \left\| \left(\sum_{i=2}^n |y_i|^{\frac{1}{\lambda_i}} + |y_1|^{\frac{1}{\lambda_1}} \right)^{-\sum_{i=2}^n \frac{\lambda_i}{r_i} - \frac{\lambda_1}{r_1}} \right\|_{r_1}$$

$$\leqslant C \left(\sum_{i=2}^n |y_i|^{\frac{1}{\lambda_i}} \right)^{-\sum_{i=2}^n \frac{\lambda_i}{r_i}}.$$

Obviously, after $n-1$ analogous steps, we arrive at the estimate

$$\| \rho^{-\mu}(\cdot, y_n) \|_{r(n)} \leqslant C_1 |y_n|^{-\frac{1}{r_n}} = C_1 |y_n|^{-\left(1 - \frac{1}{p_n} + \frac{1}{q_n}\right)}.$$

Using this estimate and the Hardy-Littlewood inequality (32) (under the condition $1 < p_n < q_n < \infty$), we finally get

$$\| \mathcal{I} \|_q \leqslant C_1 \left\| \int_{E^1} \left\| f(\cdot, y_n) \right\|_{p^{(n)}} |y_n - x_n|^{-\left(1 - \frac{1}{p_n} + \frac{1}{q_n}\right)} dy_n \right\|_{q_n}$$

$$\leqslant C_2 \| f \|_{(p^{(n)}, p_n)} = C_2 \| f \|_p.$$

This completes the proof of the theorem.

Corollary. *Suppose that* $1 \leqslant m < n$, *that, under the conditions of the theorem,*

$$f = \prod_{i=1}^{m} \delta(y_i) \varphi(y_{m+1}, \ldots, y_n),$$

where $\delta(y_i)$ *is the delta function, and that* $\mathbf{p} = (1, \ldots, 1, p_{m+1}, \ldots, p_n)$. *Then,*

$$\| f \|_{\mathbf{p},\, E^n} = \| \varphi \|_{(p_{m+1}, \ldots, p_n),\, E^{n-m}}$$

and inequality (36) *takes the form**

$$\| \mathcal{I} \|_{q,\, E^n} \leqslant C \| \varphi \|_{(p_{m+1}, \ldots, p_n),\, E^{n-m}}. \tag{37}$$

REMARK. Application of Minkowski's generalized inequality (11′), which enables us (when certain relationships involving the components of the vector are satisfied) to reverse the order in which the partial norms are taken, also enables us to generalize somewhat the formulation of the theorem by replacing condition (35) with the more general condition that, for some m such that $1 \leqslant m \leqslant n$, we have

$$\begin{gathered} 1 < p_m < q_m < \infty, \quad q_i \geqslant q_m, \\ p_i \leqslant p_m \quad (i = m+1, \ldots, n). \end{gathered} \tag{35′}$$

To see that this condition leads to the result, we first apply inequality (11′) $n - m$ times (where $q_i \geqslant q_m$ for $i = m+1, \ldots, n$), we then apply the theorem just proven, and we finally apply inequality (11′) $n - m$ times (where $p_i \leqslant p_m$ for $i = m+1, \ldots, n$). We obtain successively

$$\| \mathcal{I} \|_{q} = \| \mathcal{I} \|_{(q_1, \ldots, q_m, \ldots, q_n)} \leqslant \| \mathcal{I} \|_{(q_1, \ldots, q_{m-1}, q_{m+1}, \ldots, q_n, q_m)} \leqslant$$
$$C \| f \|_{(p_1, \ldots, p_{m-1}, p_{m+1}, \ldots, p_n, p_m)} \leqslant C \| f \|_{(p_1, \ldots, p_m, \ldots, p_n)} = C \| f \|_{\mathbf{p}}.$$

*Strictly speaking, inequality (37) must be proven by taking the limit with respect to a delta-form sequence of functions. Actually, this inequality will not be used in what follows.

The basic purpose of the next subsection is to prove Lemmas 2.22 and 2.24. As a preliminary, we prove an auxiliary lemma.

2.21. Lemma. *Suppose that* $a > 0$, $0 < \mu < 1$, $1 \leqslant r < \infty$, *and* $\mu r' > 1$. *Let* $\varphi(x)$ *denote a positive nonincreasing function defined on* $E^1_+ = (0, \infty)$ *that belongs to* $L_r(E^1_+)$. *Then, there exists a constant C independent of a and* φ *such that*

$$\mathcal{J} = \int_0^\infty \varphi(x)\, |x - a|^{-\mu}\, dx \leqslant C a^{-\mu + \frac{1}{r'}} \|\varphi\|_{r, E^1_+}. \qquad (38)$$

PROOF. Since $\varphi(x)$ is nonincreasing on $(0, \infty)$, we have

$$\int_0^a \varphi(x) |x - a|^{-\mu}\, dx \geqslant \int_a^{2a} \varphi(x) |x - a|^{-\mu}\, dx.$$

Therefore,

$$\mathcal{J} \leqslant 2 \int_0^a \varphi(x) |x - a|^{-\mu}\, dx + \int_{2a}^\infty \varphi(x) |x - a|^{-\mu}\, dx = 2\mathcal{J}_1 + \mathcal{J}_2.$$

Applying successively the inequalities of Čebyšev and Hölder, we obtain

$$\mathcal{J}_1 = \int_0^a \varphi(x) |x - a|^{-\mu}\, dx \leqslant \frac{1}{a} \int_0^a \varphi(x)\, dx \int_0^a |x - a|^{-\mu}\, dx$$
$$\leqslant \frac{1}{1-\mu}\, a^{-\mu + \frac{1}{r'}} \|\varphi\|_r.$$

Also,

$$\mathcal{J}_2 = \int_{2a}^\infty \varphi(x)(x - a)^{-\mu}\, dx \leqslant \|\varphi\|_r \left[\int_a^\infty x^{-\mu r'}\, dx\right]^{1/r'} \leqslant C a^{-\mu + \frac{1}{r'}} \|\varphi\|_r.$$

The lemma now follows when we combine these three inequalities.

2.22. Lemma. *Suppose that* $1 \leqslant p < q < \infty$, $1 \leqslant \theta \leqslant q$, *and* $\mu = 1 - \frac{1}{p} + \frac{1}{q}$. *Let* $\psi(y, t)$ *and* $h(y, t)$ *denote measurable functions defined on* E^2. *Suppose that* $\psi \in L_{(p, \theta)}(E^2)$ *and that* $h(y, t)$ *is nonnegative and symmetrically decreasing with respect to the variable* y *about the point* $y = 0$ *for almost all* $t \in E^1$. *Suppose in addition that*

$$h(y, t) \leqslant |y|^{-\mu - \alpha_1 \gamma} |t|^{-\frac{1}{\theta'} + \alpha_1} \quad \left(\frac{1}{\theta} + \frac{1}{\theta'} = 1\right), \tag{39}$$

where $\alpha_1 > 0$, $\gamma > 0$, *and* $\mu + \alpha_1\gamma < 1$, *and that*

$$h(y, t) \leqslant |y|^{-\mu + \alpha_2 \gamma} |t|^{-\frac{1}{\theta'} - \alpha_2}, \tag{40}$$

where $\alpha_2 > 0$ *and* $0 < \alpha_2 \gamma < \frac{1}{q}$. *Then, there exists a constant* C *independent of* ψ *such that*

$$\left\| \int_{E^1} \int_{E^1} \psi(y, t)\, h(y - x, t)\, dy\, dt \right\|_q \leqslant C \|\psi\|_{(p, \theta)}. \tag{41}$$

PROOF. By virtue of the converse to Hölder's inequality, to prove inequality (41) we need only show that, for any function $g \in L_{q'}(E^1)$, where $\frac{1}{q} + \frac{1}{q'} = 1$, we have

$$I = \int_{E^1} \int_{E^1} \int_{E^1} g(x)\, \psi(y, t)\, h(y - x, t)\, dy\, dt\, dx \leqslant C \|g\|_{q'} \|\psi\|_{(p, \theta)}. \tag{42}$$

Obviously, to prove this inequality we may assume that g and ψ are nonnegative.

Suppose that $g^*(x)$ is equimeasurable with $g(x)$ and that it is symmetrically decreasing about the point $x = 0$. Suppose that, for almost all fixed t, the function $\psi^*(y, t)$, as a function of y, is equimeasurable with $\psi(y, t)$ and symmetrically decreasing about the point $y = 0$. Then, on the basis of Lemma (2.17) we have

$$\int_{E^1}\int_{E^1} g(x)\,\psi(y, t)\,h(y-x, t)\,dy\,dx \leqslant \int_{E^1}\int_{E^1} g^*(x)\,\psi^*(y, t)\,h(y-x, t)\,dy\,dx.$$

Therefore, we may assume that the functions $g(x)$ and $\psi(y, t)$ in the expression for the integral I decrease symmetrically with respect to x and y respectively.

Let us break the integral I into eight parts corresponding to the eight octants of the space. It will be sufficient to obtain an estimate for the integral corresponding to the octant $y > 0$, $t > 0$, $x > 0$ since the others are estimated analogously. Consequently, we need only show that

$$\mathcal{I} = \int_0^\infty\int_0^\infty\int_0^\infty g(x)\,\psi(y, t)\,h(y-x, t)\,dy\,dt\,dx \leqslant C\,\|g\|_{q'}\,\|\psi\|_{(p,\,\theta)}. \tag{43}$$

Estimating the integral $\mathcal{I}$ with the aid of Hölder's and Minkowski's inequalities, we obtain

$$\mathcal{I} \leqslant \|g\|_{q'}\left\|\int_0^\infty\int_0^\infty \psi(y, t)\,h(y-x, t)\,dy\,dt\right\|_q \leqslant \|g\|_{q'}\left(\|\mathcal{I}_1\|_q + \|\mathcal{I}_2\|_q\right), \tag{44}$$

where

$$\mathcal{I}_1(x) = \int_0^{x^\gamma} dt \int_0^\infty \psi(y, t)\,h(y-x, t)\,dy,$$

$$\mathcal{I}_2(x) = \int_{x^\gamma}^\infty dt \int_0^\infty \psi(y, t)\,h(y-x, t)\,dy.$$

Let us estimate the function $h(y-x, t)$ in the first of these integrals by using inequality (39) and in the second by using inequality (40). Then, we have

$$\mathcal{I}_1(x) \leqslant \int_0^{x^\gamma} t^{-\frac{1}{\theta'}+\alpha_1}\, dt \int_0^\infty \psi(y, t)\, |y-x|^{-\mu-\alpha_1\gamma}\, dy,$$

$$\mathcal{I}_2(x) \leqslant \int_{x^\gamma}^\infty t^{-\frac{1}{\theta'}-\alpha_2}\, dt \int_0^\infty \psi(y, t)\, |y-x|^{-\mu+\alpha_2\gamma}\, dy.$$

In both cases, the inner integrals can be estimated with the aid of inequality (38). The conditions of Lemma 2.21 under which this inequality was proven are satisfied by virtue of the conditions on the parameters p, q, α_1, α_2, and γ.

Applying inequality (38) and then Hardy's inequality (23), in which p is replaced with θ, we obtain

$$\|\mathcal{I}_1\|_q \leqslant C_1 \left\| x^{-\frac{1}{q}-\alpha_1\gamma} \int_0^{x^\gamma} \left\|\psi(\cdot\,, t)\right\|_p \cdot t^{-\frac{1}{\theta'}+\alpha_1}\, dt \right\|_q \leqslant C_2 \|\psi\|_{(p,\,\theta)},$$

$$\|\mathcal{I}_2\|_q \leqslant C_3 \left\| x^{-\frac{1}{q}+\alpha_2\gamma} \int_{x^\gamma}^\infty \left\|\psi(\cdot\,, t)\right\|_p \cdot t^{-\frac{1}{\theta'}-\alpha_2} dt \right\|_q \leqslant C_4 \|\psi\|_{(p,\,\theta)},$$

where C_2 and C_4 are independent of ψ.

The last two inequalities together with (44) prove inequality (43) and hence the lemma.

2.23. Remark. Let us look at some examples of functions $h(y, t)$ satisfying the conditions of this last lemma. With the parameters p, q, μ, and θ as in the lemma, let us set

$$h_1(y, t) = |y|^{-\mu+\beta_1} |t|^{-\frac{1}{\theta'}+\beta_2} \left(|y|^{\frac{1}{\lambda_1}} + |t|^{\frac{1}{\lambda_2}} \right)^{-(\beta_1\lambda_1+\beta_2\lambda_2)},$$

where β_1, β_2, λ_1, and λ_2 are positive numbers, and let us set

$$h_2(y, t) = |y|^{-\mu-\beta\gamma} |t|^{-\frac{1}{\theta'}+\beta} \left[\frac{|y|^{\gamma}}{|t|}\right]_1^{2\beta},$$

where β and γ are positive numbers and $[a]_1$ denotes $\min\{a, 1\}$.

Let us show that $h_1(y, t)$ and $h_2(y, t)$ satisfy inequalities (39) and (40). On the one hand, we have

$$h_1(y, t) = |y|^{-\mu-\alpha_1\gamma} |t|^{-\frac{1}{\theta'}+\alpha_1} \times$$

$$\left(|y|^{\beta_1+\alpha_1\gamma} |t|^{\beta_2-\alpha_1} \left(|y|^{\frac{1}{\lambda_1}} + |t|^{\frac{1}{\lambda_2}}\right)^{-(\beta_1\lambda_1+\beta_2\lambda_2)}\right) \leqslant |y|^{-\mu-\alpha_1\gamma} |t|^{-\frac{1}{\theta'}+\alpha_1}$$

if $0 < \alpha_1 \leqslant \beta_2$ and $\gamma = \lambda_2/\lambda_1$. Here, we may assume that $\mu + \alpha_1\gamma < 1$ since $\mu = 1 - \frac{1}{p} + \frac{1}{q} < 1$ for $q > p$. On the other hand, we have

$$h_1(y, t) = |y|^{-\mu+\alpha_2\gamma} |t|^{-\frac{1}{\theta'}-\alpha_2} \times$$

$$\left(|y|^{\beta_1-\alpha_2\gamma} |t|^{\beta_2+\alpha_2} \left(|y|^{\frac{1}{\lambda_1}} + |t|^{\frac{1}{\lambda_2}}\right)^{-(\beta_1\lambda_1+\beta_2\lambda_2)}\right) \leqslant |y|^{-\mu+\alpha_2\gamma} |t|^{-\frac{1}{\theta'}-\alpha_2}$$

if $0 < \alpha_2\gamma \leqslant \beta_1$ and $\gamma = \lambda_2/\lambda_1$ (here we may assume that $\alpha_2\gamma < 1/q$, where $q < \infty$).

Analogously,

$$h_2(y, t) = |y|^{-\mu-\alpha_1\gamma} |t|^{-\frac{1}{\theta'}+\alpha_1} \left(\frac{|y|^{\gamma}}{|t|}\right)^{\alpha_1-\beta} \left[\frac{|y|^{\gamma}}{|t|}\right]_1^{2\beta}$$

$$\leqslant |y|^{-\mu-\alpha_1\gamma} |t|^{-\frac{1}{\theta'}+\alpha_1}$$

if $0 < \alpha_1 \leqslant \beta$ (we may assume that $\mu + \alpha_1\gamma < 1$), and

$$h_2(y, t) = |y|^{-\mu+\alpha_2\gamma} |t|^{-\frac{1}{\theta'}-\alpha_2} \left(\frac{|y|^{\gamma}}{|t|}\right)^{-(\alpha_2+\beta)} \left[\frac{|y|^{\gamma}}{|t|}\right]_1^{2\beta}$$

$$\leqslant |y|^{-\mu+\alpha_2\gamma} |t|^{-\frac{1}{\theta'}-\alpha_2}$$

if $0 < \alpha_2 \leqslant \beta$ (here, $\alpha_2\gamma < \frac{1}{q}$, where $q < \infty$).

Functions of the type $h_1(y, t)$ include, as examples, the functions

$$h_3(y, t) = |t|^\beta \left(|y|^{\frac{1}{\lambda_1}} + |t|^{\frac{1}{\lambda_2}}\right)^{-\mu\lambda_1 - \left(\frac{1}{\theta'} + \beta\right)\lambda_2}, \quad \beta > 0, \tag{45}$$

and

$$h_4(y, t) = \left(|y|^{\frac{1}{\lambda_1}} + |t|^{\frac{1}{\lambda_2}}\right)^{-\mu\lambda_1 - \frac{1}{\theta'}\lambda_2} \quad \text{if} \quad \theta' < \infty \ (\theta > 1). \tag{46}$$

The function h_3 is obtained from h_1 with $\beta_1 = \mu > 0$ and $\beta_2 = \frac{1}{\theta'} + \beta > 0$; the function h_4 is obtained from h_1 with $\beta_1 = \mu > 0$ and $\beta_2 = \frac{1}{\theta'} > 0$ if $\theta' < \infty$ (that is, if $\theta > 1$).

It follows from what has been said that inequality (41) will hold if we replace the function $h(y, t)$ with the functions $h_i(y, t)$ for $i = 1$, 2, 3, 4.

2.24. Lemma. *Suppose that* $1 < p < q \leqslant \infty$, $p \leqslant \sigma \leqslant \infty$, $\mu = 1 - \frac{1}{p} + \frac{1}{q}$, *and* $f \in L_p(E^1)$. *Define*

$$K_1(y, \tau) = |y|^{-\mu+\beta_1} |\tau|^{-\frac{1}{\sigma} + \beta_2} \left(|y|^{\frac{1}{\lambda_1}} + |\tau|^{\frac{1}{\lambda_2}}\right)^{-(\beta_1\lambda_1 + \beta_2\lambda_2)}$$
$$(\beta_1 > 0, \ \beta_2 > 0, \ \lambda_1 > 0, \ \lambda_2 > 0),$$
$$K_2(y, \tau) = |y|^{-\mu-\beta\gamma} |\tau|^{-\frac{1}{\sigma} + \beta} \left[\frac{|y|^\gamma}{|\tau|}\right]_1^{2\beta}, \quad \beta > 0, \ \gamma > 0.$$

Then, there exists a constant C independent of f such that

$$\left\| \left\| \int_{E^1} f(y) K_i(y - x, \tau) \, dy \right\|_{q, E^1_x} \right\|_{\sigma, E^1_\tau} \leqslant C \| f \|_p. \tag{47}$$

PROOF. Obviously, it will be sufficient to show that the inequality

$$\int_{E^1}\int_{E^1}\int_{E^1} f(y)\,\psi(x, \tau)\,K_i(y-x, \tau)\,dy\,d\tau\,dx \leqslant C\,\|f\|_p\,\|\psi\|_{(q', \sigma')} \tag{48}$$

holds for any function $\psi(x, \tau) \in L_{(q', \sigma')}(E^2)$, where

$$\frac{1}{q}+\frac{1}{q'}=1 \quad \text{and} \quad \frac{1}{\sigma}+\frac{1}{\sigma'}=1.$$

Since the left-hand member of inequality (48) is estimated on the basis of Hölder's inequality in terms of

$$\|f\|_p \left\| \int_{E^1}\int_{E^1} \psi(x, \tau)\,K_i(y-x, \tau)\,dx\,d\tau \right\|_{p'} \qquad \left(\frac{1}{p}+\frac{1}{p'}=1\right),$$

that inequality will be proven if we can show that

$$\left\| \int_{E^1}\int_{E^1} \psi(x, \tau)\,K_i(x-y, \tau)\,dx\,d\tau \right\|_{p'} \leqslant C\,\|\psi\|_{(q', \sigma')}. \tag{49}$$

And this follows immediately from the preceding lemma. To see this, we note that the parameters p', q', σ', and μ satisfy the conditions

$$1 \leqslant q' < p' < \infty, \qquad 1 \leqslant \sigma' \leqslant p',$$
$$\mu = 1 - \frac{1}{p} + \frac{1}{q} = 1 - \frac{1}{q'} + \frac{1}{p'}$$

and that the functions $K_i(x, \tau)$ satisfy inequalities (39) and (40) in which p is replaced with q', q with p', and θ' with σ. The validity of these estimates is shown in the remark 2.23 since, for the values indicated for the parameters, we have $K_i(y, \tau) = h_i(y, \tau)$ for $i = 1, 2$. Thus, the conditions of Lemma 2.22 are satisfied and inequality (49) follows from inequality (41). This completes the proof of the lemma.

§3. Boundedness of the convolution in L_p

In this section, we shall resolve the question of boundedness, in the norm of L_p, of the convolution operator $K * f$ when certain special anisotropic conditions on $K(x)$ are satisfied. We shall show that boundedness of the convolution in L_r for one value of r in $1 < r < \infty$ implies its boundedness in L_p for all p in $1 < p < \infty$ and also its boundedness in $L_{\mathbf{p}}$ for all $\mathbf{p}$ in $\mathbf{1} < \mathbf{p} < \mathbf{\infty}$.

The basic result for the unmixed L_p-norm is contained in Theorem 3.4 with $m = 0$; its generalization to the mixed $L_{\mathbf{p}}$-norm is contained in Theorem 3.5.

The exposition will be made according to a plan close to the one adopted for the isotropic case in the article [1] by Benedek, Calderón, and Panzone. Here, we have, in particular, a generalization of the corresponding results of Kree [1] dealing with the anisotropic case and the spaces L_p.

3.1. Quasilinear operators of weak and strong types. For a measurable function $f(x)$ defined on E^n, we denote by $\mu(f; t)$ the measure of the set of points x for which $|f(x)| > t$:

$$\mu(f; t) = \operatorname{mes}\{x : |f(x)| > t\}.$$

The function $\mu = \mu(f; t)$ is called the *distribution function* for $|f(x)|$. For $1 \leqslant p < \infty$, we have*

$$\int_{E^n} |f(x)|^p \, dx = \int_0^\infty \mu(|f|^p; t)\, dt$$

$$= \int_0^\infty \mu(|f|; t^{1/p})\, dt = p \int_0^\infty t^{p-1} \mu(f; t)\, dt,$$

*The validity of the first equation for functions of bounded support with values in an interval $[a, b] \subset (0, \infty)$ follows from the coincidence of the approximating sums of the first and second integrals. The general case is obtained with the aid of monotonic passages to the limit.

so that

$$\| f \|_p = \left\{ p \int_0^\infty t^{p-1} \mu(f;\ t)\, dt \right\}^{1/p}. \tag{1}$$

We denote by M_p the set of all functions $f(x)$ for which the following quantities are finite:

$$\begin{aligned} \| f \|_{M_p} &= \sup_{0<t<\infty} t\,[\mu(f;\ t)]^{1/p}, \qquad 1 \leqslant p < \infty, \\ \| f \|_{M_\infty} &= \| f \|_{L_\infty} = \operatorname*{ess\,sup}_{E^n} | f(x) |. \end{aligned} \tag{2}$$

If we narrow the interval of integration in (1) to $(0, h)$ and remember that $\mu(f;\ t)$ is monotonic, we see that

$$\| f \|_{M_p} \leqslant \| f \|_p, \qquad 1 \leqslant p < \infty.$$

That the converse does not hold is shown by the function $f(x) = |x|^{-1/p}$.

An operator A defined from one function space into another is said to be *quasilinear* if, for any two functions f_1 and f_2 in its domain of definition, their sum $f_1 + f_2$ is also in its domain of definition and there exists a constant $\varkappa$ independent of f_1 and f_2 such that

$$| A(f_1 + f_2) | \leqslant \varkappa (| Af_1 | + | Af_2 |).$$

If $\varkappa = 1$, the operator A is said to be *sublinear* (that is, convex downward).

We shall say that a quasilinear operator A is *of type* (p, q) or *of strong type* (p, q) if it is defined on $L_p(E^n)$ into $L_q(E^m)$ and there exists a constant K independent of f such that

$$\| Af \|_{q,\, E^m} \leqslant K \| f \|_{p,\, E^n}, \qquad \forall f \in L_p(E^n).$$

If instead of the last inequality the weaker inequality

$$\| Af \|_{M_q} \leqslant K \| f \|_{p,\, E^n}, \qquad \forall f \in L_p(E^n),$$

holds, we shall say that the operator is *of weak type* (p, q). In this case, the smallest value that can serve as K may be called the *weak* (p, q)*-norm* of A.

3.2. Marcinkiewicz's theorem [1]. *Suppose that* $1 \leqslant p_i \leqslant q_i \leqslant \infty$ *(for* $i = 1, 2$*),* $q_1 \neq q_2$, $0 < \tau < 1$, $\frac{1}{p} = \frac{1-\tau}{p_1} + \frac{\tau}{p_2}$, *and* $\frac{1}{q} = \frac{1-\tau}{q_1} + \frac{\tau}{q_2}$. *If a quasilinear operator A is simultaneously of weak type* (p_1, q_1) *and* (p_2, q_2) *with norms* K_1 *and* K_2 *respectively, then the operator A is of strong type* (p, q) *and we have*

$$\| Af \|_{q, E^m} \leqslant M K_1^{1-\tau} K_2^{\tau} \| f \|_{p, E^n},$$

where $M = M(\tau, \varkappa, p_1, q_1, p_2, q_2)$ *is independent of f and, for fixed* p_1, q_1, p_2, q_2, *and* $\varepsilon > 0$, *it is bounded when* $\tau \in [\varepsilon, 1 - \varepsilon]$.

PROOF (Zygmund [1]). We shall systematically use the notation

$$\alpha = \frac{1}{p}, \quad \alpha_1 = \frac{1}{p_1}, \quad \alpha_2 = \frac{1}{p_2}, \quad \beta = \frac{1}{q}, \quad \beta_1 = \frac{1}{q_1}, \quad \beta_2 = \frac{1}{q_2}.$$

In the $\alpha\beta$-plane, the segment connecting the points (α_1, β_1) and (α_2, β_2), where $\beta_1 \neq \beta_2$, lies in the triangle $0 \leqslant \beta \leqslant \alpha \leqslant 1$. The assertion of the theorem is that the operator A is of strong type $\left(\frac{1}{\alpha}, \frac{1}{\beta}\right)$ for any interior point (α, β) of that segment.

We shall prove the theorem only for the case in which $q_1 \neq \infty$ and $q_2 \neq \infty$. We shall assume for definiteness that $0 < \beta_1 < \beta_2 \leqslant 1$.

Case 1: $\alpha_1 < \alpha_2$. (In what follows, we shall use only this case with $p_1 = q_1$ and $p_2 = q_2$.) Suppose that $f \in L_{1/\alpha}$. For fixed $c > 0$, let us define $f_1 = f$ when $|f| \leqslant c$ and $f_1 = e^{i \arg f} c$ elsewhere and then let us define $f_2 = f - f_1$. We have

$$|f_1| = \min(|f|, c), \quad |f| = |f_1| + |f_2|.$$

Since $f \in L_{1/\alpha}$, we have $f_1 \in L_{1/\alpha_1}$ and $f_2 \in L_{1/\alpha_2}$. Consequently, there exist by assumption $g_1 = Af_1$, $g_2 = Af_2$, and $g = Af$ such that $|g| \leqslant \varkappa(|g_1| + |g_2|)$.

This last inequality and the conditions of weak type imply that

$$\mu(g;\ 2\varkappa t) \leqslant \mu(g_1;\ t) + \mu(g_2;\ t) \leqslant K_1^{q_1} t^{-q_1} \| f_1 \|_{p_1}^{q_1} + K_2^{q_2} t^{-q_2} \| f_2 \|_{p_2}^{q_2}. \tag{3}$$

The right-hand member of (3) is dependent on c. The idea of the proof consists in choosing $c > 0$ in the best manner.

Noting that

$$\begin{aligned} &\mu(f_1;\ t) = \mu(f;\ t) && \text{for} \quad 0 < t < c, \\ &\mu(f_1;\ t) = 0 && \text{for} \quad t > c, \\ &\mu(f_2;\ t) = \mu(f;\ t + c) && \text{for} \quad t > 0, \end{aligned}$$

we obtain, on the basis of (3) and (1),

$$\mu(g;\ 2\varkappa t) \leqslant K_1^{q_1} t^{-q_1} p_1^{q_1/p_1} \left\{ \int_0^c h^{p_1 - 1} \mu(f;\ h)\, dh \right\}^{q_1/p_1}$$
$$+ K_2^{q_2} t^{-q_2} p_2^{q_2/p_2} \left\{ \int_c^\infty (h - c)^{p_2 - 1} \mu(f;\ h)\, dh \right\}^{q_2/p_2}.$$

Taking $c = \left(\frac{t}{a}\right)^{1/\xi}$, where $a > 0$ and $\xi \neq 0$, we obtain, by virtue of (1) and this last estimate,

$$\| g \|_q^q = q \int_0^\infty t^{q-1} \mu(g;\ t)\, dt = (2\varkappa)^q q \int_0^\infty t^{q-1} \mu(g;\ 2\varkappa t)\, dt$$

$$\leqslant (2\varkappa)^q q \left\{ K_1^{q_1} p_1^{q_1/p_1} \int_0^\infty t^{q-1-q_1} \left[\int_0^{(t/a)^{1/\xi}} h^{p_1 - 1} \mu(f;\ h)\, dh \right]^{q_1/p_1} dt \right.$$
$$\left. + K_2^{q_2} p_2^{q_2/p_2} \int_0^\infty t^{q-1-q_2} \left[\int_{(t/a)^{1/\xi}}^\infty h^{p_2 - 1} \mu(f;\ h)\, dh \right]^{q_2/p_2} dt \right\}.$$

Let us use Minkowski's inequality for the integrals 2(10) to estimate the right-hand member:

$$\left\{\int_0^\infty t^{q-1-q_1}\left[\int_0^{(t/a)^{1/\xi}} h^{p_1-1}\mu(f;\,h)\,dh\right]^{q_1/p_1} dt\right\}^{p_1/q_1}$$

$$\leqslant \int_0^\infty h^{p_1-1}\mu(f;\,h)\left\{\int_{ah^\xi}^\infty t^{q-1-q_1}\,dt\right\}^{p_1/q_1} dh$$

$$=\left(\frac{1}{q_1-q}\right)^{p_1/q_1} a^{(q-q_1)\frac{p_1}{q_1}}\int_0^\infty h^{p_1-1+\xi p_1\frac{q-q_1}{q_1}}\mu(f;\,h)\,dh, \quad (4)$$

$$\left\{\int_0^\infty t^{q-1-q_1}\left[\int_{(t/a)^{1/\xi}}^\infty h^{p_2-1}\mu(f;\,h)\,dh\right]^{q_2/p_2} dt\right\}^{p_2/q_2}$$

$$\leqslant \int_0^\infty h^{p_2-1}\mu(f;\,h)\left\{\int_0^{ah^\xi} t^{q-1-q_2}\,dt\right\}^{p_2/q_2} dh \leqslant$$

$$=\left(\frac{1}{q-q_2}\right)^{p_2/q_2} a^{p_2\frac{q-q_2}{q_2}}\int_0^\infty h^{p_2-1+\xi p_2\frac{q-q_2}{q_2}}\mu(f;\,h)\,dh. \quad (5)$$

Let us choose ξ so that the exponents of h will be equal to $p-1$ in the last integrals of both (4) and (5). This will be the case if we choose

$$\xi=\frac{\beta(\alpha-\alpha_1)}{\alpha(\beta-\beta_1)}=\frac{\beta(\alpha-\alpha_2)}{\alpha(\beta-\beta_2)}=\frac{q_1(p_1-p)}{p_1(q_1-q)}=\frac{q_2(p_2-p)}{p_2(q_2-q)}.$$

Combining the results, we obtain

$$\|g\|_q^q\leqslant(2\varkappa)^q\,q\left\{K_1^{q_1}\left(\frac{p_1}{p}\right)^{\frac{q_1}{p_1}}\left(\frac{1}{q_1-q}\right)a^{q-q_1}\|f\|_p^{p\frac{q_1}{p_1}}\right.$$

$$\left.+K_2^{q_2}\left(\frac{p_2}{p}\right)^{\frac{q_2}{p_2}}\left(\frac{1}{q-q_2}\right)a^{q-q_2}\|f\|_p^{p\frac{q_2}{p_2}}\right\}. \quad (6)$$

When we take

$$a = K_1^{\frac{q_1}{q_1-q_2}} K_2^{\frac{q_2}{q_2-q_1}} \| f \|_p^{p\frac{\frac{q_2}{p_2}-\frac{q_1}{p_1}}{q_2-q_1}}$$

in both terms of the right-hand member of (6), the exponents of K_1, K_2, and $\| f \|_p$ will coincide and we get the assertion of the theorem with

$$M^q = (2\varkappa)^q \, q \left\{ \frac{\left(\frac{p_1}{p}\right)^{q_1/p_1}}{q_1 - q} + \frac{\left(\frac{p_2}{p}\right)^{q_2/p_2}}{q - q_2} \right\}.$$

Case 2: $\alpha_1 > \alpha_2$. This case is treated in an analogous fashion and we get the assertion of the theorem with the same constant M as for Case 1.

Case 3: $\alpha_1 = \alpha_2$. We have

$$\| g \|_q^q = (2\varkappa)^q \, q \left\{ \int_0^c + \int_c^\infty \right\} t^{q-1} \mu\,(g;\; 2\varkappa t)\, dt$$

$$\leqslant (2\varkappa)^q \, q \left\{ (\| g \|_{M_{q_2}})^{q_2} \int_0^c t^{q-q_2-1}\, dt + (\| g \|_{M_{q_1}})^{q_1} \int_c^\infty t^{q-q_1-1}\, dt \right\}$$

$$= (2\varkappa)^q \, q \left\{ \frac{1}{q - q_2} \| g \|_{M_{q_2}}^{q_2} c^{q-q_2} + \frac{1}{q_1 - q} \| g \|_{M_{q_1}}^{q_1} c^{q-q_1} \right\}.$$

If we take

$$c = (\| g \|_{M_{q_1}})^{\frac{q_1}{q_1-q_2}} (\| g \|_{M_{q_2}})^{\frac{q_2}{q_2-q_1}},$$

we obtain the assertion of the theorem with

$$M^q = (2\varkappa)^q \, q \left\{ \frac{1}{q_1 - q} + \frac{1}{q - q_2} \right\}.$$

This completes the proof.

3.3. REMARK. Everywhere in subsections 3.1 and 3.2, we may assume that the operator A is defined not on the entire space $L_p(E^n)$ but only on the space $\mathring{L}_\infty(E^n)$ of measurable essentially bounded functions of bounded support. Then, we need to change the definitions of an operator of the strong type and an operator of the weak type, assuming the operator to be defined on $\mathring{L}_\infty(E^n)$ and the corresponding inequalities to be satisfied for all $f \in \mathring{L}_\infty(E^n)$. Marcinkiewicz's Theorem 3.2 remains valid since we use in its proof only the decomposition $f = f_1 + f_2$, which is meaningful also in $\mathring{L}_\infty(E^n)$, and the estimates for Af_1 and Af_2.

3.4. Throughout this subsection, we shall use the following notation and conditions:

$$x = (x_1, \ldots, x_n) = (\bar{x}, \bar{\bar{x}}) \in E^n, \quad \bar{x} = (x_1, \ldots, x_m),$$
$$\bar{\bar{x}} = (x_{m+1}, \ldots, x_n), \quad 0 \leqslant m \leqslant n-1,$$
$$1 < \bar{p} = (p_1, \ldots, p_m) < \infty, \quad 1 \leqslant p < \infty.$$

For $f(\bar{x}, \bar{\bar{x}})$, we obtain

$$\| f \|_{\bar{p}, p} = \left\{ \int \| f(\cdot, \bar{\bar{x}}) \|_{\bar{p}}^p \, d\bar{\bar{x}} \right\}^{1/p},$$

and we denote the corresponding space of functions by $L_{\bar{p}, p}$.

In the case $m = 0$,

$$\| f(\cdot, \bar{\bar{x}}) \|_{\bar{p}} = | f(\bar{\bar{x}}) |$$

and $\| f \|_{\bar{p}, p} = \| f \|_p$ coincides with the usual norm in $L_p(E^n)$.

We shall also assume only that the convolution is defined on $\mathring{L}_\infty(E^n)$, so that the functions $f(x) = f(\bar{x}, \bar{\bar{x}})$ in question belong to $\mathring{L}_\infty(E^n)$; that is, they are *measurable, essentially bounded, and of bounded support*. Also, for $g = g(\bar{\bar{x}})$, let us assume that

$$\mu(g; t) = \operatorname{mes} \{\bar{\bar{x}} : | g(\bar{\bar{x}}) | > t\}, \quad \pi[\bar{\bar{x}}] = \max_{m+1 \leqslant i \leqslant n} | x_i |^{1/\lambda_i}, \quad \lambda_i > 0.$$

Theorem. *Let A denote the operator defined by*

$$Af = \int\int K(\bar{x}-\bar{y}, \bar{\bar{x}}-\bar{\bar{y}}) f(\bar{y}, \bar{\bar{y}})\, d\bar{y}\, d\bar{\bar{y}},$$

$$K(x) = K(\bar{x}, \bar{\bar{x}}) \in L^{\text{loc}}(E^n).$$

Suppose that the following conditions are satisfied:

1°. *For some r in* $1 < r < \infty$,

$$\| Af \|_{\bar{p}, r} \leqslant c_r \| f \|_{\bar{p}, r}.$$

2°. *For some* $N > 0$ *and all* $t > 0$,

$$\int\limits_{\pi[\bar{\bar{x}}] > Nt} \int | K(\bar{x}, \bar{\bar{x}}-\bar{\bar{y}}) - K(\bar{x}, \bar{\bar{x}}) |\, d\bar{x}\, d\bar{\bar{x}} \leqslant c \quad \text{if} \quad \pi[\bar{\bar{y}}] < t.$$

Then, for all p in $1 < p < \infty$ *there exists a constant* c_p *depending on* c_r, c, N, r, *and p such that*

$$\| Af \|_{\bar{p}, p} \leqslant c_p \| f \|_{\bar{p}, p}.$$

The proof of this theorem is based on a number of lemmas and will be given below.

Theorem 3.5. *Let A denote the operator defined by*

$$Af = \int K(x-y) f(y)\, dy, \qquad K(x) \in L^{\text{loc}}(E^n).$$

Suppose that the following conditions are satisfied:

1°. *For some r in* $1 < r < \infty$,

$$\| Af \|_r \leqslant c_r \| f \|_r.$$

2°. *For some* $N > 0$ *and all* $t > 0$,

$$\int\limits_{\pi[x] > Nt} | K(x-y) - K(x) |\, dx \leqslant c \quad \text{if} \quad \pi[y] < t.$$

Then, for all p *in* $1 < p < \infty$,

$$\| Af \|_p \leqslant c_p \| f \|_p .$$

PROOF. Theorem 3.5 follows easily from Theorem 3.4. We note that condition 2° implies for $\pi[\bar{\bar{y}}] < t$ that

$$\int\limits_{\pi[\bar{\bar{x}}] > Nt} \int | K(\bar{x}, \bar{\bar{x}} - \bar{\bar{y}}) - K(\bar{x}, \bar{\bar{x}}) | \, d\bar{x} \, d\bar{\bar{x}}$$

$$\leqslant \int\limits_{\pi[x] > Nt} | K(\bar{x}, \bar{\bar{x}} - \bar{\bar{y}}) - K(\bar{x}, \bar{\bar{x}}) | \, dx \leqslant c_1. \quad (7)$$

By virtue of Theorem 3.4, the assertion of Theorem 3.5 holds for $p = (p_1, \ldots, p_1)$. By virtue of inequality (7) and Theorem 3.4 with $m = 1$, the assertion of Theorem 3.5 holds for $p = (p_1, p_2, \ldots, p_2)$. By virtue of inequality (7) and Theorem 3.4 with $m = 2$, the assertion of Theorem 3.5 holds for $p = (p_1, p_2, p_3, \ldots, p_3)$. Proceeding in this way, we get the assertion of Theorem 3.5 in the general case.

3.6. A covering lemma.* *Suppose that*

$$u(x) \in L_1(E^n), \quad \lambda = (\lambda_1, \ldots, \lambda_n), \quad \lambda_i > 0, \quad t > 0.$$

Then, there exists a number $\lambda_0 \geqslant 1$ *(depending only on* $\lambda_1, \ldots, \lambda_n$*) and a sequence of disjoint parallelepipeds* Q_k *of the form*

$$\{x: |x_i - x_i^0| < b_i, \quad 0 < a^{\lambda_i} \leqslant b_i \leqslant (\lambda_0 a)^{\lambda_i}, \ i = 1, \ldots, n\}$$

such that

$$t \leqslant | Q_k |^{-1} \int\limits_{Q_k} | u(x) | \, dx \leqslant 2^{|\lambda|} \lambda_0^{|\lambda|} t \qquad \left(|\lambda| = \sum_1^n \lambda_i \right),$$

*F. Riesz proved this for the one-dimensional case, Zygmund and Calderón [1] for the isotropic case, and Jones [1] for a special anisotropic case.

$|u(x)| \leqslant t$ *almost everywhere outside* $\bigcup Q_k$.

PROOF. For given $\lambda = (\lambda_1, \ldots, \lambda_n)$, we shall refer to a parallelepiped of the form

$$\{x: |x_i - x_i^0| < a^{\lambda_i} \ (i = 1, \ldots, n)\} = \{x: \pi[x - x^0] < a\}$$

as a proper parallelepiped. Let us partition the space E^n into a grid of equal (proper) parallelepipeds $\{Q_{0j}\}$ with edges of half-length a^{λ_i} (for $i = 1, \ldots, n$). We shall refer to these as parallelepipeds of zeroth rank. Let us define* $s_{1i} = [2^{\lambda_i}] \leqslant 2^{\lambda_i}$. We now partition the ith edge into s_{1i} equal segments and then partition the parallelepipeds of zeroth rank into $s_{11}s_{12} \ldots s_{1n}$ smaller parallelepipeds. We shall refer to these as parallelepipeds of first rank.

We then partition each of these parallelepipeds of first rank into $s_{21}s_{22} \ldots s_{2n}$ parallelepipeds of second rank. Proceeding in this way, we obtain parallelepipeds of kth rank the edges of which have half-length

$$\frac{a^{\lambda_i}}{s_{1i}s_{2i} \ldots s_{ki}},$$

where

$$s_{ki} = \left[\frac{2^{k\lambda_i}}{s_{1i} \ldots s_{k-1\,i}}\right].$$

Since $\dfrac{2^{k\lambda_i}}{s_{1i} \ldots s_{ki}} \geqslant 1$, it follows that $s_{k+1\,i} \geqslant [2^{\lambda_i}] \geqslant 1$ and since there are arbitrarily large values of k such that $s_{ki} > 1$, the partitioning process is well defined.

The following inequalities for the half-lengths of the edges are obvious:

*The important case in our subsequent applications will be the case of rational λ_i. Then, instead of 2^{λ_i}, we can take integers $b^{\lambda_i} > 1$, which simplifies our subsequent constructions.

$$\frac{a^{\lambda_i}}{s_{1i} \ldots s_{ki}} \geqslant \left(\frac{a}{2^k}\right)^{\lambda_i},$$

$$\frac{a^{\lambda_i}}{s_{1i} \ldots s_{ki}} \leqslant \frac{a^{\lambda_i}}{s_{1i} \ldots s_{k-1\,i}(s_{ki}+1)} \left(\frac{s_{ki}+1}{s_{ki}}\right)$$

$$\leqslant \left(\frac{a}{2^k}\right)^{\lambda_i} 2 \leqslant \left(\lambda_0 \frac{a}{2^k}\right)^{\lambda_i}, \quad \min_i \lambda_0^{\lambda_i} = 2.$$

Thus, every parallelepiped Q of kth rank contains a concentric proper parallelepiped Q^* of volume $2^n (a/2^k)^{|\lambda|}$ and is in turn contained in a concentric proper parallelepiped Q^{**} of volume $|Q^{**}| = \lambda_0^{|\lambda|} |Q^*|$.

Let us now suppose that $a > 0$ is chosen by the condition $2^n a^{|\lambda|} > t^{-1} \int\limits_{E^n} |u|\, dx$, so that the mean value of $|u|$ in each parallelepiped of zeroth rank is less than t. Consider the parallelepipeds Q_{11}, Q_{12}, Q_{13}, ... of first rank in which the mean value of $|u|$ is no less than t.

Then, consider the parallelepipeds Q_{21}, Q_{22}, Q_{23}, ... of second rank for which the mean value of $|u|$ is no less than t and which are not covered by the preceding ones. Continuing the process, let us consider the parallelepipeds Q_{k1}, Q_{k2}, Q_{k3}, ... of rank k for which the mean value of $|u|$ is no less than t and which have not already been considered.

Suppose that Q_{kj} is contained in a parallelepiped $Q_{k-1(j)}$ of rank $k-1$ not yet considered. Then,

$$t|Q_{kj}| \leqslant \int\limits_{Q_k} |u|\, dx \leqslant \int\limits_{Q_{k-1\,(j)}} |u|\, dx \leqslant t|Q_{k-1\,(j)}|$$

$$\leqslant t \frac{|Q^{**}_{k-1\,(j)}|}{|Q^*_{kj}|} |Q_{kj}| \leqslant t\lambda_0^{|\lambda|} 2^{|\lambda|} |Q_{kj}|.$$

Let us arrange in a sequence $\{Q_h\}$ the parallelepipeds that we have considered. For almost all $x \notin \bigcup Q_h$, there exist parallelepipeds of all ranks with mean value of $|u| < t$ to which x belongs. Since $u(x) \in L_1 \subset L^{\mathrm{loc}}$, it follows from Theorem 1.7 that $|u(x)| \leqslant t$ at every Lebesgue point of the set $E^n \setminus \overline{\bigcup Q_k}$, that is, almost everywhere on that set. This completes the proof of the lemma.

3.7. Lemma. *Suppose that condition 2° of Theorem 3.4 is satisfied for $K(\bar{x}, \bar{\bar{x}}) \in L^{\mathrm{loc}}(E^n)$. Then, the inequality*

$$\int_{\pi[\bar{\bar{x}}-\bar{\bar{x}}^0]>Nt} \| Af \|_{\bar{p}}\, d\bar{\bar{x}} \leqslant c \| f \|_{\bar{p},1}$$

holds for all functions $f(\bar{x}, \bar{\bar{x}})$ that are concentrated in $\pi[\bar{\bar{x}}-\bar{\bar{x}}^0] < t$ and for which $\int f(\bar{x}, \bar{\bar{x}})\, d\bar{\bar{x}} = 0$ for all $\bar{x}$. Here, the constant c is the same as in condition 2° of Theorem 3.4.

PROOF.

$$\int_{\pi[\bar{\bar{x}}-\bar{\bar{x}}^0]>Nt} \| Af \|_{\bar{p}}\, dx$$

$$= \int_{\pi[\bar{\bar{x}}-\bar{\bar{x}}^0]>Nt} \left\| \int\int K(\cdot - \bar{y}, \bar{\bar{x}} - \bar{\bar{y}}) f(\bar{y}, \bar{\bar{y}})\, d\bar{y}\, d\bar{\bar{y}} \right\|_{\bar{p}} d\bar{\bar{x}}$$

$$= \int_{\pi[\bar{\bar{x}}]>Nt} \left\| \int\int [K(\cdot - \bar{y}, \bar{\bar{x}}-\bar{\bar{y}}) - K(\cdot - \bar{y}, \bar{\bar{x}})] f(\bar{y}, \bar{\bar{y}}-\bar{\bar{x}}^0)\, d\bar{y}\, d\bar{\bar{y}} \right\|_{\bar{p}} d\bar{\bar{x}}$$

$$\leqslant \int_{\pi[\bar{\bar{y}}]<t} \| f(\cdot, \bar{\bar{y}} - \bar{\bar{x}}^0) \|_{\bar{p}} \int_{\pi[\bar{\bar{x}}]>Nt} |K(\bar{x}, \bar{\bar{x}}-\bar{\bar{y}}) - K(\bar{x}, \bar{\bar{x}})|\, d\bar{x}\, d\bar{\bar{x}}\, d\bar{\bar{y}}$$

$$\leqslant c \| f \|_{\bar{p},1}.$$

3.8. Theorem. *Suppose that the following two conditions hold for all $t > 0$:*

1°. $$\mu(\| Af \|_{\bar{p}};\, t) \leqslant c t^{-r} \| f \|_{\bar{p},r}^{r}$$

for some r in $1 < r < \infty$.

2°. *For a function* $f(\bar{x}, \bar{\bar{x}})$ *concentrated in* $\pi[\bar{\bar{x}} - \bar{\bar{x}}^0] < t$ *such that* $\int f(\bar{x}, \bar{\bar{x}})\, d\bar{\bar{x}} = 0$ *for all* $\bar{x}$, *we have*

$$\int\limits_{\pi[\bar{\bar{x}}-\bar{\bar{x}}^0] > Nt} \| Af \|_{\bar{p}}\, dx \leqslant c_1 \int \| f(\cdot\,, \bar{\bar{x}}) \|_{\bar{p}}\, d\bar{\bar{x}} = c_1 \|f\|_{\bar{p},\,1}.$$

Then, for every p *in* $1 < p < r$, *there exists a constant* c_p *depending on* c, c_1, N, r, *and* p *such that*

$$\| Af \|_{\bar{p},\,p} \leqslant c_p \| f \|_{\bar{p},\,p}.$$

PROOF. Let us fix $s > 0$. Let $Q_k^* \subset Q_k \subset Q_k^{**}$ (for $k = 1, 2, \ldots$) denote parallelepipeds constructed in E^{n-m} for $\| f(\cdot\,, \bar{\bar{x}}) \|_{\bar{p}}$ in accordance with the covering Lemma, 3.6. Then,

$$f = g + h = g + \sum_1^\infty h_k,$$

where

$$g(\bar{x}, \bar{\bar{x}}) = f(\bar{x}, \bar{\bar{x}}) \qquad \text{for} \quad \bar{\bar{x}} \in E^{n-m} \setminus \bigcup Q_k,$$

$$g(\bar{x}, \bar{\bar{x}}) = | Q_k |^{-1} \int\limits_{Q_k} f(\bar{x}, \bar{\bar{y}})\, d\bar{\bar{y}} \quad \text{for} \quad \bar{\bar{x}} \in Q_k,$$

$$h_k(\bar{x}, \bar{\bar{x}}) = f(\bar{x}, \bar{\bar{x}}) - g(\bar{x}, \bar{\bar{x}}) \quad \text{for} \quad \bar{\bar{x}} \in Q_k,$$

$$h_k(\bar{x}, \bar{\bar{x}}) = 0 \qquad \text{for} \quad \bar{\bar{x}} \notin Q_k \subset Q_k^{**}.$$

We note also that, since f is of compact support, the supports of g and all the h_k are contained in some single compact set.

Thus

$$\| g(\cdot\,, \bar{\bar{x}}) \|_{\bar{p}} \leqslant 2^{|\lambda|} \lambda_0^{|\lambda|} t \qquad \text{for almost all} \quad \bar{\bar{x}},$$

$$\| g \|_{\bar{p},\,1} \leqslant \| f \|_{\bar{p},\,1}, \qquad \left\| \sum_1^\infty h_k \right\|_{\bar{p},\,1} \leqslant 2 \| f \|_{\bar{p},\,1}.$$

Also (using the first condition of the theorem),

$$\mu\left(\|Ag\|_{\bar{p}}, \frac{t}{2}\right) \leqslant c2^r t^{-r} \|g\|^r_{\bar{p},r} \leqslant c2^r t^{-r} \int (2^{|\lambda|}\lambda_0^{|\lambda|} t)^{r-1} \|g(\cdot, \bar{\bar{x}})\|_{\bar{p}} d\bar{\bar{x}}$$

$$\leqslant c' t^{-1} \| g \|_{\bar{p},1} \leqslant c_1' t^{-1} \| f \|_{\bar{p},1}.$$

Since $\sum_1^K h_k \to h$ as $K \to \infty$ in the norm $\| \cdot \|_{\bar{p},r}$, we conclude on the basis of condition 1° of the theorem that

$$\left\| Ah - A\left(\sum_1^K h_k\right)\right\|_{\bar{p}} \to 0$$

with respect to measure in E^{n-m}. Therefore, for almost all $\bar{\bar{x}}$,

$$\| Ah(\cdot, \bar{\bar{x}}) \|_{\bar{p}} \leqslant \sum_1^\infty \| Ah_k(\cdot, \bar{\bar{x}}) \|_{\bar{p}}.$$

For a parallelepiped $Q_k = \{\bar{\bar{x}}: \pi[\bar{\bar{x}} - \bar{\bar{x}}^0] < s\}$, we denote by $\tilde{Q}_k$ the parallelepiped $\tilde{Q}_k = \{x: \pi[x - x^0] < Ns\}$. By virtue of 2°,

$$\int_{E^{n-m} \setminus \cup \tilde{Q}_j^{**}} \| Ah \|_{\bar{p}} d\bar{\bar{x}} \leqslant \sum \int_{E^{n-m} \setminus \cup \tilde{Q}_j^{**}} \| Ah_k \|_{\bar{p}} d\bar{\bar{x}}$$

$$\leqslant \sum \int_{E^{n-m} \setminus \tilde{Q}_k^{**}} \| Ah_k \|_{\bar{p}} d\bar{\bar{x}} \leqslant c_1 \sum \| h_k \|_{\bar{p},1} = c_1 \| h \|_{\bar{p},1}.$$

Consequently,

$$\mu\left(\| Ah \|_{\bar{p}}, \frac{t}{2}\right) \leqslant | \cup \tilde{Q}_k^{**} | + 2t^{-1} \int_{E^{n-m} \setminus \cup \tilde{Q}_j^{**}} \| Ah \|_{\bar{p}} d\bar{\bar{x}}$$

$$\leqslant N^{n-m} \lambda_0^{|\lambda|} | \cup Q_j^* | + 2t^{-1} c_1 \| h \|_{\bar{p},1}.$$

Since $\| h \|_{\bar{p},1} \leqslant 2 \| f \|_{\bar{p},1}$ and

$$|\cup Q_i^*| \leqslant \sum |Q_i| \leqslant \sum t^{-1} \int_{Q_j} \|f(\cdot, \bar{\bar{x}})\|_{\bar{p}}\, d\bar{\bar{x}} \leqslant t^{-1} \|f\|_{\bar{p}, 1},$$

we get

$$\mu\left(\|Ah\|_{\bar{p}}, \frac{t}{2}\right) \leqslant c_2 t^{-1} \|f\|_{\bar{p}, 1}.$$

Consequently,

$$\mu(\|Af\|_{\bar{p}}, t) = \mu(\|Ag + Ah\|_{\bar{p}}, t)$$
$$\leqslant \mu\left(\|Ag\|_{\bar{p}}, \frac{t}{2}\right) + \mu\left(\|Ah\|_{\bar{p}}, \frac{t}{2}\right) \leqslant c_3 t^{-1} \|f\|_{\bar{p}, 1},$$

where c_3 is a constant depending on c, c_1, N, and r.

This inequality and condition 1° of the theorem lead to the assertion of the theorem by virtue of the following lemma:

3.9. Lemma. *Suppose that a linear operator A satisfies the following conditions for some r in $1 < r < \infty$ and all $t > 0$:*

$$\mu(\|Af\|_{\bar{p}}, t) \leqslant c_1 t^{-1} \|f\|_{\bar{p}, 1},$$
$$\mu(\|Af\|_{\bar{p}}, t) \leqslant c t^{-r} \|f\|_{\bar{p}, r}^{r}.$$

Then, for all p in $1 < p < r$, there exists a number c_p dependent on c_1, c, and p such that

$$\|Af\|_{\bar{p}, p} \leqslant c_p \|f\|_{\bar{p}, p}.$$

PROOF. Let us define

$$\varphi(x) = f(x) \|f(\cdot, \bar{\bar{x}})\|_{\bar{p}}^{-1} \quad \text{for} \quad \|f(\cdot, \bar{\bar{x}})\|_{\bar{p}} \neq 0,$$
$$\varphi(x) = 0 \quad \text{for} \quad \|f(\cdot, \bar{\bar{x}})\|_{\bar{p}} = 0.$$

Consider the operator

$$(Bg)(\overline{\overline{x}}) = \| A[g(\overline{\overline{x}})\varphi(\cdot, \overline{\overline{x}})] \|_{\overline{p}},$$

defined on the set of functions $g(\overline{\overline{x}}) \in \overset{\circ}{L}_{\infty}(E^{n-m})$. Obviously, the operator B is sublinear (see subsections 3.1 and 3.3) and

$$\mu(Bg, t) \leqslant c_1 t^{-1} \| g \|_1, \quad \mu(Bg, t) \leqslant c t^{-r} \| g \|_r^r.$$

By virtue of Marcinkiewicz's theorem 3.2 and the remark 3.3, there exists a constant c_p depending only on c_1, c, and p such that

$$\| Bg \|_p \leqslant c_p \| g \|_p.$$

If we take $g(\overline{\overline{x}}) = \| f(\cdot, \overline{\overline{x}}) \|_{\overline{p}}$, we get the assertion of the lemma.

3.10. Proof of Theorem 3.4. Suppose that the operator

$$Af = \int \int K(\bar{x} - \bar{y}, \overline{\overline{x}} - \overline{\overline{y}}) f(\bar{y}, \overline{\overline{y}})\, d\bar{y}\, d\overline{\overline{y}}, \quad f \in \overset{\circ}{L}_{\infty}(E^n),$$

satisfies the conditions of Theorem 3.4. Let us define the operator

$$A^* g = \int \int K(\bar{y} - \bar{x}, \overline{\overline{y}} - \overline{\overline{x}}) g(\bar{y}, \overline{\overline{y}})\, d\bar{y}\, d\overline{\overline{y}}, \quad g \in \overset{\circ}{L}_{\infty}(E^n).$$

Since the inequality $\left| \int f(x) g(x)\, dx \right| \leqslant C \| f \|_{\overline{p}, r}$, which holds for all $f \in \overset{\circ}{L}_{\infty}(E^n)$, implies (see 2.6) that $g \in L_{\overline{p}', r'}$, and $\| g \|_{\overline{p}', r'} \leqslant C$, where

$$\frac{1}{\overline{p}} + \frac{1}{\overline{p}'} = 1, \quad \frac{1}{r} + \frac{1}{r'} = 1,$$

we easily see that the dual inequalities

$$\| Af \|_{\overline{p}, r} \leqslant c_r \| f \|_{\overline{p}, r}, \quad \| A^* g \|_{\overline{p}', r'} \leqslant c_r \| g \|_{\overline{p}', r'}$$

hold simultaneously. To see this, suppose, for example, that

$\| Af \|_{\bar{p}, r} \leqslant c_r \| f \|_{\bar{p}, r}$ and $g \in L_{\bar{p}', r'}$. Then,

$$\left| \int f A^* g \, dx \right| = \left| \int g A f \, dx \right| \leqslant \| Af \|_{\bar{p}, r} \| g \|_{\bar{p}', r'} \leqslant c_r \| f \|_{\bar{p}, r} \| g \|_{\bar{p}', r'},$$

from which the assertion follows.

Thus, the operator A^* satisfies the conditions of Theorem 3.4 with $\bar{p}$ and r replaced with $\bar{p}'$ and r'.

By virtue of Lemma 3.7, the operator A satisfies the conditions of Theorem 3.8 and the operator A^* satisfies the conditions of Theorem 3.8 with $\bar{p}$ and r replaced with $\bar{p}'$ and r'. Therefore, we have

$$\| A^* g \|_{\bar{p}', q} \leqslant c_q \| g \|_{\bar{p}', q}$$

for all q in $1 < q < r'$. By virtue of duality and Theorem 3.8, we have

$$\| Af \|_{\bar{p}, s} \leqslant c_s \| f \|_{\bar{p}, s}$$

for all s in $1 < s < q'$. Since q' can be taken arbitrarily large, this last inequality coincides with the assertion of Theorem 3.4.

3.11. Consider a convolution operator of the form

$$A_n f = \int_{-\infty}^{\infty} K(x_n - y_n) f(\bar{x}, y_n) \, dy_n, \quad K(u) \in L^{\text{loc}}(E^1), \qquad (8)$$

where $x = (x_1, \ldots, x_{n-1}, x_n) = (\bar{x}, x_n)$. In §15, we shall use a result dealing with the boundedness of such an operator in $L_{\boldsymbol{p}}(E^n)$, $\boldsymbol{p} = (p_1, \ldots, p_{n-1}, p_n) = (\bar{\boldsymbol{p}}, p_n)$. We shall now treat this result. Since the proof in many ways repeats the proofs given above, we shall confine ourselves at times to references to analogous situations.

Lemma. *Suppose that there exist a* $\bar{\boldsymbol{p}}$ *in* $1 < \bar{\boldsymbol{p}} < \infty$, *an* r *in* $1 < r < \infty$, *and an* $N > 0$ *such that, for all* $t > 0$,

1°. $\mu(\| A_n f \|_{\bar{p}}; t) \leqslant c t^{-r} \| f \|_{\bar{p}, r}^r$.

$$2^\circ \quad \int_{-Nt}^{Nt} | K(u-v) - K(u) | du \leqslant c \text{ if } |v| < t.$$

Then, for all $\mathbf{p}$ *such that* $1 < p_n < r$,

$$\| A_n f \|_{\mathbf{p}} \leqslant c_{\mathbf{p}} \| f \|_{\mathbf{p}}, \tag{9}$$

where $c_{\mathbf{p}}$ *is a constant that depends on* c, $\mathbf{p}$, r, *and* N.

PROOF. It follows from condition 2° that the inequality

$$\int_{|x_n - \overset{\circ}{x}_n| > Nt} \| A_n f \|_{\bar{p}} \, dx_n \leqslant c \| f \|_{\bar{\mathbf{p}}, 1} \tag{10}$$

holds with the same constant as in 2° for all t and all $f(\bar{x}, x_n)$ concentrated in the strip $|x_n - \overset{\circ}{x}_n| < t$ such that $\int f(\bar{x}, x_n) \, dx_n = 0$ for almost all $\bar{x}$. Specifically,

$$\int_{|x_n - \overset{\circ}{x}_n| > Nt} \| A_n f \|_{\bar{p}} dx_n$$

$$= \int_{|x_n - \overset{\circ}{x}_n| > Nt} \left\| \int K(x_n - y_n) f(\cdot, y_n) \, dy_n \right\|_{\bar{p}} dx_n$$

$$= \int_{|x_n| > Nt} \left\| \int [K(x_n - y_n) - K(x_n)] f(\cdot, y_n - \overset{\circ}{x}_n) \, dy_n \right\|_{\bar{p}} dx_n \leqslant$$

$$\int_{|y_n| < t} \| f(\cdot, y_n - \overset{\circ}{x}_n) \|_{\bar{p}} \int_{|x_n| > Nt} | K(x_n - y_n) - K(x_n) | dx_n \leqslant c \| f \|_{\bar{\mathbf{p}}, 1}.$$

It now follows from condition 1° and inequality (10) on the basis of Theorem 3.8 that inequality (9) holds if $1 < p_n < r$.

3.12. Lemma. *Suppose that condition* 2° *of Lemma* 3.11 *holds for the convolution operator* $A_n f$ *defined by* (8) *and that, for some* $(\bar{\mathbf{p}}, r)$ *in* $\mathbf{1} < (\bar{\mathbf{p}}, r) < \infty$,

$$\| A_n f \|_{\bar{\mathbf{p}}, r} \leqslant c \| f \|_{\bar{\mathbf{p}}, r}.$$

Then, there exists a constant $\tilde{c}_{\mathbf{p}}$ *depending on* $\mathbf{p}$, *c, N, and r such that the inequality*

$$\| A_n f \|_{\mathbf{p}} \leqslant \tilde{c}_{\mathbf{p}} \| f \|_{\mathbf{p}}$$

holds for all $\mathbf{p}$ *such that* $1 < p_n < \infty$.

This is proven by shifting to the adjoint operator in a manner analogous to what was done in 3.10 in the proof of Theorem 3.4.

3.13. Theorem. *Suppose that the kernel* $K(u)$ *of the operator* (8) *is such that*

1° *for some r in* $1 < r < \infty$ *and* $n = 1$,

$$\| A_1 \varphi \|_r \leqslant c_r \| \varphi \|_r, \qquad \varphi \in \overset{\circ}{L}_\infty (E^1),$$

and

2° *for some* $N > 0$ *and all* $t > 0$,

$$\int_{-Nt}^{Nt} | K(u-v) - K(u) | \, du \leqslant c \qquad \text{if} \qquad |v| < t.$$

Then, for $\mathbf{p}$ *in* $\mathbf{1} < \mathbf{p} < \infty$ *and* $n = 1, 2, \ldots,$ *there exists a constant* $c_{n,\,\mathbf{p}}$ *depending on r and N such that*

$$\| A_n f \|_{\mathbf{p}} \leqslant c_{n,\,\mathbf{p}} \| f \|_{\mathbf{p}}, \qquad f \in \overset{\circ}{L}_\infty (E^n). \tag{11}$$

PROOF. We shall prove this by mathematical induction on the dimension n of the space. For $n = 1$, the assertion of the theorem is valid by virtue of Theorems 3.4 and 3.5. Suppose that the theorem has been proven for E^{n-1}, where $n > 1$. Then, (11) holds in E^n for those $\mathbf{p}$ for which $p_n = p_{n-1}$. To see this, it is sufficient to apply (11) for E^{n-1} to the function $f(x)$ for every fixed x_n and then take the L_{p_n}-norms with respect to x_n of both sides.

Now, it remains only to refer to Lemma 3.12.

§4. Singular integrals in L_p

In this section, we shall use the results of §3 to obtain, in the anisotropic case, estimates in L_p of the type of Mihlin-Zygmund-Calderón estimates of singular integrals and estimates of limiting operators for convolutions.* We begin with certain auxiliary considerations.

4.1. Generalized spherical coordinates. Definition. Let $\lambda = (\lambda_1, \ldots, \lambda_n)$ denote a vector with positive** coordinates, $|\lambda| = \sum_1^n \lambda_i$, and let a denote a real number. We shall say that a function $\Phi(x) = \Phi(x_1, \ldots, x_n)$ is *λ-homogeneous of degree a* if, for arbitrary positive t and nonzero x,

$$\Phi(t^\lambda x) = \Phi(t^{\lambda_1} x_1, \ldots, t^{\lambda_n} x_n) = t^{\frac{|\lambda|}{n} a} \Phi(x).$$

We shall refer to a function $\rho(x)$ that is continuous, λ-homogeneous of first degree, and positive for nonzero x as the *λ-distance* (between the point x and the coordinate origin).

In the case in which $\lambda_1 = \ldots = \lambda_n$, a λ-homogeneous function of degree a is also homogeneous of degree a in the usual sense of the word.

Examples of λ-distances are

$$\pi(x) = \max_{1 \leqslant i \leqslant n} |x_i|^{|\lambda|/n\lambda_i}, \quad \sqrt{\sum_1^n |x_i|^{2|\lambda|/n\lambda_i}}, \quad \left(\sum_1^n |x_i|^{1/\lambda_i}\right)^{|\lambda|/n},$$

as is the function $\rho(x)$ (positive-valued for $x \neq 0$) defined implicitly by the equality

*We shall generalize in particular the results of Lewis [1] dealing with the anisotropic case and the space L_p.

**We can define λ-homogeneity also for arbitrary real λ_i by taking

$$|\lambda| = \sum_1^n |\lambda_i|.$$

$$\sum_{i=1}^{n} x_i^2 \rho^{-n\lambda_i/|\lambda|} = 1. \tag{1}$$

For notational simplicity, we shall write

$$\lambda_{\max} = \max_i \lambda_i, \qquad \lambda_{\min} = \min_i \lambda_i, \qquad \lambda^{(0)} = n\lambda/|\lambda|.$$

For $|\lambda| = n$, we have $\pi(x) = \pi[x]$ (see §3.4).

Corresponding to every λ-distance $\rho(x)$ is the *unit* ρ*-sphere* $\{x : \rho(x) = 1\}$.

We shall refer to the curve $x_i = t^{\lambda_i}\xi_i$, for $i = 1, \ldots, n$ and $0 < t < \infty$, as a λ*-trajectory,* assuming that not all the ξ_i are equal to 0.

The unit π-sphere is seen to be the surface of the cube $\{x : |x_i| \leqslant 1\}$ and the unit ρ-sphere, with ρ as in (1), coincides with the usual sphere $\{x : |x| = 1\}$.

In what follows, we shall often write simply "distance," "sphere," etc., instead of "λ-distance," "ρ-sphere," etc.

Obviously, the unit sphere intersects each trajectory at one point. Since the different trajectories do not intersect and since they fill the entire space, the unit sphere surrounds the coordinate origin. The converse is also true: every closed surface surrounding the coordinate origin and intersecting each λ-trajectory at a unique point is a unit sphere for some λ-distance.

Let $\rho(x)$ denote an arbitrary λ-distance. We shall call a point $\xi = \xi_x$ having the property that $\rho(\xi) = 1$ and lying on the same λ-trajectory as a point $x \neq 0$ the (curvilinear) *projection* of the point x onto the unit sphere $\rho(\xi) = 1$. In this case $x = t^{\lambda}\xi$ and $\rho(x) = t^{|\lambda|/n}\rho(\xi) = t^{|\lambda|/n}$, that is, $x = \rho^{n\lambda/|\lambda|}\xi = \rho^{\lambda^{(0)}}\xi$.

Every λ-distance $\rho(x)$ is equivalent to the λ-distance $\pi(x)$; that is, there exist positive numbers c' and c'' such that

$$c'\pi(x) \leqslant \rho(x) \leqslant c''\pi(x). \tag{2}$$

It will be sufficient to show this for points of the sphere $\pi(x) = 1$. On this sphere, inequalities (2) follow from the positiveness and continuity of the function $\rho(x)$.

For $\pi(x)$, we have the following inequality, which we shall call the generalized triangle inequality:

$$\pi(x+y) \leqslant \max_{1\leqslant i\leqslant n} |2x_i|^{|\lambda|/n\lambda_i} + \max_{1\leqslant i\leqslant n} |2y_i|^{|\lambda|/n\lambda_i}$$
$$\leqslant 2^{|\lambda|/n\lambda_{\min}}(\pi(x)+\pi(y)). \quad (3)$$

As a consequence of (2) and (3), the generalized triangle inequality holds for an arbitrary λ-distance when written in the form

$$\rho(x+y) \leqslant c^*\rho(x) + c^*\rho(y).$$

We shall later have occasion to use the λ-distances $\pi(x)$, $\left(\sum_1^n |x_i|^{1/\lambda_i}\right)^{|\lambda|/n}$, and $\rho(x)$ (defined by (1)) although many propositions hold and are analogously formulated for more general λ-distances.

In addition to the Cartesian coordinates $(x_1, \ldots, x_n)$ of the point x, it is also convenient to use generalized spherical coordinates (ρ, ξ), where $\rho = \rho(x)$ is defined by (1) and $\xi = \xi_x$ is the curvilinear projection of the point x onto the unit sphere $\rho(x) = |x| = 1$.

Obviously, the correspondence between the coordinates $x = \rho^{n\lambda/|\lambda|}\xi = \rho^{\lambda^{(0)}}\xi$ is one-to-one on $E^n \setminus \{0\}$.

Let us express one of the coordinates of ξ, where $|\xi| = 1$ (for definiteness, let us take ξ_n) in terms of the remaining ones and let us shift from the variable vector x to the variable vector

$$(\xi_1, \ldots, \xi_{n-1}, \rho),$$

where

$$\xi_1^2 + \ldots + \xi_{n-1}^2 < 1 \text{ and } \rho > 0.$$

Here, the Jacobian of the transformation is

$$\frac{\partial(x_1, \ldots, x_{n-1}, x_n)}{\partial(\xi_1, \ldots, \xi_{n-1}, \rho)} = \rho^{|\lambda^{(0)}|-1} \left.\frac{\partial(x_1, \ldots, x_{n-1}, x_n)}{\partial(\xi_1, \ldots, \xi_{n-1}, \rho)}\right|_{\rho=1}$$

$$= \rho^{|\lambda^{(0)}|-1} |A|,$$

where the elements of the matrix A are

$$a_{ij} = \delta_{ij} \qquad (1 \leqslant i \leqslant n-1, \ 1 \leqslant j \leqslant n-1),$$

$$a_{in} = \lambda_i^{(0)} \xi_i \qquad (i = 1, \ldots, n),$$

$$a_{nj} = \frac{\partial \xi_n}{\partial \xi_j} \qquad (j = 1, \ldots, n-1).$$

If we subtract from the last column a linear combination of the preceding columns with coefficients $\lambda_j^{(0)} \xi_j$, we obtain a column of zeros except for the last element, which is equal to

$$\lambda_n^{(0)} \xi_n - \sum_{j=1}^{n-1} \lambda_j^{(0)} \xi_j \frac{\partial \xi_n}{\partial \xi_j}.$$

Therefore,

$$\frac{\partial(x_1, \ldots, x_{n-1}, x_n)}{\partial(\xi_1, \ldots, \xi_{n-1}, \rho)} = \rho^{|\lambda^{(0)}|-1} \left(\lambda_n^{(0)} \xi_n - \sum_{j=1}^{n-1} \lambda_j^{(0)} \xi_j \frac{\partial \xi_n}{\partial \xi_j}\right)$$

$$= \rho^{|\lambda^{(0)}|-1} \xi_n^{-1} \left(\lambda_n^{(0)} \xi_n^2 + \sum_{j=1}^{n-1} \lambda_j^{(0)} \xi_j^2\right).$$

Thus,

$$\left|\frac{\partial(x_1, \ldots, x_{n-1}, x_n)}{\partial(\xi_1, \ldots, \xi_{n-1}, \rho)}\right| = \rho^{n-1} |\xi_n|^{-1} \sum_{j=1}^{n} \lambda_j^{(0)} \xi_j^2.$$

Noting that a surface element of the sphere $|\xi| = 1$ is

$$d\xi = |\xi_n|^{-1} d\xi_1 \ldots d\xi_{n-1},$$

we obtain the following useful formula:

$$\int\limits_{\varepsilon<\rho(x)<\nu} K(x)\,dx = \int\limits_{\varepsilon}^{\nu}\int\limits_{|\xi|=1} K(\rho^{\lambda^{(0)}}\xi)\left(\sum_{i=1}^{n}\lambda_i^{(0)}\xi_i^2\right) d\xi\,\rho^{n-1}\,d\rho. \quad (4)$$

Suppose now that

$$\pi(x) = \max_{1\leqslant i\leqslant n} |x_i|^{|\lambda|/n\lambda_i}.$$

Then,

$$x = \pi(x)^{n\lambda/|\lambda|}\xi = \pi^{\lambda^{(0)}}\xi, \qquad \pi(\xi) = 1.$$

Let us find the Jacobian of the transformation $x = (x_1, \ldots, x_n)$ into $(\xi_1, \ldots, \xi_n)$, π, where $\max\limits_{1\leqslant i\leqslant n} |\xi_i| = 1$.

Suppose that the point ξ is in that face of the unit cube $\max\limits_{1\leqslant i\leqslant n} |\xi_i| = 1$ the equation for which is $\xi_j = 1$ (or $\xi_j = -1$). One can easily see that the Jacobian of the transformation is

$$\frac{\partial(x_1, \ldots, x_n)}{\partial(\xi_1, \ldots, \xi_{j-1}, \xi_{j+1}, \ldots, \xi_n, \pi)} = \pm\,\lambda_j^{(0)}\pi^{|\lambda^{(0)}|-1}.$$

4.2. Let $\chi(x)$ denote a function belonging to $\overset{\circ}{L}_\infty(E^n)$ (that is, it is measurable, essentially bounded, and of bounded support) such that

$$\int \chi(x)\,dx = 0 \quad (5)$$

and

$$\int |\chi(x+y) - \chi(x)|\,dx \leqslant \delta(|y|), \quad (6)$$

where $\delta(t)$ is a nonnegative nondecreasing function defined on $[0, \infty)$ and satisfying the Dini condition

$$\int_0^1 \delta(t)\,\frac{dt}{t} < \infty .$$

For $\lambda = (\lambda_1, \ldots, \lambda_n)$, where $\lambda_i > 0$ (for $i = 1, \ldots, n$), and $0 < \varepsilon < \nu \leqslant \infty$, we define

$$K_{\varepsilon\nu}(x) = \int_{\varepsilon}^{\nu} \chi(xh^{-\lambda})\, h^{-|\lambda|-1}\, dh,$$

$$f_{\varepsilon\nu}(x) = \int K_{\varepsilon\nu}(x-y)\, f(y)\, dy,$$

where $f \in L_p(E^n)$ with $1 < p < \infty$.

Let us show that $K_{\varepsilon\nu}$ satisfies the conditions of Theorem 3.5. Since, for arbitrary q in $1 < q < \infty$, we have

$$\|\chi(\cdot\, h^{-\lambda})\|_q = h^{|\lambda:\, q|}\, \|\chi\|_q < \infty,$$

where

$$|\lambda:\ q| = \frac{\lambda_1}{q_1} + \ldots + \frac{\lambda_n}{q_n},$$

it follows that

$$K_{\varepsilon\nu} \in L_q(E^n), \qquad \|K_{\varepsilon\infty} - K_{\varepsilon\nu}\|_q = \|K_{\nu\infty}\|_q \to 0 \quad \text{as} \quad \nu \to \infty. \tag{7}$$

4.3. Lemma. *Suppose that*

$$\hat{K}_{\varepsilon\nu}(\xi) = \int e^{-2\pi i\,(x,\ \xi)} K_{\varepsilon\nu}(x)\, dx, \quad 0 < \varepsilon < \nu < \infty.$$

Then, there exists a number C independent of ε *and* ν *such that*

$$|\hat{K}_{\varepsilon\nu}(\xi)| \leqslant C$$

for every $\xi \in E^n$.

PROOF. It follows from (5) that $\hat{K}_{\varepsilon\nu}(0) = 0$. We note that, for $t > 0$,

$$\hat{K}_{\varepsilon\nu}(\xi) = \int\int_{\varepsilon}^{\nu} e^{-2\pi i (x, \xi)} \chi(xh^{-\lambda}) h^{-|\lambda|-1} \, dh \, dx$$

$$= \int\int_{\varepsilon/t}^{\nu/t} e^{-2\pi i (t^\lambda x, \xi)} \chi(xh^{-\lambda}) h^{-|\lambda|-1} \, dh \, dx = \hat{K}_{\varepsilon/t, \nu/t}(t^\lambda \xi).$$

Therefore, it will be sufficient to prove the lemma for arbitrary ε and ν such that $0 < \varepsilon < \nu < \infty$ and $|\xi| = 1$. Since $\hat{K}_{\varepsilon\nu}(\xi) = \hat{K}_{\varepsilon\mu}(\xi) + \hat{K}_{\mu\nu}(\xi)$, we need consider only the two cases $0 < \varepsilon < \nu \leqslant 1$ and $1 \leqslant \varepsilon < \nu < \infty$.

For $0 < \varepsilon < \nu \leqslant 1$ and $|\xi| = 1$, equation (5) and the fact that $\chi(x)$ is of bounded support imply that

$$|\hat{K}_{\varepsilon\nu}(\xi)| = \left| \int\int_{\varepsilon}^{\nu} (e^{-2\pi i (x, \xi)} - 1) \chi(xh^{-\lambda}) h^{-|\lambda|-1} \, dh \, dx \right|$$

$$\leqslant C_1 \int_{\varepsilon}^{\nu}\int |x| |(\chi(xh^{-\lambda})| h^{-|\lambda|-1} dx \, dh$$

$$\leqslant C_2 \int_0^1 \int_{\operatorname{supp} \chi} |h^\lambda x| h^{-1} \, dx \, dh \leqslant C_3.$$

For $1 \leqslant \varepsilon < \nu < \infty$, $|\xi| = 1$, and $y = \frac{\xi}{2}$, we have

$$2|\hat{K}_{\varepsilon\nu}(\xi)| = |\hat{K}_{\varepsilon\nu}(\xi)[1 - e^{2\pi i (y, \xi)}]|$$

$$= \left| \int\int_{\varepsilon}^{\nu} e^{-2\pi i (x, \xi)} [\chi(xh^{-\lambda}) - \chi((x - y) h^{-\lambda})] h^{-|\lambda|-1} \, dh \, dx \right|$$

$$\leqslant C_4 \int_{\varepsilon}^{\nu}\int |\chi(x) - \chi(x - yh^{-\lambda})| h^{-1} \, dx \, dh$$

$$\leqslant C_5 \int_1^\infty \delta(|yh^{-\lambda}|) h^{-1} dh \leqslant C_5 \int_1^\infty \delta\left(\frac{1}{2h^{\lambda_{\min}}}\right) h^{-1} dh \leqslant C_6 \int_0^{1/2} \frac{\delta(t)}{t} \, dt.$$

4.4. Let us show that conditions 1° and 2° of Theorem 3.5 are satisfied by $K_{\varepsilon\nu}$, where $0<\varepsilon<\nu<\infty$. Since the Fourier transform of a convolution is equal to the product of the Fourier transforms, it follows from Lemma 4.3 that

$$\| f_{\varepsilon\nu} \|_2 \leqslant C \| f \|_2, \qquad f \in \overset{\circ}{L}_{\infty}.$$

Let us show that condition 2° of Theorem 3.5 is satisfied. For $t>0$, $\pi[y]<t$, $\tilde{y}=yt^{-\lambda}$, and $\pi[\tilde{y}]<1$, we have

$$\begin{aligned}
&\int_{\pi[x]>Nt} | K_{\varepsilon\nu}(x-y) - K_{\varepsilon\nu}(x) | \, dx \\
&\quad \leqslant \int_{\varepsilon}^{\nu} \int_{\pi[x]>Nt} |\chi((x-y)h^{-\lambda}) - \chi(xh^{-\lambda})| \, h^{-|\lambda|-1} \, dx \, dh \\
&\quad = \int_{\varepsilon/t}^{\nu/t} \int_{\pi[\tilde{x}]>N} |\chi((\tilde{x}-\tilde{y})h^{-\lambda}) - \chi(\tilde{x}h^{-\lambda})| \, h^{-|\lambda|-1} \, d\tilde{x} \, dh \\
&\quad \leqslant \int_{a(N)}^{\infty} \delta(\tilde{y}h^{-\lambda}) \frac{dh}{h} \leqslant C_1 \int_0^{b(N)} \frac{\delta(t)}{t} \, dt,
\end{aligned}$$

where $a(N)>0$ (because of the boundedness of the support of χ) if $\inf|\tilde{x}-\tilde{y}|>0$, $\pi[\tilde{x}]>N$, and $\pi[\tilde{y}]<1$, that is, for all sufficiently large N. Also $b(N)\to 0$ as $N\to\infty$.

Thus, the conditions of Theorem 3.5 are satisfied by $K_{\varepsilon\nu}$, where $0<\varepsilon<\nu<\infty$. Therefore, for all p in $1<p<\infty$ and all f in $\overset{\circ}{L}_{\infty}$, there exists a number c_p independent of ε, ν, and f such that

$$\| K_{\varepsilon\nu} * f \|_p \leqslant c_p \| f \|_p. \tag{8}$$

Let us show that this estimate remains valid for $f\in L_p$ and $0<\varepsilon<\nu\leqslant\infty$.

Since the set $\overset{\circ}{L}_{\infty}$ is dense in L_p, inequality (8) remains valid with the same constant c_p for $f\in L_p$.

Suppose that $f\in L_p$, where $1<p<\infty$. It follows from (8) on the basis of Young's inequality 2(18) that, for $1 < p < q \leqslant \infty$,

$$\| K_{\varepsilon\nu} * f \|_q \leqslant C(\varepsilon, p, q) \| f \|_p, \quad 0 < \varepsilon < \nu \leqslant \infty,$$

where

$$C(\varepsilon, p, q) = \| \chi \|_{1-\frac{1}{p}+\frac{1}{q}} \int_{\varepsilon}^{\infty} h^{|\lambda: q|-|\lambda: p|-1} dh \to 0$$

as $\varepsilon \to \infty$. In particular, the convergence $K_{\varepsilon\nu} * f \to K_{\varepsilon\infty} * f$ as $\nu \to \infty$ is uniform on E^n, so that we can take the limit in the inequality

$$\| K_{\varepsilon\nu} * f \|_{p, |x| \leqslant R} \leqslant c_p \| f \|_p$$

(see (8)) first as $\nu \to \infty$, then as $R \to \infty$. This shows that (8) holds for $f \in L_p$ and $0 < \varepsilon < \nu \leqslant \infty$.

4.5. Theorem. *Suppose that* $1 < p < \infty$, $0 < \varepsilon < \nu \leqslant \infty$, *and* $f \in L_p$. *Then,* $f_{\varepsilon\nu}$ *converges in* L_p *to some function* $f_{0\nu}$ *as* $\varepsilon \to 0$, *and there exists a constant* c_p *independent of* f *and* ν *such that*

$$\| f_{0\nu} \|_p \leqslant c_p \| f \|_p. \tag{9}$$

PROOF. Let us suppose first that $f \in C_0^\infty$. Then,

$$\| f(\cdot - y) - f(\cdot) \|_p \leqslant M_f |y|.$$

Using property (5) of a kernel, we obtain for $0 < \varepsilon_1 < \varepsilon_2$

$$f_{\varepsilon_1\nu}(x) - f_{\varepsilon_2\nu}(x) = f_{\varepsilon_1\varepsilon_2}(x)$$

$$= \int_{\varepsilon_1}^{\varepsilon_2} \int [f(x-y) - f(x)] \chi(yh^{-\lambda}) h^{-|\lambda|-1} dy\, dh.$$

By letting ε_1 and ε_2 approach 0, we get

$$\|f_{\varepsilon_1\nu}-f_{\varepsilon_2\nu}\|_p \leqslant \int\limits_{\varepsilon_1}^{\varepsilon_2} \int \|f(\cdot-y)-f(\cdot)\|_p |\chi(yh^{-\lambda})| h^{-|\lambda|-1}\, dy\, dh$$

$$\leqslant M_f \int\limits_{\varepsilon_1}^{\varepsilon_2} \int |\chi(y)|\, |yh^{\lambda}| h^{-1}\, dy\, dh \to 0.$$

The assertion of the theorem for $f \in C_0^\infty$ follows from this and inequality (8).

Now, let f denote an arbitrary function in L_p for $1 < p < \infty$ and let δ denote a positive number. Since C_0^∞ is dense in L_p, there exists a $\varphi \in C_0^\infty$ such that $\|f-\varphi\|_p < \delta$. By using (8), we obtain, for $0 < \varepsilon_1 < \varepsilon_2 < \nu \leqslant \infty$,

$$\|f_{\varepsilon_1\nu}-f_{\varepsilon_2\nu}\|_p = \|f_{\varepsilon_1\varepsilon_2}\|_p \leqslant \|(f-\varphi)_{\varepsilon_1\varepsilon_2}\|_p + \|\varphi_{\varepsilon_1\varepsilon_2}\|_p \leqslant c_p\delta + \|\varphi_{\varepsilon_1\varepsilon_2}\|_p.$$

The right-hand member of this inequality can be made arbitrarily small by appropriate choice of the number δ and the function φ (given δ) by virtue of the smallness of the last term for fixed φ and all sufficiently small ε_1 and ε_2. This fact and the completeness of L_p imply that $f_{\varepsilon\nu}$ converges to some function $f_{0\nu} \in L_p$ as $\varepsilon \to 0$.

Now, by taking the limit in (8) with $\nu = \infty$ as $\varepsilon \to 0$, we obtain (9).

4.6. Let us now derive an analogue of Theorem 4.5 for a different kernel. This analogue is a generalization of the corresponding classical result of Mihlin [1] (with $p = 2$) and Zygmund and Calderón [1] (with $1 < p < \infty$) to the anisotropic case. However, the result obtained will not be used in what follows. Let us do some preliminary work.

Suppose that $K(x) \in L^{\text{loc}}(E^n \setminus \{0\})$ and that the following conditions hold for certain constants $M > 0$ and $N \geqslant 1$ and for all $t > 0$, $\mu > 0$, $\varepsilon > 0$, and $\nu > \varepsilon$:

$$\int\limits_{\pi[x] > Nt} |K(x-y)-K(x)|\, dx \leqslant M \quad \text{if} \quad \pi[y] < t, \tag{10}$$

$$\int_{t<\pi[x]<2t} |K(x)|\,dx \leqslant M, \tag{11}$$

$$\left|\int_{U_\nu \setminus U_\varepsilon} K(x)\,dx\right| \leqslant M \quad \text{if} \quad 0<\varepsilon<\nu<\infty, \tag{12}$$

and a finite limit

$$\lim_{\varepsilon\to 0} \int_{U_1 \setminus U_\varepsilon} K(x)\,dx \tag{13}$$

exists for some increasing sequence of neighborhoods U_ν such that

$$\left\{x:\ \pi[x] < \frac{\nu}{N}\right\} \subset U_\nu \subset \{x:\ \pi[x] < \nu\}. \tag{14}$$

For $0 < \varepsilon < \nu < \infty$, let us define

$$K_\varepsilon(x) = \begin{cases} K(x) & \text{for} \quad x \in E^n \setminus U_\varepsilon, \\ 0 & \text{for} \quad x \in U_\varepsilon, \end{cases}$$
$$K_\varepsilon^\nu(x) = K_\varepsilon(x) - K_\nu(x),$$

and let us show that the conditions of Theorem 3.5 hold for K_ε.

To show that condition 2° of Theorem 3.5 holds for K_ε, let us estimate the integral

$$\int_{\pi[x] > N_1 t} |K_\varepsilon(x-y) - K_\varepsilon(x)|\,dx$$

$$\leqslant \int_Q |K(x-y) - K(x)|\,dx + \int_{Q_1 \cup Q_2} |K(x)|\,dx = \mathcal{J} + \mathcal{J}',$$

where

$$Q=\{x:\ \pi[x]>N_1t,\ x-y\in U_\varepsilon\},$$
$$Q_1=\{x:\ \pi[x]>N_1t,\ x\notin U_\varepsilon,\ x-y\in U_\varepsilon,\ \pi[y]<t\},$$
$$Q_2=\{x:\ \pi[x]>N_1t,\ x\in U_\varepsilon,\ x-y\notin U_\varepsilon,\ \pi[y]<t\}.$$

The estimate of $\mathcal{J}$ follows from (10). From the definition (14) of U_ε, we get

$$Q_1\subset Q_1'=\{x:\ \pi[x]>N_1t,\ \pi[x]>\varepsilon/N,\ \pi[x-y]<\varepsilon,\ \pi[y]<t\},$$
$$Q_2\subset Q_2'=\{x:\ \pi[x]>N_1t,\ \pi[x]<\varepsilon,\ \pi[x-y]>\varepsilon/N,\ \pi[y]<t\}$$
$$\subset\{x:\ \pi[x]<\varepsilon,\ \pi[x-y]>\varepsilon/N,\ \pi[y]<\varepsilon/N_1\}.$$

On the basis of inequality (3), we conclude that, for a sufficiently large number $N_1=N_1(N,\ \lambda)$, there exists a positive number d such that

$$Q_1'\subset Q_1''=\left\{x:\ \frac{\varepsilon+t}{2N}<\pi[x]<2^{1/\lambda_{\min}}(\varepsilon+t)\right\},$$
$$Q_2'\subset Q_2''=\{x:\ d\varepsilon<\pi[x]<\varepsilon\}.$$

Now, for the estimate of $\mathcal{J}'$ it remains only to increase the region of integration to $Q_1''\cup Q_2''$ and apply (11).

Let us show that condition 1° of Theorem 3.5 holds for $K_\varepsilon(x)$ with $r=2$. To do this, it will be sufficient to show that there exists a constant C independent of ε, ν, and ξ such that

$$|\widehat{K}_\varepsilon^\nu(\xi)|\leqslant CM,\qquad 0<\varepsilon<\nu<\infty,\quad \xi\in E^n.$$

This is true because, for $|x|\leqslant R<\infty$, $f\in\overset{\circ}{L}_\infty$, and sufficiently large $\nu=\nu(R,\ f)$, we have

$$(K_\varepsilon*f)(x)=(K_\varepsilon^\nu*f)(x),$$

so that

$$\|K_\varepsilon*f\|_{2\,(|x|\leqslant R)}\leqslant\sup_\xi|\widehat{K}_\varepsilon^\nu(\xi)|\,\|f\|_2\leqslant CM\|f\|_2,$$

and it only remains to take the limit as $R\to\infty$.

Thus, proof that condition 1° of Theorem 3.5 holds reduces to the following lemma:

4.7. Lemma. *Suppose that conditions* (10)-(12) *are satisfied for* $K(x) \in L_1(E^n)$. *Then, there exists a constant C independent of K such that*

$$|\hat{K}(\xi)| \leqslant CM.$$

Proof. That the assertion holds at the point $\xi = 0$ follows from the continuity of $\hat{K}(\xi)$ since it is the Fourier transform of a summable function or from (12) if we take the limit as ε approaches 0 and ν approaches ∞. Let us define

$$K^{(r)}(x) = r^{|\lambda|}K(r^{\lambda}x) \qquad (r > 0).$$

Obviously, $K^{(r)}(x)$ possesses properties (10)-(12) if we replace ε, ν, and t with ε/r, ν/r, and t/r. Replacing x with $r^{\lambda}x$, we obtain

$$\hat{K}(\xi) = \int e^{-2\pi i(x,\,\xi)}K(x)\,dx = \int e^{-2\pi i(r^{\lambda}x,\,\xi)}K^{(r)}(x)\,dx = \hat{K}^{(r)}(r^{\lambda}\xi).$$

It follows that the lemma will be proven if we can find a common bound for $|\hat{K}^{(r)}(\xi)|$ for $|\xi| = 1$ and all $r > 0$.

Omitting the superscript (r), we see that it will be sufficient to prove the lemma for $|\xi| = 1$. We have

$$|\hat{K}(\xi)| \leqslant \left|\int_{\pi[x]<1} (e^{-2\pi i(x,\,\xi)} - 1)\,K(x)\,dx\right| + \left|\int_{\pi[x]<1} K(x)\,dx\right|$$

$$+ \left|\int_{\pi[x]>1} e^{-2\pi i(x,\,\xi)}K(x)\,dx\right| = |\mathcal{J}_1| + |\mathcal{J}_2| + |\mathcal{J}_3|.$$

It obviously follows from (11) and (12) that $|\mathcal{J}_2| \leqslant C_1M$.

We obtain an estimate for $|\mathcal{J}_1|$ by using the inequality $|1 - e^{ib}| \leqslant |b|$ and the obvious inequality $|x| \leqslant \sqrt{n}\,\pi[x]^{\lambda_{\min}}$ for $\pi[x] \leqslant 1$, where $\lambda_{\min} = \min_i \lambda_i$:

$$|\mathcal{J}_1| \leqslant C_2 \int\limits_{\pi[x]<1} |x|\,|K(x)|\,dx$$

$$\leqslant C_2 \sqrt{n} \sum_{j=0}^{\infty} 2^{-j\lambda_{\min}} \int\limits_{2^{-j-1} \leqslant \pi[x] < 2^{-j}} K(x)\,|\,dx \leqslant C_3 M.$$

Finally, let us estimate $|\mathcal{J}_3|$ for $|\xi| = 1$. We choose δ from the conditions $0 < \delta < 1/2$ and

$$\pi[\delta\xi] < \min\left\{\frac{1}{N},\ 2^{\frac{1}{\lambda_{\min}}-1}\right\}.$$

Suppose that $y = \delta\xi$. Then,

$$|1 - e^{-2\pi i (y,\,\xi)}| > \gamma = \gamma(\delta) > 0, \qquad \pi[y] < \frac{1}{N}.$$

By a translation of the variables of integration, we obtain

$$(1 - e^{-2\pi i (y,\,\xi)})\,\mathcal{J}_3 = \int\limits_{\pi[x]>1} e^{-2\pi i (x,\,\xi)}[K(x) - K(x-y)]\,dx$$

$$+ \mathcal{J}_5 + \mathcal{J}_6 = \mathcal{J}_4 + \mathcal{J}_5 + \mathcal{J}_6,$$

where

$$|\mathcal{J}_5| \leqslant \int\limits_{\substack{\pi[x]>1\\ \pi[x-y]<1}} |K(x-y)|\,dx = \int\limits_{\substack{\pi[x+y]>1\\ \pi[x]<1}} |K(x)|\,dx,$$

$$|\mathcal{J}_6| \leqslant \int\limits_{\substack{\pi[x]<1\\ \pi[x-y]>1}} |K(x-y)|\,dy = \int\limits_{\substack{\pi[x+y]<1\\ \pi[x]>1}} |K(x)|\,dx.$$

By virtue of (10), we have $|\mathcal{J}_4| \leqslant M$, and it remains to estimate $\mathcal{J}_5$ and $\mathcal{J}_6$. By virtue of (3) and the choice of y, we have

$$\pi[x] \leqslant 2^{1/\lambda_{\min}}[\pi[x+y] + \pi[-y]] \leqslant 2^{1/\lambda_{\min}}\,\pi[x+y] + \frac{1}{2},$$

$$2^{1/\lambda_{\min}}\pi[x] \geqslant \pi[x+y] - 2^{1/\lambda_{\min}}\pi[y] \geqslant \pi[x+y] - \frac{1}{2}.$$

Enlarging the region of integration, we see that

$$|\mathcal{J}_5| \leqslant \int\limits_{2^{-1-\frac{1}{\lambda_{\min}}} < \pi[x] < 1} |K(x)|\,dx, \quad |\mathcal{J}_6| \leqslant \int\limits_{1 < \pi[x] < 2^{-1+\frac{1}{\lambda_{\min}}}} |K(x)|\,dx.$$

From these results and (11), we see that $|\mathcal{J}_5| + |\mathcal{J}_6| \leqslant C_4 M$, and hence

$$|\mathcal{J}_3| < \gamma^{-1}(1 + C_4)M.$$

This completes the proof of the lemma.

4.8. Theorem. *Suppose that conditions* (10)-(12) *are satisfied for* $K(x) \in L^{\mathrm{loc}}(E^n \setminus \{0\})$. *Then, for* $\mathbf{1} < \mathbf{p} < \infty$, *there exists a constant* $c_{\mathbf{p}}$ *independent of* ε *and* f *such that*

$$\| K_\varepsilon * f \|_{\mathbf{p}} \leqslant c_{\mathbf{p}} \| f \|_{\mathbf{p}}, \qquad f \in \mathring{L}_\infty.$$

The proof consists in showing, as we have just done, that the conditions of Theorem 3.5 hold for $K_\varepsilon(x)$.

4.9. Theorem. *Suppose that conditions* (10)-(13) *hold for* $K(x) \in L^{\mathrm{loc}}(E^n \setminus \{0\})$. *Let* $\mathbf{p}$ *denote a vector in* $\mathbf{1} < \mathbf{p} < \infty$ *and let* $f(x)$ *denote a function in* $L_{\mathbf{p}}$. *Define*

$$f_\varepsilon^\nu(x) = \left(K_\varepsilon^\nu * f\right)(x)$$

for $0 < \varepsilon < \nu < \infty$. *Then, for every compact set of the form* $|x| \leqslant R$, *the restriction of the function* $f_\varepsilon^\nu(x)$ *to that set converges in* $L_{\mathbf{p}}(|x| \leqslant R)$ *to some function* $f_0(x)$ *as* $\varepsilon \to 0$ *and* $\nu \to \infty$. *Also, there exists a constant* $c_{\mathbf{p}}$ *independent of* f *such that*

$$\| f_0 \|_{\mathbf{p}} \leqslant c_{\mathbf{p}} \| f \|_{\mathbf{p}}.$$

PROOF. By virtue of Theorem 4.8,

$$\left\| f_\varepsilon^\nu \right\|_{\mathbf{p}} \leqslant 2c_{\mathbf{p}} \| f \|_{\mathbf{p}}, \qquad f \in C_0^\infty.$$

It will be sufficient to prove the assertion for functions in the set C_0^∞, which is dense in L_p. For $f \in C_0^\infty$ and $|x| \leqslant R$, the functions $f_\varepsilon^\nu(x)$ are independent of ν for all sufficiently large ν. The conclusion of the theorem then follows as we let ν approach ∞. Now, we need only show that

$$\left\| f_{\varepsilon_1}^\nu - f_{\varepsilon_2}^\nu \right\|_p = \left\| f_{\varepsilon_1}^{\varepsilon_2} \right\|_p \to 0 \quad \text{for} \quad 0 < \varepsilon_1 < \varepsilon_2 \to 0, \quad f \in C_0^\infty.$$

We have

$$f_{\varepsilon_1}^{\varepsilon_2}(x) = \int_{U_{\varepsilon_2} \setminus U_{\varepsilon_1}} K(y) f(x-y)\, dy$$

$$= \int_{y \in U_{\varepsilon_2} \setminus U_{\varepsilon_1}} K(y) [f(x-y) - f(x)]\, dy + f(x) \int_{y \in U_{\varepsilon_2} \setminus U_{\varepsilon_1}} K(y)\, dy.$$

Since $|f(x-y) - f(x)| \leqslant \varphi(x) |y|$, where $\varphi(x) \in C_0^\infty$, the result follows from (11) and (13).

4.10. REMARK. Condition (13) does not follow from conditions (10)-(12). Thus, for $n=1$, $K(x) = 1/x$, and

$$U_\varepsilon = \{x: \; -\varepsilon < x < 2^m\} \quad \text{for} \quad 2^{m-1} < \varepsilon \leqslant 2^m,$$

conditions (10)-(12) are satisfied whereas condition (13) is not. The situation changes if we take $U_\varepsilon = \{x: |x| < \varepsilon\}$ as our neighborhood of zero.

4.11. Theorem. *Suppose that $\lambda = (\lambda_1, \ldots, \lambda_n)$, where $\lambda_i > 0$ for $i = 1, \ldots n$. Suppose that the following conditions hold for $K(x) \in L^{\mathrm{loc}}(E^n \setminus \{0\})$:*

1°. *$K(x)$ is λ-homogeneous of degree $-n$; that is, $K(t^\lambda x) = t^{-|\lambda|} K(x)$ for arbitrary positive t and nonzero x.*

2°.
$$\int_{|\xi|=1} K(\xi) \sum_{j=1}^n \lambda_j \xi_j^2 \, d\xi = 0.$$

3°.
$$\int_0^1 \frac{\omega(t)}{t}\, dt < \infty;$$

where $\omega(t) = \sup |K(\xi) - K(\eta)|$ *for* $|\xi| = |\eta| = 1$ *and* $|\xi - \eta| \leqslant t$. *Suppose also that* $f \in L_p$, *where* $1 < p < \infty$, *and that, for* $\varepsilon > 0$,

$$f_\varepsilon(x) = (K_\varepsilon * f)(x) = \int_{\sum \frac{(x_j - y_j)^2}{\varepsilon^{2\lambda_j}} > 1} K(x-y) f(y)\, dy.$$

Then, f_ε *converges in* L_p *to some function* f_0 *as* $\varepsilon \to 0$, *and there exists a constant* c_p *independent of* ε *and* f *such that*

$$\| f_\varepsilon \|_p \leqslant c_p \| f \|_p, \qquad \varepsilon \geqslant 0.$$

PROOF. We note first of all that, by virtue of 2°, $\omega(t) \to 0$ as $t \to 0$; that is, $K(x)$ is continuous on the sphere $|x| = 1$. Consequently, by virtue of 1°, $K(x)$ is continuous on $E^n \setminus \{0\}$.

If we can show that $K_\varepsilon \in L_q$ for $\varepsilon > 0$ and $q > 1$, this will, on the basis of Hölder's inequality, imply that $(K_\varepsilon * f)(x)$, where $f L_p \in$, exists at every x. Let us show that K_ε does indeed belong to L_q.

The kernel $K(x)$ is dominated by the function $C\left(\sum_1^n | x_i |^{1/\lambda_i}\right)^{-|\lambda|}$.

This follows from their λ-homogeneity and the boundedness of $|K(x)|$ on the unit π-sphere. Consequently,

$$| K_\varepsilon(x) | \leqslant C_\varepsilon \left(1 + \sum_1^n | x_i |^{1/\lambda_i}\right)^{-|\lambda|}, \qquad x \in E^n,$$

so that it will be sufficient to prove the finiteness of the L_q-norm of the right-hand member of the last inequality:

$$\left\{ \int_{-\infty}^{\infty} \left(1 + \sum_1^n | x_i |^{\frac{1}{\lambda_i}}\right)^{-|\lambda| q_1} dx_1 \right\}^{\frac{1}{q_1}}$$

$$\leqslant C_1\left\{\int_{-\infty}^{\infty}\left[\left(1+\sum_2^n |x_l|^{\frac{1}{\lambda_l}}\right)^{\lambda_1}+|x_1|\right]^{-\frac{|\lambda|}{\lambda_1}q_1}dx_1\right\}^{\frac{1}{q_1}}$$

$$\leqslant C_2\left(1+\sum_2^n |x|^{\frac{1}{\lambda_l}}\right)^{-\sum_2^n \lambda_l-\lambda_1\left(1-\frac{1}{q_1}\right)}\leqslant C_2\left(1+\sum_2^n |x_l|^{\frac{1}{\lambda_l}}\right)^{-\sum_2^n \lambda_l}.$$

If we then estimate the L_{q_2}-norm with respect to x_2 in the same way and continue the process, we obviously obtain the desired result.

The proof now reduces to Theorem 4.9. Let us show that conditions (10)–(13) are satisfied in the present case. For our neighborhood of zero we take $U_\varepsilon = \{x\colon \rho(x) < \varepsilon\}$, where $\rho(x)$ is the nonnegative-valued function defined by (1).

Condition (11) is established with the aid of 1° by replacing x with $t^\lambda x'$:

$$\int_{t<\pi|x|<2t} |K(x)|\,dx = \int_{1<\pi|x|<2} |K(x')|\,dx'.$$

To show that conditions (12) and (13) hold, the shift from x to the variables (ρ, ξ), where $\rho \geqslant 0$, $|\xi| = 1$, $x = \rho^{\lambda^{(0)}}\xi$, $\rho(x) = \rho$, $\lambda^{(0)} = \frac{n}{|\lambda|}\lambda$. For simplicity, let us assume that $|\lambda| = n$, that is, $\lambda^{(0)} = \lambda$. (No generality is lost by this.)

In accordance with (4) and 1°, we obtain

$$\int_{U_\nu \setminus U_\varepsilon} K(x)\,dx = \int_{\varepsilon\leqslant\rho<\nu} K(x)\,dx$$

$$= \int_\varepsilon^\nu \int_{|\xi|=1} K(\xi)\sum_1^n \lambda_j\xi_j^2\,d\xi\,\frac{d\rho}{\rho} = 0$$

by virtue of 2°.

Finally, let us show that condition (10) holds. Since the π- and ρ-distances are equivalent, it will, by virtue of condition 1°, be sufficient to verify (10) for $t = 1/N$ with π replaced with the ρ of (1); that is, it will be sufficient to estimate, for $\rho(y) < 1/N$, the integral

$$\mathcal{J} = \int\limits_{\rho(x)>1} |K(x-y) - K(x)|dx \leqslant \int\limits_{\rho(x)>1} |K(z) - K(x)|dx + \int\limits_{\rho(x)>1} |K(x-y) - K(z)|dz = \mathcal{J}_1 + \mathcal{J}_2.$$

We have introduced the point $z = \left(\frac{\rho(x-y)}{\rho(x)}\right)^{\lambda} x$ on the intersection of the λ-trajectory of the point x with the ρ-sphere of the point $x - y$; that is, $\rho(z) = \rho(x-y)$.

Shifting to generalized spherical coordinates, we obtain

$$\mathcal{J}_1 = \int\limits_{\rho(\xi)=|\xi|=1} |K(\xi)| \sum \lambda_i \xi_i^2 \, d\xi \int\limits_{\rho(x)>1} \left| \frac{1}{\rho^{|\lambda|}(z)} - \frac{1}{\rho^{|\lambda|}(x)} \right| \times \rho^{|\lambda|-1}(x) \, d\rho(x).$$

Since $\rho(y) < 1/N$ and N can be taken sufficiently large, we have $\rho(z) = \rho(x-y) \sim \rho(x)$ for $\rho(x) > 1$. More precisely, for $\xi = \rho(x)^{-\lambda} x$, $|\xi| = 1$, and $\eta = \rho(x)^{-\lambda} y$, we have by virtue of the smoothness of $\rho(x)$

$$\begin{aligned} |\rho(x) - \rho(x-y)| &= \rho(x)|\rho(\xi) - \rho(\xi-\eta)| \leqslant C_1\rho(x)|\eta| \\ &= C_1\rho(x)\left|\rho(x)^{-\lambda} y\right| \leqslant C_2\rho^{1-\lambda_{\min}}(x)\,\rho^{\lambda_{\min}}(y) \\ &\leqslant C_2 N^{-\lambda_{\min}}\rho^{1-\lambda_{\min}}(x). \end{aligned} \tag{15}$$

Then, by the theorem on finite increments,

$$\left| \frac{1}{\rho^{|\lambda|}(z)} - \frac{1}{\rho^{|\lambda|}(x)} \right| = \frac{|\rho^{|\lambda|}(x) - \rho^{|\lambda|}(z)|}{\rho^{|\lambda|}(z)\,\rho^{|\lambda|}(x)} \leqslant C_3 \frac{|\rho(x) - \rho(z)|}{\rho^{|\lambda|+1}(x)} \leqslant C_4 \rho^{-|\lambda|-\lambda_{\min}},$$

so that $\mathcal{J}_1 \leqslant C_5$. Let us estimate $\mathcal{J}_2$:

$$\mathcal{I}_2 = \int\limits_{\rho(x)>1} \int\limits_{|\xi_x|=1} |K(\xi_{x-y}) - K(\xi_x)| \sum_1^n \lambda_i \xi_{xi}^2 \, d\xi_x \rho^{-|\lambda|}(x-y) \times$$

$$\rho^{|\lambda|-1}(x)\, d\rho(x) \leqslant C_6 \int\limits_{\rho(x)>1} \int\limits_{|\xi_x|=1} \omega(|\xi_{x-y} - \xi_x|)\, d\xi_x \rho^{-1}(x)\, d\rho(x).$$

We note that

$$|\xi_{x-y} - \xi_x| = |\xi_{x-y} - \xi_z|$$
$$\leqslant C_6 \pi^{\lambda_{\min}}(\xi_{x-y} - \xi_z) \leqslant C_7 \rho^{\lambda_{\min}}(\xi_{x-} - \xi_z).$$

Then, by virtue of the relation $\rho(x-y) = \rho(z)$ and the generalized triangle inequality, we have

$$\rho(\xi_{x-y} - \xi_z) = \rho\left(\frac{x-y}{\rho^\lambda(x-y)} - \frac{z}{\rho^\lambda(z)}\right) = \rho^{-1}(x-y)\,\rho(x-y-z)$$

$$\leqslant C^* \rho^{-1}(x-y)[\rho(x-z) + \rho(-y)]$$
$$\leqslant C^* \rho^{-1}(x)[\rho(x-z) + N^{-1}].$$

But

$$\rho(x-z)$$
$$= \rho\left(x - \left(\frac{\rho(x-y)}{\rho(x)}\right)^\lambda x\right) \leqslant C'' \max_i \left|1 - \left(\frac{\rho(x-y)}{\rho(x)}\right)^{\lambda_i}\right|^{1/\lambda_i} |x_i|^{1/\lambda_i}$$
$$\leqslant C''C' \rho(x) \max_i \left|1 - \left(\frac{\rho(x-y)}{\rho(x)}\right)^{\lambda_i}\right|^{1/\lambda_i}$$

$$= C''C' \max_i |\rho^{\lambda_i}(x) - \rho^{\lambda_i}(x-y)|^{1/\lambda_i}.$$

From the theorem on finite increments and inequality (15), we have

$$\rho(x-z) \leqslant C_8 \max_i \rho^{1-\frac{1}{\lambda_i}}(x)\,|\rho(x) - \rho(x-y)|^{\frac{1}{\lambda_i}}$$

$$\leqslant C_9 \rho(x) \max_i \rho(x)^{-\frac{\lambda_{\min}}{\lambda_i}} = C_9 \rho(x)^{1-\frac{\lambda_{\min}}{\lambda_{\max}}}.$$

Combining these inequalities, we obtain

$$|\xi_{x-y}-\xi_x|\leqslant C_{10}\rho^{\lambda_{\min}}(\xi_{x-y}-\xi_z)$$

$$\leqslant C_{11}\rho^{-\lambda_{\min}}(x)\left[\rho(x-z)+N^{-1}\right]^{\lambda_{\min}}\leqslant C_{12}\rho(x)^{-\frac{\lambda_{\min}^2}{\lambda_{\max}}}.$$

Then, on the basis of 3°, we have

$$\mathcal{I}_2\leqslant C_{13}\int_{\rho>1}\omega\left(C_{12}\rho^{-\frac{\lambda_{\min}^2}{\lambda_{\max}}}\right)\rho^{-1}\,d\rho\leqslant C_{14}\int_0^1\frac{\omega(t)}{t}\,dt<\infty,$$

which completes the proof.

Chapter 2

INTEGRAL REPRESENTATIONS OF DIFFERENTIABLE FUNCTIONS

The prime purpose of the present chapter is to obtain various kinds of integral representations of functions of several real variables. These representations express the value of a function at an arbitrary point of some region in terms of integrals of that function and its derivatives or finite differences.

In what follows, each class of functions that is defined by certain differential or differential-difference properties will be given an integral representation specific for that particular class. Here, we should emphasize that, if the form of the representation is determined by the nature of the class of functions in question, the "carrier" of the representation, that is, the form of the region of integration in the representation, is determined by the parameters of the particular class of domains of definition of the functions. This enables us to investigate the relationships between the properties of the class of functions in question and the properties of their domains of definition. The methods of investigation used in the present book are based on integral representations of functions.

In the present chapter, we shall also look at the question of approximation of functions with averages, we shall introduce the concept of a generalized Sobolev derivative and present some of its properties, and we shall define the basic classes of regions to be treated later.

We point out that all the questions considered here in connection with the method of integral representations of functions originated in the well-known works of Sobolev [1], [2].

In what follows, we shall use the following notation:

For any vector $k=(k_1, \ldots, k_n)$ with nonnegative components, we define

$$|k|=\sum_{i=1}^{n} k_i,$$

$$k-\mathbf{1}=(k_1-1, \ldots, k_n-1),$$

where $\mathbf{1}=(1, \ldots, 1)$. If k_i is an integer for $i=1, \ldots, n$, we define

$$k!=k_1! \ldots k_n!.$$

Suppose that x is a vector $(x_1, \ldots, x_n)$ belonging to E^n. Then, we define

$$x^k=x_1^{k_1} \ldots x_n^{k_n},$$

$(-x)^k=(-1)^{|k|}x^k$ (for nonnegative integers k_i), and

$$x^{(i)}=(x_1, \ldots, x_{i-1}, x_{i+1}, \ldots, x_n).$$

Suppose that $\lambda=(\lambda_1, \ldots, \lambda_n)$, where $\lambda_i>0$ for $i=1, \ldots, n$. Let v denote a positive number. Then,

$$v^\lambda=(v^{\lambda_1}, \ldots, v^{\lambda_n}), \quad xv^\lambda=(x_1v^{\lambda_1}, \ldots, x_nv^{\lambda_n}),$$

$$x\colon v^\lambda=\left(\frac{x_1}{v^{\lambda_1}}, \ldots, \frac{x_n}{v^{\lambda_n}}\right)=(x_1v^{-\lambda_1}, \ldots, x_nv^{-\lambda_n}).$$

Let us also write

$$(k, \lambda) = \sum_{i=1}^{n} k_i \lambda_i,$$

$$I_{v^\lambda}(a) = \{x: |x_i - a_i| \leqslant v^{\lambda_i} \ (i = 1, \ldots, n)\} \qquad (a = (a_1, \ldots, a_n)),$$

$$I_{v^\lambda} = I_{v^\lambda}(\mathbf{0}) \qquad (\mathbf{0} = (0, \ldots, 0)),$$

$$e_i = (0, \ldots, 0, \overset{i}{1}, 0, \ldots, 0)$$

(the unit vector directed along the x_i axis), and

$$\prod_j^{(i)} t_j = \prod_{j \neq i} t_j.$$

We denote the derivative with respect to x_i by D_{x_i} and we denote the derivative with respect to the ith argument by D_i. Thus,

$$D_{x_i} K(x: v^\lambda) = \frac{\partial}{\partial x_i} K(x: v^\lambda) = D_i K(x: v^\lambda) v^{-\lambda_i}.$$

We also define

$$D_x = (D_{x_1}, \ldots, D_{x_n}), \ D_{x_i}^{k_i} = (D_{x_i})^{k_i} = \frac{\partial^{k_i}}{\partial x_i^{k_i}}, \ D_x^k = D_{x_1}^{k_1} \ldots D_{x_n}^{k_n},$$

$$D = (D_1, \ldots, D_n), \qquad D_i^{k_i} = (D_i)^{k_i}, \qquad D^k = D_1^{k_1} \ldots D_n^{k_n}.$$

Obviously, $D_{x_i} f(x) = D_i f(x)$ and $D_x^k f(x) = D^k f(x)$.

Consider two sets $\Omega_1 \subset E^n$ and $\Omega_2 \subset E^n$. We denote their arithmetic sum* (resp. difference) by $\Omega_1 + \Omega_2$ (resp. $\Omega_1 - \Omega_2$). The notation $x + \Omega$ or $x - \Omega$, for a point x in E^n and a set Ω, has an analogous meaning.

*Translation editor's note. Arithmetic sum as opposed to logical sum or union. Thus, $\Omega_1 + \Omega_2 = \{y: y = x_1 + x_2, x_i \in \Omega_i, i = 1, 2\}$.

§5. Averaging of functions

5.1. We introduce a process of averaging a given function f that will lead to a sequence of infinitely differentiable functions converging in a certain sense to a function f.

Let $K(x)$ denote an infinitely differentiable function whose support is bounded and contained in E^n (that is, $\left(K \in C_0^\infty(E^n)\right)$. Suppose that $K(x)$ also satisfies the condition

$$\int_{E^n} K(x)\,dx = 1. \tag{1}$$

For purposes of simplicity only, we shall assume in what follows that $\operatorname{supp} K = S(K) \subset I_1$.

An example of such a function is the function

$$K(x) = \begin{cases} \mu e^{\frac{\rho^2(x,\,a)}{\rho^2(x,\,a) - r^2}} & \text{for} \quad 0 \leqslant \rho(x, a) < r, \\ 0 & \text{for} \quad \rho(x, a) \geqslant r, \end{cases}$$

where $a = (a_1, \ldots, a_n)$, $\rho(x, a) = \left(\sum_1^n |x_i - a_i|^2\right)^{1/2}$, and the constant μ is chosen so that (1) will be satisfied. We note that $\operatorname{supp} K$ is a ball of radius r with center at a. For suitable choice of r and a, we can arrange, for example, for $\operatorname{supp} K$ to be contained in one of the coordinate angles [a coordinate angle being the n-dimensional analogue of an octant in E^3]. Functions with supports of this kind will be used in what follows.

Suppose that $\lambda = (\lambda_1, \ldots, \lambda_n)$, where $\lambda_i > 0$ for $i = 1, \ldots, n$. Let v denote a positive number. The function

$$K_{v^\lambda}(x) = v^{-|\lambda|} K\left(x : v^\lambda\right) \tag{2}$$

is infinitely differentiable on E^n, its support is the set $S_{v^\lambda}(K) \subset I_{v^\lambda}$, where

$$S_{v^\lambda}(K) = \{x:\ (x : v^\lambda) \in S(K)\}, \tag{3}$$

and, by virtue of (1),

$$\int_{E^n} K_{v^\lambda}(x)\,dx = \int_{E^n} v^{-|\lambda|} K(x : v^\lambda)\,dx = \int_{E^n} K(x)\,dx = 1. \tag{4}$$

Let G denote a measurable subset of E^n and let f denote a function defined on G. We define $f = 0$ on $E^n \setminus G$ and we assume that the function f now defined on all E^n belongs to $L^{\text{loc}}(E^n)$.

We define the mean function corresponding to the function f with averaging kernel K and with averaging parameter v^λ by

$$f_{v^\lambda}(x) = v^{-|\lambda|} \int_{E^n} f(x+y)\,K(y : v^\lambda)\,dy$$

$$= v^{-|\lambda|} \int_{E^n} f(y)\,(K((y-x) : v^\lambda)\,dy. \tag{5}$$

One can easily show that $f_{v^\lambda}(x)$ is continuous and has continuous derivatives of all orders on E^n and that, for arbitrary $\alpha = (\alpha_1, \ldots, \alpha_n)$, where the α_i are nonnegative integers,

$$D_x^\alpha f_{v^\lambda}(x) = (-1)^{|\alpha|}\, v^{-|\lambda|-(\alpha,\,\lambda)} \int_{E^n} f(y)\,D^\alpha K\big((y-x) : v^\lambda\big)\,dy. \tag{6}$$

5.2. Lemma. If $f \in L_p(G)$, where $p = (p_1, \ldots, p_n)$, then

$$\| f_{v^\lambda} \|_p \leqslant \| K \|_1 \| f \|_p \qquad (1 \leqslant p \leqslant \infty), \tag{7}$$

$$\lim_{v \to 0} \| f_{v^\lambda} - f \|_p = 0 \qquad (1 \leqslant p < \infty). \tag{8}$$

PROOF. Inequality (7) follows immediately from (5) with the aid of Young's inequality 2(19):

$$\| f_{v^\lambda} \|_p \leqslant v^{-|\lambda|} \| K(\cdot : v^\lambda) \|_1 \| f \|_p = \| K \|_1 \| f \|_p.$$

To prove (8), we note that, by virtue of (4),

$$f_{v^\lambda}(x) - f(x) = v^{-|\lambda|} \int_{E^n} [f(x+y) - f(x)]\, K(y : v^\lambda)\, dy.$$

Then, on the basis of the Minkowski's generalized inequality 2(12), we have

$$\| f_{v^\lambda} - f \|_p \leqslant v^{-|\lambda|} \int_{S_{v^\lambda}(K)} | K(y : v^\lambda) | \, \| f(\cdot + y) - f \|_p \, dy$$

$$\leqslant \sup_{y \in I_{v^\lambda}} \| f(\cdot + y) - f \|_p \| K \|_1.$$

If we now apply Theorem 1.5 regarding the continuity in the wide sense of a function $f \in L_p(G)$, where $1 \leqslant p < \infty$, we obtain (8).

REMARK. If $f \in L_p^{\text{loc}}(G)$, where $1 \leqslant p < \infty$, then $f_{v^\lambda} \to f$ in the sense of $L_p^{\text{loc}}(G)$.

Let us show this: Let F denote a compact subset of G. Then, for sufficiently small v, the set $F + I_{v^\lambda}$ is contained in a compact subset T of G. We note that $f \in L_p(T)$. Then, on the basis of the inequality

$$\| f_{v^\lambda} - f \|_{p,F} \leqslant \sup_{y \in I_{v^\lambda}} \| f(\cdot + y) - f \|_{p,F} \| K \|_1$$

and the theorem on continuity in mean, we see that $\| f_{v^\lambda} - f \|_{p,F} \to 0$ as $v \to 0$.

Corollary. Equation (8) shows that the set of functions that are infinitely differentiable on E^n is dense in $L_p(G)$ if $1 \leqslant p < \infty$.

Equation (8) does not in general hold if any of the components of the vector $\boldsymbol{p}$ are infinite. On the other hand, we do have the following: If f is uniformly continuous on an open set G, then

$$\lim_{v \to 0} \| f_{v^\lambda} - f \|_{\infty, U} = 0, \tag{9}$$

where $U = \{x\colon x \in G,\ x + S_{v^\lambda}(K) \subset G,\ 0 < v < \delta\}$ is the set of those x in G such that $x+S_{v^\lambda}(K)\subset G$ for every v in $0<v<\delta$.

To see this, in analogy with what we did above, we obtain

$$\|f_{v^\lambda}-f\|_{\infty,\,U}\leqslant \sup_{y\in S_{v^\lambda}(K)}\|f(\cdot+y)-f\|_{\infty,\,U}\|K\|_1.$$

This inequality and the uniform continuity of f on G imply (9).

5.3. Lemma. Suppose that G is an open subset of E^n and that f belongs to $L_p^{\text{loc}}(G)$, where $p\geqslant 1$. Then,

$$\lim_{v\to 0} f_{v^\lambda}(x)=f(x) \tag{10}$$

for almost all $x\in G$.

PROOF. Let x denote a member of G. Then, for sufficiently small positive v, the set $x+I_{v^\lambda}$ is contained in G and, by virtue of (5) and (4), we have

$$|f_{v^\lambda}(x)-f(x)|=\left|v^{-|\lambda|}\int_{E^n}[f(y)-f(x)]\,K((y-x):v^\lambda)\,dy\right|$$

$$\leqslant\|K\|_\infty v^{-|\lambda|}\int_{I_{v^\lambda}(x)}|f(y)-f(x)|\,dy. \tag{11}$$

Noting that $v^{-|\lambda|}=2^n|I_{v^\lambda}(x)|^{-1}$ and applying Theorem 1.7 (Jessen-Marcinkiewicz-Zygmund) with $\psi_i(v)=2v^{\lambda_i}$ for $i=1,\ldots,n$, we see that the right-hand member of (11) approaches 0 for almost all $x\in G$ as $v\to 0$. This proves our assertion.

REMARK. If f is continuous on G, then $f_{v^\lambda}(x)\to f(x)$ at every point x in G as $v\to 0$. This follows immediately from inequality (11).

5.4. The purpose of the present subsection is to prove Theorem 1.6. We repeat it: Suppose that G is an open subset of E^n. Then the

Suppose that G is an open subset of E^n. Then the set $C_0^\infty(G)$ is dense in $L_p(G)$ if $1\leqslant p<\infty$.

PROOF. Let f denote a member of $L_p(G)$ and let ε denote a positive number. By virtue of property 1(6) of functions in the space

$L_p(G)$, there exists an open set Ω such that $\Omega \subset \overline{\Omega} \subset G$ and

$$\|f - f_\Omega\|_{p,G} < \frac{\varepsilon}{2},$$

where $f_\Omega = f\chi_\Omega$, where in turn χ_Ω is the characteristic function of the set Ω.

We denote by δ the distance from Ω to the boundary of the set G. Obviously, $\delta > 0$. For the function f_Ω, we construct the mean function $f_{\Omega, v^1} = f_{\Omega, v}$ (where $\lambda_i = 1$ for $i = 1, \ldots, n$), assuming that $v < \frac{\delta}{\sqrt{n}}$. Obviously, $f_{\Omega, v} \in C_0^\infty(G)$. By virtue of (8), we have, for sufficiently small v,

$$\|f_\Omega - f_{\Omega, v}\|_{p,G} < \frac{\varepsilon}{2},$$

so that

$$\|f - f_{\Omega, v}\|_{p,G} \leqslant \|f - f_\Omega\|_{p,G} + \|f_\Omega - f_{\Omega, v}\|_{p,G} < \frac{\varepsilon}{2} + \frac{\varepsilon}{2} = \varepsilon.$$

This proves the assertion.

§6. Generalized derivatives

We now introduce the concept of a generalized derivative in the sense of Sobolev, which will play a highly important role in what follows. In particular, we shall use this concept to define basic classes of functions that we shall study.

6.1. Suppose that the functions f and χ are locally summable on an open subset G of E^n. Suppose that, for every infinitely differentiable function φ whose support is bounded and contained in G, we have

$$\int_G \chi(x)\varphi(x)\,dx = (-1)^{|k|} \int_G f(x)\varphi^{(k)}(x)\,dx, \qquad (1)$$

where $k = (k_1, \ldots, k_n)$ (the k_i being nonnegative integers). Then χ is called a generalized derivative of the function f of the form $f^{(k)} = D^k f = \frac{\partial^{|k|} f}{\partial x_1^{k_1} \ldots \partial x_n^{k_n}}$ in G.

One can easily see that a given function f can have only one derivative of this kind. Specifically, if χ and χ_0 are two generalized derivatives, then (1) is also valid for χ_0. By subtracting one form of (1) from the other, we obtain

$$\int_G (\chi - \chi_0)\varphi \, dx = 0,$$

and then the arbitrariness of the function φ implies that χ and χ_0 are equivalent on G.

If $f(x)$ has continuous derivatives in G of orders up through $|k|$, when we apply the formula for integration by parts, we obtain equation (1) for $f(x)$. Consequently, the generalized derivative coincides with a continuous derivative. It also follows from the definition that a generalized derivative $f^{(k)}$ does not depend on the order of the differentiation since it is possible to perform in any order the differentiation of a function φ with continuous derivatives.

We mention a few other properties of a generalized derivative. If f_1 and f_2 have generalized derivatives $f_1^{(k)}$ and $f_2^{(k)}$, then the linear combination $C_1 f_1 + C_2 f_2$, where C_1 and C_2 are constants, has as its generalized derivative $C_1 f_1^{(k)} + C_2 f_2^{(k)}$.

If f has a generalized derivative $\frac{\partial f}{\partial x_1} = \chi$ and χ has a generalized derivative $\frac{\partial \chi}{\partial x_2}$, then f has the generalized derivative

$$\frac{\partial^2 f}{\partial x_1 \partial x_2} = \frac{\partial \chi}{\partial x_2}.$$

The situation is analogous with derivatives of other kinds. Furthermore, if f has the generalized derivatives $\frac{\partial f}{\partial x_2}$ and $\frac{\partial^2 f}{\partial x_1 \partial x_2}$, then

$\frac{\partial^2 f}{\partial x_1 \partial x_2}$ is the generalized derivative of $\frac{\partial f}{\partial x_2}$ with respect to x_1.

It follows from the definition of a generalized derivative that, if $f^{(k)}$ is the generalized derivative of f on G, it is also the generalized derivative of f on any open subset G' of G.

The following assertion is to some extent a converse of this: if f_1 has a generalized derivative $f_1^{(k)}$ on a region G_1 and f_2 has a generalized derivative $f_2^{(k)}$ on a region G_2 and if $f_1 \equiv f_2$ on $G_1 \cap G_2$, then the function f defined on $G_1 \cup G_2$, by

$$f = \begin{cases} f_1 & \text{in } G_1, \\ f_2 & \text{in } G_2 \end{cases}$$

has a generalized derivative $f^{(k)}$.*

We mention that by a region** we mean an open connected subset of E^n.

We shall frequently have occasion to use both this proposition and the following lemma:

6.2. Lemma. Let G denote a region. Let f denote a function defined on G and belonging to $L_p^{\text{loc}}(G)$, where $1 \leqslant p \leqslant \infty$. Let $\{f_j\}$, for $j = 1, 2, \ldots$, denote a sequence of functions f_j defined on G, belonging to $L_p^{\text{loc}}(G)$, and possessing generalized derivatives $f_j^{(k)} \in L_q^{\text{loc}}(G)$, where $1 \leqslant q \leqslant \infty$. Suppose that $f_j \to f$ in the sense of $L_p^{\text{loc}}(G)$ as $j \to \infty$ and that $(f_j^{(k)} - f_i^{(k)}) \to 0$ in the sense of $L_q^{\text{loc}}(G)$ as $j, i \to \infty$. Then, the function f has a generalized derivative $f^{(k)} \in L_q^{\text{loc}}(G)$ on G and $f_j^{(k)} \to f^{(k)}$ in the sense of $L_q^{\text{loc}}(G)$ as $j \to \infty$.

PROOF. Since $(f_j^{(k)} - f_i^{(k)}) \to 0$ in $L_q^{\text{loc}}(G)$ as $i, j \to \infty$, it follows that there exists a function χ defined on G such that $\chi \in L_q^{\text{loc}}(G)$ and $f_j^{(k)} \to \chi$ in $L_q^{\text{loc}}(G)$.

*This result, generalizing the familiar result of Sobolev, was obtained in [1] by M. A. Galahov. See also the article [2] by V. I. Burenkov.

**Translation editor's footnote. The word "domain" is often used interchangeably with "region" for this notion of an open connected subset of E^n.

Since f_j has a generalized derivative $f_j^{(k)}$ in G, we have, for any function $\varphi \in C_0^\infty(G)$,

$$\int_G f_j^{(k)}\varphi\,dx = (-1)^{|k|}\int_G f_j\varphi^{(k)}\,dx \qquad (j = 1, \ldots).$$

Since the support of φ is bounded and contained in G, the integrals have the same value as if they were over some compact subset of G.

Furthermore, by using the facts that $f_j \to f$ in $L_p^{\text{loc}}(G)$ and $f_j^{(k)} \to \chi$ in $L_q^{\text{loc}}(G)$ and then taking the limit in the equation shown above as $j \to \infty$, we obtain

$$\int_G \chi\varphi\,dx = (-1)^{|k|}\int_G f\varphi^{(k)}\,dx.$$

It follows that $\chi = f^{(k)}$. This completes the proof of the lemma.

REMARK. Suppose that the conditions of the lemma hold, that $f_j^{(k)} \in L_q(G)$ for $j = 1, 2, \ldots$, and that $\|f_j^{(k)} - f_i^{(k)}\|_{q,G} \to 0$ as i, $j \to \infty$. Then, $f^{(k)} \in L_q(G)$ and $\|f_j^{(k)} - f^{(k)}\|_{q,G} \to 0$ as $j \to \infty$.

6.3. Let us pause to show the connection between generalized differentiation and averaging.

Suppose that a function f belonging to $L^{\text{loc}}(E^n)$ defined on an open set G has a generalized derivative $f^{(k)}$. Let us set up the mean function

$$f_{v^\lambda}(x) = v^{-|\lambda|}\int_{E^n} f(y)\,K((y-x) : v^\lambda)\,dy$$

and let us find $D_x^k f_{v^\lambda}(x)$. (Since $f_{v^\lambda} \in C^\infty(E^n)$, this derivative is the usual one.) We have

$$D_x^k f_{v^\lambda}(x) = v^{-|\lambda|}\int_{E^n} f(y)\,D_x^k K((y-x) : v^\lambda)\,dy$$

$$= (-1)^{|k|} v^{-|\lambda|} \int_{E^n} f(y) D_y^k K((y-x) : v^\lambda)\, dy.$$

The function $K((y-x) : v^\lambda)$ treated as a function of y vanishes outside the closed set $x + S_{v^\lambda}(K)$. If this set is contained in G, then $K((y-x) : v^\lambda)$ can be taken as the function φ of bounded support contained in G in formula (1). Then, since f has a generalized derivative $f^{(k)}$, we have on the basis of formula (1)

$$D_x^k f_{v^\lambda}(x) = v^{-|\lambda|} \int_{E^n} D^k f(y) \cdot K((y-x) : v^\lambda)\, dy,$$

that is,

$$D^k (f_{v^\lambda}(x)) = (D^k f)_{v^\lambda}(x). \tag{2}$$

This last equation means that the mean function of the generalized derivative coincides with the derivative of the same form of the mean function on the set

$$U = \{x : x \in G,\ x + S_{v^\lambda}(K) \subset G\}.$$

6.4. There are other definitions of a generalized derivative that are equivalent to the one given above. We shall not stop to give all of them* and mention only a connection between the existence of a generalized derivative in the sense of Sobolev and the absolute continuity of a function. This will prove useful to us later. We give only the results and refer the reader to Nikol'skiĭ's monograph [9] for the proof.

If f has a generalized derivative

$$D_{x_i}^{k_i} f = \frac{\partial^{k_i} f}{\partial x_i^{k_i}}$$

*See, for example, the book [9] by Nikol'skiĭ, §4.1.

on an open set G, then there exists a function ψ equivalent to it on G that has generalized derivatives with respect to x_i of orders through k_i almost everywhere on G and

$$\psi, \frac{\partial \psi}{\partial x_i}, \ldots, \frac{\partial^{k_i - 1} \psi}{\partial x_i^{k_i - 1}}$$

are, for almost all fixed $x^{(i)} = (x_1, \ldots, x_{i-1}, x_{i+1}, \ldots, x_n) \in G^{(i)}$ (where $G^{(i)}$ denotes the orthogonal projection of G onto the hyperplane $x_i = 0$), absolutely continuous functions on any closed interval $[a, b]$ of the variable x_i contained in G.

This implies in particular that, if $D_{x_i}^{k_i} f$ exists (we shall assume that f is already modified on a set of n-dimensional measure zero as was indicated above), if φ belongs to $C_0^\infty(G)$, and if the intersection of the line $x^{(i)} =$ const with the support of φ is contained in $[a, b]$, then

$$\int_a^b f D_{x_i}^{k_i} \varphi \, dx_i = (-1)^{k_i} \int_a^b \left(D_{x_i}^{k_i} f \right) \varphi \, dx_i; \tag{3}$$

that is, the integration by parts is valid.

We mention also that, if the segment $[x_i, x_i + k_i t] \times x^{(i)}$ is contained in G, then we have

$$\Delta_i(t) D_i^{s_i} f(x) = D_i^{s_i} f(x + te_i) - D_i^{s_i} f(x) = \int_0^t D_i^{s_i + 1} f(x + \tau e_i) \, d\tau$$

for any derivative of order s_i, where $0 \leqslant s_i \leqslant k_i - 1$. By successive application of this equation, we obtain the formula

$$\Delta_i^{k_i}(t) f(x) =$$

$$\overbrace{\int_0^t \cdots \int_0^t}^{k_i} D_i^{k_i} f(x + (\tau_1 + \ldots + \tau_{k_i}) e_i) \, d\tau_1 \ldots d\tau_{k_i}, \tag{4}$$

where

$$\Delta_i^{k_i}(t) f(x) = \overbrace{\Delta_i(t) \ldots \Delta_i(t)}^{k_i} f(x) = \sum_{j=0}^{k_i} C_{k_i}^j (-1)^{k_i - j} f(x + jte_i). \tag{5}$$

From (4) we get the following useful inequality:

$$\left|\Delta_i^{k_i}(t) f(x)\right| \leqslant t^{k_i - 1} \int_0^{k_i t} \left|D_i^{k_i} f(x + \tau e_i)\right| d\tau, \qquad t > 0. \tag{6}$$

Formula (4) and inequality (6) will be used later.

6.5. Not only can we introduce the concept of an individual generalized differentiable function but we can also introduce the concept of a generalized linear differential operator.

Let $\alpha = (\alpha_1, \ldots, \alpha_n)$ denote a vector with nonnegative-integer-valued coordinates. Let D denote $(D_1, \ldots, D_n)$, where

$$D_i = \frac{\partial}{\partial x_i}.$$

Let us define

$$D^\alpha = D_1^{\alpha_1} \ldots D_n^{\alpha_n},$$

$$(-D)^\alpha = (-1)^{|\alpha|} D^\alpha,$$

$$P(D) = \sum_\alpha c_\alpha D^\alpha,$$

this last being a differential polynomial with real or complex constant coefficients.

We shall say that a locally summable function f defined on an open set G admits application of a generalized linear differential operator $P(D)$ on G if there exists a locally summable function χ on G such that, for every function $\varphi \in C_0^\infty(G)$,

$$\int_G \chi(x)\varphi(x)\,dx = \int_G f(x)P(-D)\varphi(x)\,dx.$$

In this case, we set $P(D)f = \chi$. Accordingly, we can write the preceding equation in the form

$$\int_G P(D)f(x)\cdot\varphi(x)\,dx = \int_G f(x)P(-D)\varphi(x)\,dx. \tag{7}$$

We note that there is no assumption regarding the existence of the individual derivatives appearing in $P(D)f$.

§7. Integral representations of differentiable functions

The derivation of the integral representations of functions that we shall give in the present section is based on the following simple idea:

For a given locally summable function f, we construct its mean function $f_{v^\lambda}(x) = f(x, v)$ with some kernel Ω and averaging parameter v^λ, where λ is a fixed vector. The average $f(x, v)$ can also be regarded as a continuously differentiable function of the parameter v for $v > 0$.

Therefore, on the basis of the Newton-Leibnitz formula, we have, for arbitrary ε and h such that $0 < \varepsilon < h$,

$$f_{\varepsilon^\lambda}(x) = f_{h^\lambda}(x) - \int_\varepsilon^h \frac{\partial}{\partial v} f_{v^\lambda}(x)\,dv. \tag{1}$$

This equation is the starting point in the derivation of all the representations that we shall consider here.

With suitable choice of averaging kernel Ω, we can express the last term in (1) in terms of integrals either of the generalized derivatives of the function f or of finite differences of that function. Therefore, to obtain a particular representation, the construction of the corresponding kernel Ω is quite important.

If the function f satisfies the conditions of Lemma 5.2 or 5.3, then $f_{\varepsilon\lambda}$ approaches f in some sense or other as $\varepsilon \to 0$. If we let ε approach 0 in equation (1), we obtain the needed integral representation of f.

7.1. The kernel $\Omega(x)$. First of all, let us construct the averaging kernel $\Omega(x)$, which will be useful for obtaining an integral representation of the function in question in terms of its derivatives.

Let $K(x)$ denote a function satisfying the conditions of subsection 5.1; that is, suppose that $K \in C_0^\infty(E^n)$ and

$$\int_{E^n} K(x)\, dx = 1. \tag{2}$$

For the moment, we make no assumptions regarding the set $S(K) = \operatorname{supp} K$ although the choice of the support of K will play an important role later.

From the given function $K(x)$ we construct a function $\Omega(x)$ that possesses the listed properties of the function K and also certain new properties that will be of importance to us. Specifically, let us define

$$\Omega(x) = D_x^k \left[\frac{x^{k-1}}{(k-1)!} \int_{E^n} K(z)\, \theta(x-z)\, dz \right], \tag{3}$$

where $k = (k_1, \ldots, k_n)$, the k_i being sufficiently large natural numbers, $\mathbf{1} = (\overbrace{1, \ldots, 1}^{n})$, and

$$\theta(x) = \prod_{j=1}^{n} \theta(x_j),$$

where $\theta(x_j)$ is Heaviside's function; that is, $\theta(t) = 1$ for $t > 0$ and $\theta(t) = 0$ for $t < 0$. One can easily see that $\Omega \in C_0^\infty(E^n)$ and $\operatorname{supp} \Omega \subset S(K)$. Let us show that

$$\int_{E^n} \Omega(x)\, dx = 1. \tag{4}$$

First of all, we note that, on the basis of Leibnitz's formula for the derivative of a product,

$$D_{x_i}^{k_i-1}\left[\frac{x_i^{k_i-1}}{(k_i-1)!}\int_{E^n} K(z)\,\theta(x-z)\,dz\right]$$

$$=\int_{E^n} K(z)\,\theta(x-z)\,dz+T_i(x),$$

where

$$T_i(x)=\sum_{s=1}^{k_i-1} C_s x_i^s D_{x_i}^s\left(\int_{E^n} K(z)\,\theta(x-z)\,dz\right)$$

$$=\sum_{s=1}^{k_i-1} C_s x_i^s D_{x_i}^{s-1}\left[\int_{E^n} K(z)\,\delta(x_i-z_i)\left(\prod_j^{(i)}\theta(x_j-z_j)\right)dz\right]$$

$$=\sum_{s=1}^{k_i-1} C_s x_i^s\int_{E^n} D_{x_i}^{s-1}K(z_1,\,\ldots,\,x_i,\,\ldots,\,z_n)\left(\prod_j^{(i)}\theta(x_j-z_j)\right)dz^{(i)},$$

δ denoting the delta function.

Since supp $K\subset I_{\boldsymbol{a}}$ for some $\boldsymbol{a}$, we have $T_i(x)=0$ for $|x_i|>a_i$. Therefore,

$$\int_{E^1} D_{x_i}^{k_i}\left[\frac{x_i^{k_i-1}}{(k_i-1)!}\int_{E^n} K(z)\,\theta(x-z)\,dz\right]dx_i$$

$$=\left[\int_{E^n} K(z)\,\theta(x-z)\,dz+T_i(x)\right]\Bigg|_{x_i=-\infty}^{x_i=+\infty}$$

$$=\int_{E^n} K(z)\left(\prod_j^{(i)}\theta(x_j-z_j)\right)dz. \qquad (5)$$

If we now represent the integral in the left-hand member of (4) in the form of iterated integrals and then apply formula (5) successively n times, we get

$$\int_{E^n} \Omega(x)\,dx = \int_{E^n} K(x)\,dx = 1.$$

Thus, the function Ω satisfies all the conditions imposed on the averaging kernel in §5. Consequently, we take it for the averaging kernel.

Suppose that $\lambda = (\lambda_1, \ldots, \lambda_n)$, where $\lambda_i > 0$ for $i = 1, \ldots, n$. Let v denote a positive number. Obviously, the function

$$\Omega_{v^\lambda}(x) = v^{-|\lambda|}\Omega(x : v^\lambda) \tag{6}$$

is infinitely differentiable with respect to x on E^n and with respect to v for $v > 0$. Also, supp $\Omega_{v^\lambda} \subset S_{v^\lambda}(K)$, where $S_{v^\lambda}(K)$ is defined by 5(3).

We note also that

$$\int_{E^n} \Omega_{v^\lambda}(x)\,dx = \int_{E^n} v^{-|\lambda|}\Omega(x : v^\lambda)\,dx = \int_{E^n} \Omega(x)\,dx = 1. \tag{7}$$

Let us find the derivative of $\Omega_{v^\lambda}(x)$ with respect to the parameter v. We have

$$\begin{aligned}
\frac{\partial}{\partial v}\Omega_{v^\lambda}(x) &= \frac{\partial}{\partial v}[v^{-|\lambda|}\Omega(x : v^\lambda)] \\
&= D_x^k\left[\frac{x^{k-1}}{(k-1)!}\int_{E^n} K(z)\frac{\partial}{\partial v}\theta(x : v^\lambda - z)\,dz\right] \\
&= -\sum_{i=1}^n \lambda_i v^{-1-\lambda_i} D_x^k\left[\frac{x^{k-1}x_i}{(k-1)!}\int_{E^n} K(z)\,\delta(x_i v^{-\lambda_i} - z_i)\right. \\
&\quad \times\left.\left(\prod_j^{(i)} \theta(x_j v^{-\lambda_j} - z_j)\right)dz\right] = -\sum_{i=1}^n \lambda_i v^{-1-|\lambda|} D_i^{k_i}\tilde{\mathscr{L}}_i(x : v^\lambda),
\end{aligned} \tag{8}$$

where

$$\tilde{\mathscr{L}}_i(x) = D^{k-k_ie_i}\left[\frac{x^{k-1}x_i}{(k-1)!}\int\limits_{E^{n-1}} K(z_1, \ldots, x_i, \ldots, z_n)\right.$$

$$\left.\times\left(\prod_j^{(i)}\theta(x_j - z_j)\right)dz^{(i)}\right]. \quad (9)$$

It is obvious from formula (8) that the derivative of $\Omega_{v^\lambda}(x)$ with respect to v can be represented as the sum of n functions each of which has zero moments of the first $k_i - 1$ orders with respect to the corresponding variable x_i. This property will be used most often in the following situation:

Suppose that a function f is defined on a region G containing the support of the function $\tilde{\mathscr{L}}_i(x)$ and has a generalized derivative $D_i^{l_i}f$ on G. Then,

$$\int\limits_G f(x)\, D_{x_i}^{k_i}\tilde{\mathscr{L}}_i(x)\, dx$$

$$= (-1)^{l_i}\int\limits_G D_{x_i}^{l_i}f(x)\, D_{x_i}^{k_i-l_i}\tilde{\mathscr{L}}_i(x)\, dx \quad (l_i \leqslant k_i). \quad (10)$$

This follows from the definition 6(1) of a generalized derivative with $D_i^{k_i-l_i}\tilde{\mathscr{L}}_i(x)$ taken as φ.

This property of the function $\Omega_{v^\lambda}(x)$ is also the basis of our examination in subsequent averagings with the kernel Ω.

7.2. Integral representations of functions in terms of their derivatives. Let f denote a function defined on a region G contained in E^n. In what follows, we shall assume that G is such that the transformations in what follows make sense. As indicated above, by a region we mean an open connected set.

In addition, we shall assume that f belongs to $L^{\text{loc}}(G)$ and that it has, on G, the generalized derivatives referred to.

Let us look at an average of the function f, taking the function Ω introduced above as averaging kernel and taking v^λ as averaging parameter; that is, let us consider the function

$$f_{v^\lambda}(x) = v^{-|\lambda|} \int_{E^n} f(x+y)\,\Omega(y:\ v^\lambda)\,dy, \tag{11}$$

where $\lambda = (\lambda_1, \ldots, \lambda_n)$, with $\lambda_i > 0$ for $i = 1, \ldots, n$, and $v > 0$.

Let us apply formula (1) to this function. By virtue of (8), we have

$$\frac{\partial}{\partial v} f_{v^\lambda}(x) = \int_{E^n} f(x+y)\,\frac{\partial}{\partial v}\,[v^{-|\lambda|}\Omega(y:v^\lambda)]\,dy$$

$$= -\sum_{i=1}^{n} \lambda_i v^{-1-|\lambda|} \int_{E^n} f(x+y)\,D_i^{k_i}\tilde{\mathscr{L}}_i(y:v^\lambda)\,dy,$$

so that

$$f_{\varepsilon^\lambda}(x) = f_{h^\lambda}(x) + \int_{\varepsilon}^{h} \sum_{i=1}^{n} \lambda_i v^{-1-|\lambda|}\,dv \int_{E^n} f(x+y)\,D_i^{k_i}\tilde{\mathscr{L}}_i(y:v^\lambda)\,dy. \tag{12}$$

Let l_i, for $i = 1, \ldots, n$, denote arbitrary nonnegative integers. We may assume that $l_i \leqslant k_i$ for $i = 1, \ldots, n$ since the numbers k_i appearing in the kernel Ω can be chosen arbitrarily large. Let us use equation (10) to make a transformation of the integral with respect to y in (12). We note that

$$D_i^{k_i}\tilde{\mathscr{L}}_i(y:v^\lambda) = v^{\lambda_i l_i} D_{y_i}^{l_i} D_i^{k_i - l_i}\tilde{\mathscr{L}}_i(y:v^\lambda).$$

Then, we obtain

$$f_{\varepsilon^\lambda}(x)$$
$$= f_{h^\lambda}(x) + \int_{\varepsilon}^{h} \sum_{i=1}^{n} v^{-1-|\lambda|+l_i\lambda_i}\,dv \int_{E^n} D_i^{l_i} f(x+y)\,\mathscr{L}_i(y:v^\lambda)\,dy, \tag{13}$$

where

$$\mathscr{L}_i(x) = (-1)^{l_i} \lambda_i D_i^{k_i - l_i} \tilde{\mathscr{L}}_i(x) \qquad (i = 1, \ldots, n). \qquad (14)$$

Equation (13) can be regarded as a representation of the difference of values of the mean functions with parameters ε^λ and h^λ at the point x in terms of integrals of the generalized derivatives of the function f in the coordinate directions.

If we let ε approach 0 in (13), we obtain the following identity:

$$f(x) = f_{h^\lambda}(x) + \int_0^h \sum_{i=1}^n v^{-1-|\lambda|+l_i\lambda_i}\, dv \int_{E^n} D_i^{l_i} f(x+y)\, \mathscr{L}_i(y : v^\lambda)\, dy, \qquad (15)$$

which is valid for almost every x in the set for which the right-hand member of formula (15) is meaningful.

Let us clarify this. First of all, we note that, since supp $\mathscr{L}_i \subset S(K)$, the integral in the right-hand member of (15) is the same as if it were over the set

$$S(K, \lambda, h) = \bigcup_{0 < v \leqslant h} S_{v^\lambda}(K), \qquad (16)$$

that is, over the union of the sets $S_{v^\lambda}(K)$ defined by equation 5(3). Consequently, in the right-hand member of (15), we use the values of the function f and its derivatives at points of the set $x + S(K, \lambda, h)$, which it is natural to call the support of the representation (15).*

As usual, let G denote the domain of definition of the function f. Let us define

$$U = \{x\colon\ x \in G,\ x + S(K, \lambda, h) \subset G\}. \qquad (17)$$

Then, at least formally, the right-hand member of (15) is meaningful

*We recall that $x + S(K, \lambda, h)$ denotes the arithmetic sum of x and $S(K, \lambda, h)$ (see notation at the beginning of Chapter II.)

for all x in the set U. Furthermore, since we assume on the basis of Lemma 5.3 (with $p=1$) that $f \in L^{\mathrm{loc}}(G)$, we may assert that $f_{\varepsilon^\lambda}(x) \to f(x)$ almost everywhere on G and hence almost everywhere on U as $\varepsilon \to 0$. It follows that the identity (15) obtained by letting ε approach zero in (13) holds for almost all points x in the set U.

In applications of (15), we shall usually assume that the support of the function K is contained in one of the coordinate angles.

Suppose that $b > 0$, that $a_i \neq 0$ for $i = 1, \ldots, n$, and that

$$S(K) \subset \left\{x\colon \frac{x_i}{a_i} > 0,\ 1 < \left(\frac{x_i}{a_i}\right)^{1/\lambda_i} < 1+b \ (i=1, \ldots, n)\right\}. \tag{18}$$

Then,

$$S(K, \lambda, h) \equiv V\left(\frac{1}{\lambda}\right) = \bigcup_{0<v\leqslant h} \left\{x\colon \frac{x_i}{a_i} > 0,\ v < \left(\frac{x_i}{a_i}\right)^{1/\lambda_i} < (1+b)\,v \ (i=1, \ldots, n)\right\}. \tag{19}$$

We shall call the region $V\left(\frac{1}{\lambda}\right)$ a $\frac{1}{\lambda}$*-horn of radius h and opening b.*

In the most important cases of application of formula (15), we shall assume that $\lambda = \frac{1}{l}$ (where $\lambda_i = \frac{1}{l_i}$ for $i = 1, \ldots, n$). Then, the displaced l-horn $x + V(l)$ will serve as support of the representation (15). Obviously, if $l_1 = \ldots = l_n$, an l-horn $V(l)$ is a cone.

We mention also that the right-hand member of formula (15) is the same at any point x for all equivalent functions $f(x)$. Therefore, we can treat it as a representative of the entire class of equivalent functions.

Finally, we note that (15) is the analogue of the well-known identity of Sobolev (see [1], [2]), giving a representation of the function in terms of integrals of the function itself and all its derivatives of a specified order.*

*See also, Kantorovič [1].

7.3. Formulas (13) and (15) can be generalized. Suppose that $\alpha=(\alpha_1, \ldots, \alpha_n)$ and $l^i=(l_1^i, \ldots, l_n^i)$ (for $i=1, \ldots, n$) are vectors with nonnegative-integer-valued components satisfying the conditions

$$l_j^i \leqslant \alpha_j \quad (j=1, \ldots, n;\ j \neq i), \quad l_i^i \leqslant \alpha_i + k_i \quad (i=1, \ldots, n), \tag{20}$$

where the k_i are the numbers appearing in the formula (12).

Let us suppose that f has on G generalized derivatives $D^{l^i}f \in L^{\text{loc}}(G)$ for $i=1, \ldots, n$. If we apply the differential operator D_x^α to both sides of (12), we get

$$D_x^\alpha f_{\varepsilon^\lambda}(x) = D_x^\alpha f_{h^\lambda}(x) + \int_\varepsilon^h \sum_{i=1}^n \lambda_i v^{-1-|\lambda|}\, dv \int_{E^n} f(y)\, D_x^\alpha D_i^{k_i} \tilde{\mathscr{L}}_i((y-x):v^\lambda)\, dy. \tag{21}$$

Furthermore,

$$\int_{E^n} f(y)\, D_x^\alpha D_i^{k_i} \tilde{\mathscr{L}}_i((y-x):v^\lambda)\, dy = (-1)^{|\alpha|} v^{-(\alpha,\lambda)} \int_{E^n} f(x+y)\, D^{\alpha+k_i e_i} \tilde{\mathscr{L}}_i(y:v^\lambda)\, dy$$

$$= (-1)^{|\alpha|+|l^i|} v^{-(\alpha,\lambda)+(l^i,\lambda)} \int_{E^n} D^{l^i} f(x+y)\, D^{\alpha+k_i e_i - l^i} \tilde{\mathscr{L}}_i(y:v^\lambda)\, dy.$$

We note that the last of these equations follows from the definition 6(1) of the generalized derivative $D^{l^i}f$ if we take for φ the function $D^{\alpha+k_i e_i - l^i}\tilde{\mathscr{L}}_i(y:v^\lambda)$. Using the last transformation, we can put equation (21) in the form

$$D_x^\alpha f_{\varepsilon^\lambda}(x) = D_x^\alpha f_{h^\lambda}(x) + \int_\varepsilon^h \sum_{i=1}^n v^{-1-\varkappa_i}\, dv \int_{E^n} D^{l^i} f(x+y)\, M_i(y:v^\lambda)\, dy, \tag{22}$$

where

$$\varkappa_i = |\lambda| + (\alpha, \lambda) - (l^i, \lambda),$$
$$M_i(x) = (-1)^{|\alpha|+|l^i|}\lambda_i D^{\alpha+k_i e_i - l^i}\tilde{\mathscr{L}}(x) \qquad (j = 1, \ldots, n). \tag{23}$$

We note that the set of points y at which we use the values of the function f and its derivatives in the right-hand member of (22) (for arbitrary ε in $0 < \varepsilon < h$) is contained in the set $x + S(K, \lambda, h)$, where $S(K, \lambda, h)$ is defined by (16).

Let us now show that if the inequalities

$$\mu_i = (l^i, \lambda) - (\alpha, \lambda) > 0 \tag{24}$$

hold for $(i = 1, \ldots, n)$, then there exists on G the generalized derivative $D^\alpha f \in L^{\text{loc}}(G)$ and let us find an integral representation for it.

Let us show first that $D^\alpha f_{\varepsilon\lambda} - D^\alpha f_{h\lambda} \to 0$ in the sense of $L^{\text{loc}}(G)$ as ε and h approach 0 through combinations of values such that $0 < \varepsilon < h$. Let F denote a compact subset of G. Then, for all sufficiently small positive h, the set $F + S(K, \lambda, h) = F + S$ is contained in some compact subset of G. By virtue of equation (22), Minkowski's generalized inequality 2(12), and the estimate 2(19) for a convolution, we have

$$\left\| D^\alpha f_{\varepsilon\lambda} - D^\alpha f_{h\lambda} \right\|_{1,F} \leqslant \int_0^h \sum_{i=1}^n v^{-1-\varkappa_i} \left\| D^{l^i} f \right\|_{1,F+S} \left\| \mathscr{L}_i\left(\frac{\cdot}{v^\lambda}\right) \right\|_1 dv$$
$$\leqslant \sum_{i=1}^n \left\| D^{l^i} f \right\|_{1,F+S} \| \mathscr{L}_i \|_1 \frac{h^{\mu_i}}{\mu_i}.$$

Since $\mu_i > 0$ for $i = 1, \ldots, n$, it follows that $D^\alpha f_{\varepsilon\lambda} - D^\alpha f_{h\lambda} \to 0$ in $L(F)$ and hence in $L^{\text{loc}}(G)$ as ε and h approach zero through combinations of values such that $0 < \varepsilon < h$.

Furthermore, by virtue of the remark to Lemma 5.2, $f_{\varepsilon\lambda} \to f$ in $L^{\text{loc}}(G)$ as $\varepsilon \to 0$. We conclude from this on the basis of Lemma 6.2

that there exists $D^\alpha f \in L^{\mathrm{loc}}(G)$. Let U denote the set (17). Formula 6(2) tells us that, on this set,

$$D_x^\alpha f_{\varepsilon\lambda}(x) = (D^\alpha f)_{\varepsilon\lambda}(x).$$

It then follows on the basis of Lemma 5.3 that $D_x^\alpha f_{\varepsilon\lambda}(x) \to D^\alpha f(x)$ almost everywhere on U as $\varepsilon \to 0$.

Therefore, when we let ε approach 0 in (22), we obtain the formula

$$D^\alpha f(x) = D^\alpha f_{h\lambda}(x) + \int_0^h \sum_{i=1}^n v^{-1-\varkappa_i}\, dv \int_{E^n} D^{l^i} f(x+y)\, M_i(y : v^\lambda)\, dy, \tag{25}$$

which is valid for almost all $x \in U$. We note that the integral representation (25) was obtained under the assumption that inequalities (24) hold.

For these last estimates, it is significant that the kernels $\mathscr{L}_i$ and M_i in formulas (13), (15), (22), and (25) may be assumed to satisfy the conditions

$$\int_{E^n} D^\beta \mathscr{L}_i(x)\, dx = 0, \qquad \int_{E^n} D^\beta M_i(x)\, dx = 0 \qquad (i = 1, \ldots, n) \tag{26}$$

for arbitrary $\beta = (\beta_1, \ldots, \beta_n)$, where $\beta_i > 0$ for $i = 1, \ldots, n$.

It follows from (14) that the first of equations (26) will hold if $l_i < k_i$ and it follows from (23) that the second equation will hold if $l_i^i < k_i$. By a suitable choice of the numbers k_i, we can always arrange for these two inequalities to hold.

7.4. Representation of a function in terms of differential polynomials. We can use a theorem of Hilbert on the roots of polynomials to obtain from equation (12) more general formulas than (13) and (15). Let us review some of our notation. If $(\alpha = \alpha_1, \ldots, \alpha_n)$ is a vector with nonnegative integer-valued components, $\xi = (\xi_1,$

$\ldots, \xi_n)$, D_i is the derivative with respect to the ith argument, and $D = (D_1, \ldots, D_n)$, then $\xi^\alpha = \xi_1^{\alpha_1} \ldots \xi_n^{\alpha_n}$, $(-\xi)^\alpha = (-1)^{|\alpha|}\xi^\alpha$, and $D^\alpha = D_1^{\alpha_1} \ldots D_n^{\alpha_n}$.

The theorem of Hilbert mentioned above is as follows (see van der Waerden [1]): Let $Q(\xi)$ denote a polynomial that vanishes at all common (complex) roots of the polynomials $P_1(\xi), \ldots, P_N(\xi)$. Then, there exist polynomials $a_1(\xi), \ldots, a_N(\xi)$ such that, for some natural number m,

$$Q^m(\xi) = \sum_{j=1}^{N} a_j(\xi) P_j(\xi).$$

Suppose that we are given a vector $l = (l_1, \ldots, l_n)$ with positive-integer-valued components. Let us define $\lambda_i = \frac{1}{l_i}$ for $i = 1, \ldots, n$ and then let us define $\lambda = (\lambda_1, \ldots, \lambda_n)$.

We shall call a polynomial of the form

$$P(\xi) = \sum_{|\alpha : l|=1} c_\alpha \xi^\alpha \qquad \left(|\alpha : l| = \sum_{i=1}^{n} \frac{\alpha_i}{l_i}\right)$$

an *l-polynomial.* If all the l_i have a common value m, the polynomial $P(\xi)$ will obviously be a homogeneous polynomial of degree m.

Let $P_j(\xi)$, for $j = 1, \ldots, N$, be a collection of l-polynomials with constant (complex) coefficients. Let us suppose that the $P_j(\xi)$ have no common complex root other than $\xi = 0$. Then, by Hilbert's theorem, for a sufficiently large natural number m there exist polynomials $a_{ij}(\xi)$ (for $i = 1, \ldots, n$ and $j = 1, \ldots, N$) such that

$$\xi_i^{l_i m} = \sum_{j=1}^{N} a_{ij}(\xi) P_j(-\xi) \qquad (i = 1, \ldots, n). \tag{27}$$

Since the monomials $\xi_i^{l_i m}$ in the left-hand members of equations (27) are ml-polynomials, we may assume that the $a_{ij}(\xi)$ are $(m-1)l$-polynomials.

This is true because, if the $a_{ij}(\xi)$ satisfy the identities (27), their $(m-1)l$-polynomial parts also satisfy those identities.

Now, let us set $k_i = ml_i$ (for $i = 1, \ldots, n$) in formula (12) and let us assume that the function f admits the application of the generalized differential operators $P_j(D)$ (for $j = 1, \ldots, n$) on G.

By virtue of the identities (27), the integral with respect to y in the right-hand member of (12) can be written in the form

$$\int_{E^n} f(x+y)\, D_i^{k_i} \tilde{\mathscr{L}}_i(y : v^\lambda)\, dy$$
$$= \sum_{j=1}^{N} \int_{E^n} f(x+y)\, P_j(-D)\, a_{ij}(D)\, \tilde{\mathscr{L}}_i(y : v^\lambda)\, dy.$$

Remembering that

$$P_j(-D) = \sum_{(\alpha,\lambda)=1} c_\alpha v^{(\alpha,\lambda)} (-D_y)^\alpha = v P_j(-D_y)$$

and applying formula 6(7), we obtain

$$\int_{E^n} f(x+y)\, D_i^{k_i} \tilde{\mathscr{L}}_i(y : v^\lambda)\, dy$$
$$= \sum_{j=1}^{N} v \int_{E^n} P_j(D) f(x+y)\, a_{ij}(D)\, \tilde{\mathscr{L}}_i(y : v^\lambda)\, dy.$$

When we substitute this expression into formula (12) and define

$$T_j(x) = \sum_{i=1}^{n} \lambda_i a_{ij}(D)\, \tilde{\mathscr{L}}_i(x), \tag{28}$$

we get

$$f_{\varepsilon^\lambda}(x) = f_{h^\lambda}(x)$$
$$+ \int_{\varepsilon}^{h} \sum_{j=1}^{N} v^{-|\lambda|}\, dv \int_{E^n} P_j(D) f(x+y)\, T_j(y : v^\lambda)\, dy. \tag{29}$$

Obviously, formula (29) generalizes formula (13), which follows from (29) for $N=n$ and $P_j(\xi)=\xi_j^{l_j}$ for $j=1,\ldots,n$.

If we apply the operator D_x^ν to both sides of (29), we get

$$D_x^\nu f_{\varepsilon\lambda}(x)=D_x^\nu f_{h\lambda}(x)$$

$$+(-1)^{|\nu|}\int_\varepsilon^h\sum_{j=1}^N v^{-|\lambda|-(\nu,\lambda)}\,dv\int_{E^n} P_j(D)f(x+y)\,D^\nu T_j(y:v^\lambda)\,dy. \tag{30}$$

If we now take the limit as $\varepsilon\to 0$, we get (under the assumption that there exists $D^\nu f\in L^{\mathrm{loc}}(G)$) a representation of $D^\nu f(x)$ in terms of $P_j(D)f(x)$.

Since $\operatorname{supp} T_j\subset\operatorname{supp} K$ for $j=1,\ldots,N$, the set $x+S(K,\lambda,h)$ or the displaced horn $x+V(l)$ (where $\lambda=1/l$) is the support of the representation (30).

For the kernels T_j, we may assume that

$$\int_{E^n} D^\alpha T_j(x)\,dx=0 \qquad (j=1,\ldots,N) \tag{31}$$

for any vector α with nonnegative-integer-valued components.

We note that the question of integral representations of functions in terms of the result of the application of certain differential operators to them was examined in the works of Smith [1], Besov [8], Uspenskiĭ [5], and Rešetnyak [1].

7.5. Another representation of functions in terms of their derivatives. In this subsection, we shall show that every function f that has generalized derivatives of the form $D_i^{l_i}f$ (for $i=1,\ldots,n$) can be represented as the sum of some polynomial of degree not exceeding l_i-1 in x_i (for $i=1,\ldots,n$) and the integrals of the derivatives $D_i^{l_i}f$. Anticipating somewhat, we shall say that this representation will also be treated as an expansion of functions in the space $W_p^l(G)$ (see definition in §9) with the aid of the projection operator, which maps $W_p^l(G)$ into the space of the polynomials mentioned above. For isotropic spaces, this representation was first obtained by Sobolev [1], [2].

Let $K(x)$ and $\theta(x)$ denote the functions introduced in subsection 7.1. Let us assume that $S(K) = \text{supp } K \subset I_1$. (We recall that $I_1 = \{x: |x_i| \leqslant 1 \ (i = 1, \ldots, n)\}$.) Suppose that $l = (l_1, \ldots, l_n)$ is a vector with positive-integer-valued components.

Let us define

$$\Omega(x, y) = D_y^l \left[\frac{y^{l-1}}{(l-1)!} \int_{E^n} K(x+z)\,\theta(y-z)\,dz \right]. \quad (32)$$

One can easily see that $\Omega(x, y)$ is an infinitely differentiable function of its arguments in $E^n \times E^n$ and that, for fixed x, it is a function of y of bounded support in E^n such that $\text{supp } \Omega(x, \cdot) \subset \text{supp } K(x + \cdot)$.

We can show, just as in the proof of (4), that

$$\int_{E^n} \Omega(x, y)\,dy = \int_{E^n} K(x+y)\,dy = 1. \quad (33)$$

Suppose that f satisfies the conditions of subsection 7.2, that $\lambda = (\lambda_1, \ldots, \lambda_n)$, where $\lambda_i = \frac{1}{l_i}$ for $i = 1, \ldots, n$, and that $v > 0$. Let us define

$$f(x, v) = v^{-|\lambda|} \int_{E^n} f(x+y)\,\Omega(x, y : v^{\lambda})\,dy$$

$$= v^{-|\lambda|} \int_{E^n} f(y)\,\Omega(x, (y-x) : v^{\lambda})\,dy. \quad (34)$$

The function $f(x, v)$ is an infinitely differentiable function of x and v for $v > 0$.

Let ε denote a number such that $0 < \varepsilon < 1$. Then,

$$f(x, \varepsilon) = f(x, 1) - \int_{\varepsilon}^{1} \frac{\partial}{\partial v} f(x, v)\, dv. \tag{35}$$

The first term in the right-hand member is a polynomial of degree not exceeding $l_i - 1$ in x_i (for $i = 1, \ldots, n$). Specifically,

$$\begin{aligned} f(x, 1) &= \int_{E^n} f(x+y)\,\Omega(x, y)\, dy \\ &= \int_{E^n} f(x+y)\, D_y^l \left[\frac{y^{l-1}}{(l-1)!} \int_{E^n} K(x+z)\,\theta(y-z)\, dz \right] dy \\ &= \int_{E^n} f(y)\, D_y^l \left[\frac{(y-x)^{l-1}}{(l-1)!} \int_{E^n} K(z)\,\theta(y-z)\, dz \right] dy = P_{l-1}(f; x). \end{aligned} \tag{36}$$

We note that the coefficients in the polynomial P_{l-1} are expressed in terms of integrals of the function f over the set $S(K)$.

Let us look at the second term in the right-hand member of (35). First of all, we note that, in analogy with formula (8),

$$\frac{\partial}{\partial v} [v^{-|\lambda|}\Omega(x, y : v^{\lambda})] = - \sum_{i=1}^{n} \lambda_i v^{- -|\lambda|+l_i\lambda_i} D_{y_i}^{l_i} \tilde{\mathscr{L}}_i(x, y : v^{\lambda}),$$

where

$$\tilde{\mathscr{L}}_i(x, y) = D_y^{l-l_ie_i} \left[\frac{y^{l-1} y_i}{(l-1)!} \int_{E^{n-1}} K(x_1 + z_1, \ldots, x_i + y_i, \ldots \right.$$
$$\left. \ldots, x_n + z_n) \left(\prod_j^{(i)} \theta(y_j - z_j) \right) dz^{(i)} \right]. \tag{37}$$

Therefore, remembering that $l_i\lambda_i = 1$ (for $i = 1, \ldots n$), we have

$$\frac{\partial}{\partial v} f(x, v) = \int_{E^n} f(x+y) \frac{\partial}{\partial v} [v^{-|\lambda|}\Omega(x, y : v^{\lambda})]\, dy$$

$$= -\sum_{i=1}^{n} \lambda_i v^{-|\lambda|} \int_{E^n} f(x+y) D_{y_i}^{l_i} \tilde{\mathscr{L}}_i(x, y : v^{\lambda})\, dy$$

$$= -\sum_{i=1}^{n} v^{-|\lambda|} \int_{E^n} D_i^{l_i} f(x+y) \mathscr{L}_i(x, y : v^{\lambda})\, dy, \quad (38)$$

where

$$\mathscr{L}_i(x, y) = \lambda_i (-1)^{l_i} \tilde{\mathscr{L}}_i(x, y). \quad (39)$$

By virtue of (36) and (38), formula (35) takes the form

$$f(x, \varepsilon) =$$

$$P_{l-1}(f; x) + \int_{\varepsilon}^{1} \sum_{i=1}^{n} v^{-|\lambda|} dv \int_{E^n} D_i^{l_i} f(y) \mathscr{L}_i(x, (y-x) : v^{\lambda})\, dy. \quad (40)$$

Let us show that, if $f \in L_p^{\text{loc}}(G)$, where $p \geqslant 1$, then $f(x, v) \to f(x)$ almost everywhere on G as $v \to 0$. On the basis of (34) and (33), we have

$$f(x, v) = v^{-|\lambda|} \int_{E^n} [f(y) - f(x)] \Omega(x, (y-x) : v^{\lambda})\, dy$$

$$+ f(x) \int_{E^n} v^{-|\lambda|} \Omega(x, (y-x) : v^{\lambda})\, dy = \mathscr{I}_{v\lambda}(x) + f(x), \quad (41)$$

where

$$|\mathscr{I}_{v\lambda}(x)| \leqslant v^{-|\lambda|} \int_{E^n} |f(y) - f(x)|\, |\Omega(x, (y-x) : v^{\lambda})|\, dy.$$

It follows immediately from the definition of $\Omega(x, y)$ that the integral on the right is the same as if it were over the set of those y

such that $K(x+(y-x): v^\lambda) \neq 0$. Denoting the arithmetic sum and difference of two sets by plus and minus signs as before and defining $S(K) = \operatorname{supp} K$,

$$(S(K)-x)v^\lambda = \{y\colon\ y = \bar{y}v^\lambda = (\bar{y}_1 v^{\lambda_1}, \ldots, \bar{y}_n v^{\lambda_n}),\quad \bar{y} \in S(K)-x\},$$

we easily conclude that the value of the integral is as if it were over the set

$$S(K, \lambda, v; x) = \{y\colon\ y \in x + (S(K)-x)v^\lambda\}. \tag{42}$$

Since $S(K) \subset I_1$, the set $S(K, \lambda, v; x)$ is contained in the rectangular parallelepiped

$$\overset{*}{I}_{v\lambda}(x) = \{y\colon\ |y_i - x_i| \leqslant (1+|x_i|)v^{\lambda_i}\ \ (i=1, \ldots, n)\}.$$

Let G' denote a subset of G of the form

$$G' = \{x\colon\ x \in G,\ |x_i| < R\ \ (i=1, \ldots, n)\},$$

where R is a positive number. Then, for every $x \in G'$, we have

$$S(K, \lambda, v; x) \subset I^*_{v\lambda}(x) \subset \tilde{I}_{v\lambda}(x),$$

where

$$\tilde{I}_{v\lambda}(x) = \{y\colon\ |y_i - x_i| \leqslant (1+R)v^{\lambda_i}\ \ (i=1, \ldots, n)\}.$$

Obviously, $|\tilde{I}_{v\lambda}(x)| = 2^n(1+R)^n v^{|\lambda|} = C_1(R)v^{|\lambda|}$. If we now use the estimate

$$|\Omega(x, (y-x): v^\lambda)| \leqslant C_2(R),$$

which holds for $y \in \tilde{I}_{v\lambda}(x)$, we obtain, for $x \in G'$,

$$|\mathcal{I}_{v\lambda}(x)| \leqslant v^{-|\lambda|} C_2(R) \int\limits_{\tilde{I}_{v\lambda}(x)} |f(y) - f(x)|\, dy$$

$$\leqslant \frac{C(R)}{|\tilde{I}_{v\lambda}(x)|} \int\limits_{\tilde{I}_{v\lambda}(x)} |f(y) - f(x)|\, dy,$$

where $C(R) = C_1(R)\, C_2(R)$. We then obtain on the basis of Theorem 1.7 the result that $|\mathcal{I}_{v\lambda}(x)| \to 0$ almost everywhere on G' as $v \to 0$. By virtue of equation (41) and the arbitrariness of the subregion G', this proves our assertion.

If we now let ε approach 0 in (40), we obtain

$$f(x) = P_{l-1}(f; x) + \int\limits_0^1 \sum_{i=1}^{n} v^{-|\lambda|}\, dv \int\limits_{E^n} D_i^{l_i} f(y)\, \mathcal{L}_i\left(x, (y-x) : v^{\lambda}\right) dy, \tag{43}$$

which is valid for almost all x in the set for which the right-hand member is meaningful.

The chief purpose of the present subsection was the obtaining of the identity (43).

Since supp $\mathcal{L}_i(x, \cdot) \subset$ supp $\Omega(x, \cdot)$ and $\lambda = \frac{1}{l}$, we easily conclude that the support of the representation, that is, the set of those y making a contribution to the integration in the right-hand member of (43), is the set

$$V(l; x) = \bigcup_{0 < v \leqslant 1} S\left(K, \frac{1}{l}, v; x\right), \tag{44}$$

where $S(K, \lambda, v; x)$ is defined by formula (42).

Since $S(K, \lambda, 1; x) = S(K)$ for $v = 1$ and since $S(K, \lambda, v; x) \to x$ as $v \to 0$, the set $V(l; x)$, whose form depends on $S(K)$ and the vector l, extends from the point x to the set $S(K)$. Therefore, the following definition is natural: *a region G containing $S(K)$ is called an l-star* *region with respect to* $S(K)$ *if* $V(l; x) \subset G$ *for every point x in G.*

Thus, the identity (43) holds for almost every point x of an l-star region G with respect to $S(K)$ If f is continuous on G, that identity holds at every point of G.

Finally, we emphasize that the operator

$$\mathcal{P}f = P_{l-1}(f; x) = \int_{E^n} f(y)\,\Omega(x, y - x)\,dy$$

has the property that it maps every polynomial in x_i of degree not exceeding $l_i - 1$ for $i = 1, \ldots, n$ into itself. This follows from formula (43) since the second term of such a polynomial disappears upon substitution into that formula.

We conclude the subsection by noting that the integral representations (15) and (43) are suitable for investigation of the properties of functions in the Sobolev classes W_p^l which are characterized by vectors l with positive-integer-valued components. In the following subsection, we shall obtain integral representations typical of the classes W_p^l and L_p^l for arbitrary nonnegative values of the components of the vector l. For integral values of the components of the vector l, these classes coincide with Sobolev's classes.

7.6. Generalization of the integral representation (15). First of all, let us construct the corresponding averaging kernel. We begin with the function Ω defined in subsection 7.1. However, we shall assume for simplicity that it can be represented as the product of functions of a single variable.

Let K_i denote a member of $C_0^\infty(E^1)$ such that

$$\int_{E^1} K_i(t)\,dt = 1 \qquad (i = 1, \ldots, n). \tag{45}$$

Let us define

$$\Omega_i(t) = D_t^{k_i}\left[\frac{t^{k_i - 1}}{(k_i - 1)!}\int_{E^1} K_i(\tau)\,\theta(t - \tau)\,d\tau\right] \qquad (i = 1, \ldots, n). \tag{46}$$

One can easily show that $\Omega_i \in C_0^\infty(E^1)$, supp $\Omega_i \subset$ supp K_i, and

$$\int_{E^1} \Omega_i(t)\,dt = \int_{E^1} K_i(t)\,dt = 1 \qquad (i = 1, \ldots, n). \tag{47}$$

Let $\Psi(\xi)$ denote a nonnegative function of a single variable defined on E^1 such that supp $\Psi \subset [1, 1+\sigma]$, where $\sigma > 0$, and

$$\int_{E^1} \Psi(\xi)\,d\xi = 1. \tag{48}$$

Let us define

$$\tilde{\Omega}_i(t) = \int_{E^1} \Psi(\xi)\,\Omega_i(t\xi^{-\lambda_i})\,\xi^{-\lambda_i}\,d\xi \qquad (i = 1, \ldots, n), \tag{49}$$

where, as usual, $\lambda_i > 0$ for $i = 1, \ldots, n$.

Let us now set

$$\Omega_i^*(t) = \int_{E^1} \tilde{\Omega}_i(t-\tau)\,\tilde{\Omega}_i(\tau)\,d\tau \qquad (i = 1, \ldots, n). \tag{50}$$

Obviously, $\Omega_i^* \in C_0^\infty(E^1)$ and, by virtue of (47-50),

$$\int_{E^1} \Omega_i^*(t)\,dt = \left(\int_{E^1} \tilde{\Omega}_i(t)\,dt\right)^2 = \left(\int_{E^1} \Psi(\xi)\,d\xi \int_{E^1} \Omega_i(t\xi^{-\lambda_i})\,\xi^{-\lambda_i}\,dt\right)^2 = 1. \tag{51}$$

Thus, the functions $\Omega_i^*(t)$ can then be taken as averaging kernels.

Let us suppose also that the function f satisfies the conditions imposed on it in subsection 7.2; that is, let us assume that it is defined on any region in E^n that is necessary for a discussion to be meaningful: that it is summable on G and that it has all the generalized derivatives referred to in that discussion.

Consider the average

$$f^*_{v\lambda}(x) = \int\limits_{E^n} f(x+y) \prod_{j=1}^{n} v^{-\lambda_j} \Omega^*_j \left(y_j v^{-\lambda_j}\right) dy. \tag{52}$$

With an eye to application of the identity (1) to the function $f^*_{v\lambda}(x)$, let us find $\frac{\partial}{\partial v} f^*_{v\lambda}(x)$. It follows from (52) that

$$\frac{\partial}{\partial v} f^*_{v\lambda}(x) = \sum_{i=1}^{n} \int\limits_{E^n} f(x+y) \left(\prod_j^{(i)} v^{-\lambda_j} \Omega^*_j \left(y_j v^{-\lambda_j}\right)\right)$$
$$\times \frac{\partial}{\partial v} \left[v^{-\lambda_i} \Omega^*_i \left(y_i v^{-\lambda_i}\right)\right] dy, \tag{53}$$

where, by virtue of (50) and (49),

$$\frac{\partial}{\partial v}\left[v^{-\lambda_i}\Omega^*_i\left(y_i v^{-\lambda_i}\right)\right] = \frac{\partial}{\partial v} \int\limits_{E^1} v^{-\lambda_i}\tilde{\Omega}_i\left(y_i v^{-\lambda_i} - t\right)\tilde{\Omega}_i(t)\,dt$$
$$= \frac{\partial}{\partial v} \int\limits_{E^1} \left[v^{-\lambda_i}\tilde{\Omega}_i\left((y_i - t)\,v^{-\lambda_i}\right)\right]\left[v^{-\lambda_i}\tilde{\Omega}_i\left(tv^{-\lambda_i}\right)\right] dt$$
$$= 2 \int\limits_{E^1} v^{-\lambda_i}\tilde{\Omega}_i\left((y_i - t)\,v^{-\lambda_i}\right) \frac{\partial}{\partial v}\left[v^{-\lambda_i}\tilde{\Omega}_i\left(tv^{-\lambda_i}\right)\right] dt, \tag{54}$$

$$\frac{\partial}{\partial v}\left[v^{-\lambda_i}\tilde{\Omega}_i\left(tv^{-\lambda_i}\right)\right] = \int\limits_{E^1} \Psi(\xi)\,\xi \frac{\partial}{\partial (v\xi)}\left[(v\xi)^{-\lambda_i}\Omega_i\left(t\,(v\xi)^{-\lambda_i}\right)\right] d\xi$$
$$= \int\limits_{E^1} \Psi(\xi)\,\xi D_t^{k_i}\left[\frac{t^{k_i-1}}{(k_i-1)!}\int\limits_{E^1} K_i(\tau)\frac{\partial}{\partial (v\xi)}\theta\left(t\,(v\xi)^{-\lambda_i} - \tau\right) d\tau\right] d\xi$$
$$= -v^{-1-\lambda_i}\int\limits_{E^1}\Psi(\xi)\,\xi^{-\lambda_i} D_t^{k_i}\tilde{\mathscr{L}}_i\left(t\,(v\xi)^{-\lambda_i}\right) d\xi, \tag{55}$$

$$\tilde{\mathscr{L}}_i(t) = \frac{\lambda_i t^{k_i}}{(k_i-1)!} K_i(t). \tag{56}$$

By virtue of (54) and (55), when we replace y_i with $y_i + t$ in formula (53), that formula takes the form

$$\frac{\partial}{\partial v} f^*_{v\lambda}(x) = -2\sum_{i=1}^{n} v^{-1-|\lambda|-\lambda_i} \int_{E^1} \Psi(\xi)\,\xi^{-\lambda_i}\,d\xi \int_{E^1} \tilde{\Omega}_i\left(y_i v^{-\lambda_i}\right) dy_i$$

$$\times \int_{E^n} f(x + y + te_i)\left(\prod_j^{(i)} \Omega^*_j\left(y_j v^{-\lambda_j}\right)\right) D_i^{k_i}\tilde{\mathscr{L}}_i\left(t(v\xi)^{-\lambda_i}\right) dy^{(i)}\,dt,$$

(57)

where $|\lambda| = \sum_1^n \lambda_i$ and $dy^{(i)} = dy_1 \ldots dy_{i-1}\,dy_{i+1} \ldots dy_n$.

If we use the definition 6(1) of a generalized derivative to transform the inner integral in (57), we obtain

$$\frac{\partial}{\partial v} f^*_{v\lambda}(x) = -\sum_{i=1}^{n} v^{-1-|\lambda|+l_i\lambda_i}\mathscr{I}_i, \tag{58}$$

where

$$\mathscr{I}_i = v^{-\lambda_i} \int_{E^n}\int_{E^1}\int_{E^1} D_i^{l_i} f(x + y + te_i)\left(\prod_j^{(i)} \Omega^*_j\left(y_j v^{-\lambda_j}\right)\right)$$

$$\times\, \tilde{\Omega}_i\left(y_i v^{-\lambda_i}\right) \Psi_i(\xi)\, \mathscr{L}_i\left(t(v\xi)^{-\lambda_i}\right) dy\,d\xi\,dt, \tag{59}$$

$$\mathscr{L}_i(t) = 2(-1)^{l_i} D^{k_i - l_i}\tilde{\mathscr{L}}_i(t), \tag{60}$$

$$\Psi_i(\xi) = \Psi(\xi)\,\xi^{-\lambda_i + l_i\lambda_i} \qquad (i = 1, \ldots, n), \tag{61}$$

and the l_i (for $i = 1, \ldots, n$) are natural numbers such that $l_i \leqslant k_i$.

We note that, if $l_i < k_i$, then the functions $\mathscr{L}_i(t)$ satisfy the condition

$$\int_{E^1} \mathscr{L}_i(t)\,dt = 0 \qquad (i = 1, \ldots, n). \tag{62}$$

Keeping in mind the special features of normalization of the spaces W_p^l and L_p^l for integer and noninteger values of the coordinates of the vector l that will be introduced, we can represent $\mathcal{J}_i$ in two different forms.

The first of these representations is easily obtained: We set $y_i + t = w_i$ in (59) and then replace w_i with y_i. This gives us

$$\mathcal{J}_i = \int_{E^n} D_i^{l_i} f(x+y) M_i(y : v^\lambda)\, dy, \tag{63}$$

where

$$M_i(y) = \prod_j^{(i)} \Omega_j^*(y_j) \int_{E^1} \int_{E^1} \tilde{\Omega}_i(y_i - t) \mathcal{L}_i(t\xi^{-\lambda_i}) \Psi_i(\xi)\, d\xi\, dt. \tag{64}$$

We note that, by virtue of (62),

$$\int_{E^n} D^\alpha M_i(y)\, dy = 0, \qquad (i = 1, \ldots, n) \tag{65}$$

for arbitrary $\alpha = (\alpha_1, \ldots, \alpha_n)$, where $\alpha_i \geqslant 0$ for $i = 1, \ldots, n$.

It is somewhat more complicated to get the other representation of $\mathcal{J}_i$. Let m_i denote a natural number. We may assume that $l_i + m_i < k_i$, where k_i is the number appearing in the definition of the kernel $\Omega_i(t)$. Let us define

$$S_i(t) = 2(-1)^{l_i} D^{k_i - l_i - m_i} \tilde{\mathcal{L}}_i(t), \quad \mathcal{L}_i(t) = D^{m_i} S_i(t) \tag{66}$$

and let us represent $\mathcal{J}_i$ in the form

$$\mathcal{J}_i = v^{-\lambda_i} \int_{E^{n-1}} \int_{E^1} \left(\prod_j^{(i)} \Omega_j^*(y_j v^{-\lambda_j}) \right) \Psi_i(\xi)\, d\xi\, dy^{(i)}$$
$$\times \int_{E^1} \int_{E^1} D_i^{l_i} f(x+y) \tilde{\Omega}_i((y_i - t) v^{-\lambda_i}) D^{m_i} S_i(t(v\xi)^{-\lambda_i})\, dy_i\, dt. \tag{67}$$

The iterated integral in the last factor can be transformed in accordance with the following lemma:

Lemma. Let $\varphi(\tau)$ denote a locally summable function defined on E^1. Let $\Omega(\tau)$ and $S(\tau)$ denote functions belonging to $C_0^\infty(E^1)$. Let u and η denote positive numbers and let m denote a natural number. Then,

$$\mathcal{J} = \int_{E^1}\int_{E^1} \varphi(\tau)\,\Omega\left(\frac{\tau - t}{u}\right) D^m S\left(\frac{t}{u\eta}\right) dt\, d\tau =$$

$$\int_{E^1}\int_{E^1} \Delta^m(\delta t)\,\varphi(\tau)\left(\sum_{j=0}^{m} A_j(\delta)\,\eta^{m-j} D^{m-j}\Omega\left(\frac{\tau - t}{u}\right) D^j S\left(\frac{t}{u\eta}\right)\right) dt\, d\tau, \tag{68}$$

where

$$\Delta^m(\delta t)\,\varphi(\tau) = \sum_{r=0}^{m} C_m^r (-1)^{m-r} \varphi(\tau + r\,\delta t), \tag{69}$$

δ is any nonzero number, and the $A_j(\delta)$ are coefficients to be determined below.

PROOF. Suppose that $\delta \neq 0$. We have

$$1 = \frac{1}{m!\,\delta^m}[(\delta + 1) - 1][(2\delta + 1) - 1]\ldots[(m\delta + 1) - 1]$$

$$= \frac{1}{m!\,\delta^m}\sum_{j=0}^{m} C_j(\delta), \tag{70}$$

where

$$C_0(\delta) = \prod_{i=1}^{m}(i\delta + 1),\ C_1(\delta)$$

$$= -\sum_{s=1}^{m}\prod_i^{(s)}(i\delta + 1), \ldots, C_m(\delta) = (-1)^m.$$

One can easily see that, for any r in $1 \leqslant r \leqslant m$, there exist numbers

$B_{j,r}$ (for $j = 0, 1, \ldots, m-1$) satisfying the conditions

$$\begin{aligned} -B_{0,r}(r\delta+1) &= C_0, \\ B_{0,r} - B_{1,r}(r\delta+1) &= C_1, \\ B_{1,r} - B_{2,r}(r\delta+1) &= C_2, \\ \cdots\cdots\cdots & \\ B_{m-2,r} - B_{m-1,r}(r\delta+1) &= C_{m-1}, \\ B_{m-1,r} &= C_m, \end{aligned} \tag{71}$$

where $C_j = C_j(\delta)$ are numbers to be determined below.

Noting that the equation

$$\int_{E^1} \Omega\left(\frac{\tau-t}{u}\right) D^m S\left(\frac{t}{u\eta}\right) dt = \int_{E^1} D^{m-j}\Omega\left(\frac{\tau-t}{u}\right) D^j S\left(\frac{t}{u\eta}\right) \eta^{m-j}\, dt,$$

holds for every j in $0 \leqslant j \leqslant m$ and using the partition of unity (70), we obtain the following expression for the integral $\mathcal{J}$ constituting the left-hand member of (68):

$$\mathcal{J} = \frac{1}{m!\,\delta^m} \int_{E^1}\int_{E^1} \varphi(\tau) \left(\sum_{j=0}^{m} C_j(\delta)\, D^{m-j}\Omega\left(\frac{\tau-t}{u}\right) D^j S\left(\frac{t}{u\eta}\right) \eta^{m-j} \right) dt\, d\tau. \tag{72}$$

Let us show that, for every r in $1 \leqslant r \leqslant m$,

$$J_r = \int_{E^1}\int_{E^1} \varphi(\tau + r\,\delta t) \left(\sum_{j=0}^{m} C_j(\delta)\, D^{m-j}\Omega\left(\frac{\tau-t}{u}\right) \times D^j S\left(\frac{t}{u\eta}\right) \eta^{m-j} \right) dt\, d\tau = 0. \tag{73}$$

If we set $\tau + r\,\delta t = \tau'$, then replace τ' with τ, and use equations

(71), we obtain

$$J_r=\int_{E^1}\varphi(\tau)\,d\tau\int_{E^1}\sum_{j=0}^{m-1}B_{j,r}\Big[-(r\delta+1)\,D^{m-j}\Omega\Big(\frac{\tau-(r\delta+1)\,t}{u}\Big)$$

$$\times D^jS\Big(\frac{t}{u\eta}\Big)\eta^{m-j}+D^{m-(j+1)}\Omega\Big(\frac{\tau-(r\delta+1)\,t}{u}\Big)\,D^{j+1}S\Big(\frac{t}{u\eta}\Big)\eta^{m-(j+1)}\Big]dt$$

$$=\sum_{j=0}^{m-1}B_{j,r}u\int_{E^1}\varphi(\tau)\,d\tau\int_{E^1}D_t\Big[D^{m-(j+1)}\Omega\Big(\frac{\tau-(r\delta+1)\,t}{u}\Big)D^jS\Big(\frac{t}{u\eta}\Big)\eta^{m-j}\Big]dt=0$$

since Ω and S are functions belonging to $C_0^\infty(E^1)$.

By virtue of (69) and (73), we can write equation (72) in the form

$$\mathcal{J}=\frac{(-1)^m}{m!\,\delta^m}\int_{E^1}\int_{E^1}\Delta^m(\delta t)\,\varphi(\tau)\Bigg(\sum_{j=0}^{m}C_j(\delta)\,D^{m-j}\Omega\Big(\frac{\tau-t}{u}\Big)$$

$$\times D^jS\Big(\frac{t}{u\eta}\Big)\eta^{m-j}\Bigg)\,dt\,d\tau.$$

This last equation coincides with equation (68) for $A_j(\delta)=\frac{(-1)^m}{m!\,\delta^m}C_j(\delta)$ (for $j=0, 1, \ldots, m$). This completes the proof of the lemma.

If we now use equation (68) to transform the iterated integral in the last factor in (67), making the substitution $y_i-t=w_i$, and then replace w_i with y_i, we obtain

$$\int_{E^1}\int_{E^1}D_i^{l_i}f(x+y)\,\tilde{\Omega}_i\left((y_i-t)\,v^{-\lambda_i}\right)D^{m_i}S_i\left(t\,(v\xi)^{-\lambda_i}\right)dy_i\,dt$$

$$=\sum_{j=0}^{m_i}A_j(\delta)\,\xi^{\lambda_i(m_i-j)}\int_{E^1}\int_{E^1}\Delta_i^{m_i}(\delta t)\,D_i^{l_i}f(x+y+te_i)$$

$$\times D^{m_i-j}\,\tilde{\Omega}_i\left(y_iv^{-\lambda_i}\right)D^jS_i\left(t\,(v\xi)^{-\lambda_i}\right)dy_i\,dt. \qquad (74)$$

Let us define

$$\Phi_{ij}(y) = A_j(\delta)\, D^{m_i - j}\tilde{\Omega}_i(y_i) \prod_j^{(i)} \Omega_j^*(y_i), \tag{75}$$

$$\Psi_{ij}(\xi) = \Psi_i(\xi)\, \xi^{\lambda_i(m_i - j)} = \Psi(\xi)\, \xi^{\lambda_i(m_i + l_i - j - 1)}, \tag{76}$$

$$N_{ij}(t) = D^j S_i(t). \tag{77}$$

Then, it follows from (67) and (74)-(77) that

$$\mathcal{I}_i = \sum_{j=0}^{m_i} v^{-\lambda_i} \int_{E^n} \int_{E^1} \int_{E^1} \Delta_i^{m_i}(\delta t)\, D^{l_i} f(x + y + te_i)\, \Phi_{ij}(y : v^{\lambda})$$

$$\times \Psi_{ij}(\xi)\, N_{ij}\left(t\,(v\xi)^{-\lambda_i}\right) dy\, d\xi\, dt. \tag{78}$$

We note that we can assume that the functions $N_{ij}(t)$ satisfy the condition

$$\int_{E^1} N_{ij}(t)\, dt = 0 \qquad (j = 1, \ldots, m_i). \tag{79}$$

(This follows from (77) and (66) if $k_i > l_i + m_i$.)

If we now apply formula (1) and equations (58), (63), and (78), we obtain the identity

$$f_{\varepsilon\lambda}^*(x) = f_{h\lambda}^*(x) + \int_\varepsilon^h \sum_{i=1}^n v^{-1-|\lambda|+l_i\lambda_i} \mathcal{I}_i(x, v)\, dv, \tag{80}$$

where $\mathcal{I}_i(x, v) = \mathcal{I}_i$ can be represented either in the form (63) or in the form (78).

If $f \in L_p^{\mathrm{loc}}(G)$ for $p \geqslant 1$, it follows on the basis of equations (52) and (54) and Lemma (5.3) that $f_{\varepsilon\lambda}^*(x) \to f(x)$ almost everywhere on G as $\varepsilon \to 0$. Therefore, by letting ε approach zero in (80), we get

$$f(x) = f_{h\lambda}^*(x) + \int_0^h \sum_{i=1}^n v^{-1-|\lambda|+l_i\lambda_i} \mathcal{I}_i(x, v)\, dv, \tag{81}$$

which is valid for almost all $x \in G$ for which the support of the representation (81) is contained in G.

Let us now see just what the carrier of that representation is; that is, let us ascertain the set of values of the function f and its derivatives that contribute to the right-hand member of (81).

As in the case of formula (15), let us assume that

$$S(K_i) = \operatorname{supp} K_i \subset \left\{ t\colon \frac{t}{a_i} > 0,\ 1 < \left(\frac{t}{a_i}\right)^{1/\lambda_i} < 1 + b \right\}$$
$$(i = 1, \ldots, n),$$

where $b > 0$ and $a_i \neq 0$ for $i = 1, \ldots, n$. Since $S(\Omega_i) \subset S(K_i)$, we have, by virtue of equations (49) and (50) and the definition of $\Psi(\xi)$,

$$S(\tilde{\Omega}_i) \subset \left\{ t\colon \frac{t}{a_i} > 0,\ 1 < \left(\frac{t}{a_i}\right)^{1/\lambda_i} < (1 + b)(1 + \sigma) \right\},$$
$$S(\Omega_i^*) \subset \left\{ t\colon \frac{t}{a_i} > 0,\ 2^{1/\lambda_i} < \left(\frac{t}{a_i}\right)^{1/\lambda_i} < 2^{1/\lambda_i}(1 + b)(1 + \sigma) \right\}$$
$$(i = 1, \ldots, n).$$

From this we conclude on the basis of (64) that

$$\operatorname{supp} M_i(\cdot\colon v^\lambda) \subset \left\{ y\colon \frac{y_j}{a_j} > 0,\ 2^{1/\lambda_j} v \right.$$
$$\left. < \left(\frac{y_j}{a_j}\right)^{1/\lambda_j} < 2^{1/\lambda_j}(1 + b)(1 + \sigma) v \quad (j = 1, \ldots, n) \right\}.$$

We note that the representation of $\mathcal{I}_i(x, v)$ by formula (63) uses the values of the function f and its derivatives only at points of the set $x + \operatorname{supp} M_i(\cdot\colon v^\lambda)$.

Furthermore, from equations (75)-(77), (66), and (65) and the definition of the function $\Psi_{ij}(\xi)$ we get the following estimates for the supports of the functions $\Phi_{ij}(y\colon v^\lambda)$, $N_{ij}\left(t(v\xi)^{-\lambda_i}\right)$, and $\Psi_{ij}(\xi)$:

$$2^{1/\lambda_s}v < \left(\frac{y_s}{a_s}\right)^{1/\lambda_s} < 2^{1/\lambda_s}(1+b)(1+\sigma)v \ (s=1, \ldots, n; s \neq i),$$

$$v < \left(\frac{y_i}{a_i}\right)^{1/\lambda_i} < (1+b)(1+\sigma)v,$$

$$v\xi < \left(\frac{t}{a_i}\right)^{1/\lambda_i} < (1+b)v\xi, \qquad 1 \leqslant \xi \leqslant 1+\sigma.$$

From these estimates, we see that the representation of $\mathcal{I}_i(x, v)$ by formula (78) uses the values of the function f at points of a set contained in $x + F_{v^\lambda}$, where

$$F_{v^\lambda} = \left\{ y \colon \frac{y_j}{a_j} > 0,\ 2^{1/\lambda_j}v < \left(\frac{y_j}{a_j}\right)^{1/\lambda_j} \right.$$

$$\left. < 2^{1/\lambda_j}(1+b)(1+\sigma)(1+m_j\delta)^{1/\lambda_j}v \ (j=1, \ldots, n) \right\}. \quad \delta > 0.$$

Noting that $M_i(\cdot\colon v^\lambda) \subset F_{v^\lambda}$, we easily conclude that the support of the representation (81) is the displaced horn

$$x + V\left(\frac{1}{\lambda}\right) = x + \bigcup_{0<v\leqslant h} F_{v^\lambda}$$

(see (19)). Here, the opening of the horn $V\left(\frac{1}{\lambda}\right)$ can be made arbitrarily small by suitable choice of the parameters a_i, $b > 0$, $\sigma > 0$, and $\delta > 0$.

Let U denote the set of $x \in G$ such that $x + V\left(\frac{1}{\lambda}\right) \subset G$. Then, we may assert that the identity (81) holds for almost all $x \in U$ (for *all* $x \in U$ if f is continuous on G).

We note that, if the functions $\mathcal{I}_i(x, v)$ for $i = 1, \ldots, n$ are all represented by formula (63), then the identity (81) becomes the identity (15). If all of them are represented by formula (78), we obtain a representation of the function f in terms of the integrals of f and of the finite differences of the derivatives of f with respect to the coordinate directions. In the particular case in which all the l_i for $i = 1, \ldots, n$ are equal to zero, the right-hand member of (81) will involve only f and its finite differences.

As was mentioned in the introduction to the present chapter, the different integral representations of the functions correspond to different normalizations of those classes of functions under consideration. We give separately a representation that is a special case of the representation (81) corresponding to the basic norm of the space L_p^l.

Suppose that we are given a vector $l = (l_1, \ldots, l_n)$ with positive components. Let us set $\bar{l}_i = [l_i]$ so that $l_i = \bar{l}_i + \alpha_i$, where $0 \leqslant \alpha_i < 1$ for $i = 1, \ldots, n$.

If $l_i = \bar{l}_i$ is an integer, we represent $\mathcal{G}_i(x, v)$ according to formula (63). If $l_i > \bar{l}_i$, we represent $\mathcal{G}_i(x, v)$ according to (78), taking $m_i = 1$ and replacing l_i everywhere in that formula with $\bar{l}_i$. We then obtain

$$f(x) = f^*_{h^\lambda}(x) + \int_0^h \sum_{i=1}^n v^{-1-|\lambda|+l_i\lambda_i} \mathcal{G}_{i,\,l_i}(x, v)\, dv, \tag{82}$$

where

$$\mathcal{G}_{i,\,l_i}(x, v) = \int_{E^n} D_i^{l_i} f(x+y)\, M_i(y\colon v^\lambda)\, dy \tag{83}$$

if l_i is an integer or

$$\mathcal{G}_{i,\,l_i}(x, v) = \sum_{j=0}^{1} v^{-\lambda_i(1+\alpha_i)} \int_{E^n} \int_{E^1} \int_{E^1} \Delta_i(\delta t)\, D_i^{\bar{l}_i} f(x+y+te_i)$$
$$\times\, \Phi_{ij}(y\colon v^\lambda)\, \Psi_{ij}(\zeta)\, N_{ij}\left(t(v\xi)^{-\lambda_i}\right) dy\, d\xi\, dt \tag{84}$$

if $l_i = \bar{l}_i + \alpha_i$ is not an integer.

We conclude this subsection by noting that, if in all the preceding formulas we set $\psi(\xi) = \delta(\xi - 1)$, where $\delta(\xi - 1)$ is the translated delta function (which is equivalent to taking $\tilde{\Omega}_j = \Omega_j$ for $j = 1, \ldots, n$), then formula (82) retains its form and $\mathcal{G}_i$ is written in the form (83) in the case of integer-valued l_i but assumes the simpler form

$$\mathcal{G}_{i,l_i}(x, v) = \sum_{j=0}^{1} v^{-\lambda_i(1+\alpha_i)} \int_{E^n} \int_{E^1} \Delta_i(\delta t)\, D_i^{\bar{l}_i} f(x + y + te_i) \times \Phi_{ij}(y: v^{\lambda})\, N_{ij}(tv^{-\lambda_i})\, dy\, dt \qquad (84')$$

in the case of non-integer-valued l_i. The representation characterized by formulas (82), (83), and (84′) is used for investigating the properties of functions in the class W_p^l.

It is easy to show that the identity (82) always holds for functions belonging to L_p^l or W_p^l (in this connection, see Remark 7.8.1).

7.7. Let us look at another identity, one that follows directly from formula (12).

To derive it, we apply to both sides of (12) the differential operator D_x^α, where $\alpha = (\alpha_1, \ldots, \alpha_n)$, where in turn the α_i are nonnegative integers. We obtain

$$D^\alpha f_{\varepsilon\lambda}(x) = D^\alpha f_{h\lambda}(x) + (-1)^{|\alpha|} \int_\varepsilon^h \sum_{i=1}^n \lambda_i v^{-1-(\alpha,\lambda)-|\lambda|}\, dv \int_{E^n} f(y)\, D^\alpha D_i^{k_i} \tilde{\mathcal{L}}_i((y-x) : v^\lambda)\, dy, \qquad (85')$$

where the parameters ε, h, and $\lambda = (\lambda_1, \ldots, \lambda_n)$ and the functions $f(x)$, $f_{v\lambda}(x)$, and $\tilde{\mathcal{L}}_i(x)$ have the same meaning as in subsection 7.2.

Let b denote a positive number and let $j = (j_1, \ldots, j_n)$ denote a vector with nonnegative-integer-valued coordinates. We set

$$P_{b,\lambda}(y; x, v) = \sum_{\substack{j \geqslant 0 \\ 0 \leqslant (j,\lambda) \leqslant b}} \frac{A_j(x, v)}{j!} y^j \qquad (y^j = y_1^{j_1} \ldots y_n^{j_n}),$$

where the $A_j(x, v)$ are arbitrary measurable functions of x and v

defined wherever they need to be for a discussion to be meaningful. Thus, $P_{b,\lambda}(y;\ x, v)$ is a polynomial in y formed by the sum of powers y^j characterized by the inequality $(j, \lambda) \leqslant b$ with coefficients depending on x and v.

One can easily see that, if the numbers k_i for $i = 1, \dots, n$ are sufficiently great, we have for all $\alpha \geqslant 0$

$$\int_{E^n} P_{b,\lambda}(y;\ x,\ v)\, D^\alpha D_i^{k_i} \tilde{\mathscr{L}}_i (y : v^\lambda)\, dy = 0 \qquad (i = 1, \dots, n).$$

Therefore, we can write formula (85′) in the form

$$D^\alpha f_{\varepsilon\lambda}(x) = D^\alpha f_{h\lambda}(x) + (-1)^{|\alpha|} \int_\varepsilon^h v^{-1-(\alpha,\lambda)-|\lambda|}\, dv \int_{E^n} [f(y) - P_{b,\lambda}(y - x;\ x,\ v)] \times D^\alpha M((y - x) : v^\lambda)\, dy,$$

$$M(y) = \sum_{i=1}^n \lambda_i D^{k_i} \tilde{\mathscr{L}}_i (y). \tag{85}$$

Obtaining of this formula was the chief purpose of the present subsection. In addition, we shall use the following modification of it:

$$D^\alpha \tilde{f}_{\varepsilon\lambda}(x) = D^\alpha \tilde{f}_{h\lambda}(x) + (-1)^{|\alpha|} \int_\varepsilon v^{-1-(\alpha,\lambda)-|\lambda|}\, dv \times \int_{E^n} D^\alpha \Omega((y - x) : v^\lambda)\, dy \int_{E^n} [f(z) - P_{b,\lambda}(z - y;\ y,\ v)] \times M((z - y) : v^\lambda)\, dz, \tag{86}$$

where

$$D^\alpha \tilde{f}_{v\lambda}(x) = (-1)^{|\alpha|} v^{-(\alpha,\lambda)-|\lambda|} \int_{E^n} f(y)\, D^\alpha \tilde{\Omega}((y - x) : v^\lambda)\, dy,$$

$$\tilde{\Omega}(y) = \int_{E^n} \Omega(y - w)\Omega(w)\, dw,$$

and Ω and M are the same functions as in formula (85). Formula (86) can also be obtained by replacing $\Omega(y)$ everywhere in formula (12) with $\tilde{\Omega}(y)$.

If there exists $D^{\alpha}f \in L^{\text{loc}}(G)$, we get the corresponding integral representations for $D^{\alpha}f$ by letting ε approach zero in the formulas that we have obtained.

Remembering that supp K, where K is the function appearing in the definition of Ω, is characterized by (18), we easily conclude that for arbitrary ε such that $0 < \varepsilon < h$ the right-hand members of formulas (85) and (86) involve the values of the function f only on some horn $x + V\left(\frac{1}{\lambda}\right)$.

We shall use these identities later to investigate the properties of those classes of functions that are characterized by integral estimates of a certain type for $|f - P_{b,\lambda}|$. In addition to the identities shown, we shall also use a representation of the coefficients of an arbitrary polynomial in terms of integrals of the polynomial itself. These formulas follow easily from formula (85′).

Suppose that

$$P(x) = \sum_{0 \leqslant j \leqslant k-1} \frac{A_j}{j!} x^j$$

is a polynomial in $x_1, \ldots, x_n$ of degree not exceeding $k_i - 1$ with respect to x_i for $i = 1, \ldots, n$. Let us set $f(x) = P(x)$ in (85′). Integrating by parts, we can easily see that the second term in this formula is equal to zero. Therefore, for all $\alpha \geqslant 0$, we have

$$D^{\alpha} P_{\varepsilon\lambda}(x) = D^{\alpha} P_{h\lambda}(x).$$

By letting ε approach zero, we get

$$D^{\alpha} P(x) = (-1)^{|\alpha|} h^{-(\alpha,\lambda)-|\lambda|} \int_{E^n} P(y)\, D^{\alpha}\Omega\left((y-x) : h^{\lambda}\right) dy.$$

This equation is valid at every point x in E^n. If we now set $x = 0$

and $\alpha = j$, we get

$$A_j = (-1)^{|j|} h^{-(j,\lambda)-|\lambda|} \int_{E^n} P(y)\, D^j \Omega(y : h^\lambda)\, dy \quad (\mathbf{0} \leqslant j \leqslant \boldsymbol{k} - 1).$$

Let us replace $P(y)$ in this formula with the polynomial $P_{b,\lambda}(y; x, v)$ defined above. Then, we obtain the following representation of the coefficients in the polynomial $P_{b,\lambda}(y; x, v)$:

$$A_j(x, v) = (-1)^{|j|} h^{-(j,\lambda)-|\lambda|} \int_{E^n} P_{b,\lambda}(y; x, v)\, D^j \Omega(y : h^\lambda)\, dy$$

$$(0 \leqslant (j, \lambda) \leqslant b). \tag{87}$$

We note that formula (87) holds for arbitrary positive h and arbitrary v, in particular, for $v = h$.

7.8. Integral representation of functions in terms of differences. As was mentioned above, a special case of formula (81) yields a representation of the function f in terms of integrals of f and its differences with respect to the coordinate directions. However, we shall indicate here a simple method for obtaining such a representation, one that is based on a different choice of the averaging kernels Ω_i.

Let $K_i(t)$ for $i = 1, \ldots, n$ denote the same functions as in subsection 7.6; that is, all the K_i belong to $C_0^\infty(E^1)$ and satisfy equations (45). Let us define

$$\breve{\Omega}_i(t) = \frac{1}{A_i} \sum_{j=0}^{k_i} \frac{(-1)^{k_i - j}}{(1 + j\delta)^2} C_{k_i}^j K_i\left(\frac{t}{1 + j\delta}\right) \qquad (i = 1, \ldots, n), \tag{88}$$

where the k_i are natural numbers, $0 < \delta < 1$, and

$$A_i = (-1)^{k_i} \sum_{j=0}^{k_i} \frac{(-1)^j}{1 + j\delta} C_{k_i}^j = (-1)^{k_i} \int_0^1 (1 - t^\delta)^{k_i}\, dt \neq 0.$$

Obviously, $\breve{\Omega}_i \in C_0^\infty(E^1)$ and

$$\int_{E^1} \breve{\Omega}_i(t)\,dt = \frac{1}{A_i}\sum_{j=0}^{k_i} \frac{(-1)^{k_i-j}}{1+j\delta} C_{k_i}^j \int_{E^1} K\left(\frac{t}{1+j\delta}\right)\frac{dt}{1+j\delta}$$

$$= \frac{1}{A_i}\sum_{j=0}^{k_i} \frac{(-1)^{k_i-j}}{1+j\delta} C_{k_i}^j = 1 \qquad (i = 1, \ldots, n). \qquad (89)$$

Let us define

$$\bar{\Omega}_i(t) = \int_{E^1} \breve{\Omega}_i(t-\tau)\,\breve{\Omega}_i(\tau)\,d\tau \qquad (i = 1, \ldots, n). \qquad (90)$$

Obviously, $\bar{\Omega}_i \in C_0^\infty(E^1)$ and, by virtue of (89),

$$\int_{E^1} \bar{\Omega}_i(t)\,dt = 1 \qquad (i = 1, \ldots, n). \qquad (91)$$

Consequently, the $\bar{\Omega}_i(t)$ can be taken as averaging kernels.

Suppose that $\lambda = (\lambda_1, \ldots, \lambda_n)$, where $\lambda_i > 0$ for $i = 1, \ldots, n$. We define

$$\bar{f}_{v^\lambda}(x) = \int_{E^n} f(x+y) \prod_{j=1}^{n} v^{-\lambda_j}\bar{\Omega}_j(y_j v^{-\lambda_j})\,dy. \qquad (92)$$

We have

$$\frac{\partial}{\partial v}\bar{f}_{v^\lambda}(x) =$$

$$\sum_{i=1}^{n}\int_{E^n} f(x+y)\left(\prod_j^{(i)} v^{-\lambda_j}\bar{\Omega}_j(y_j v^{-\lambda_j})\right)\frac{\partial}{\partial v}\left[v^{-\lambda_i}\bar{\Omega}_i(y_i v^{-\lambda_i})\right]dy, \qquad (93)$$

where

$$\frac{\partial}{\partial v}[v^{-\lambda_i}\bar{\Omega}_i(y_i v^{-\lambda_i})] = \frac{\partial}{\partial v}\int_{E^1} v^{-\lambda_i}\breve{\Omega}_i(y_i v^{-\lambda_i} - t)\,\breve{\Omega}_i(t)\,dt$$

$$= \frac{\partial}{\partial v}\int_{E^1}[v^{-\lambda_i}\breve{\Omega}_i((y_i - t)v^{-\lambda_i})][v^{-\lambda_i}\breve{\Omega}_i(tv^{-\lambda_i})]\,dt$$

$$= 2\int_{E^1} v^{-\lambda_i}\breve{\Omega}_i((y_i - t)v^{-\lambda_i})\frac{\partial}{\partial v}[v^{-\lambda_i}\breve{\Omega}_i(tv^{-\lambda_i})]\,dt$$

$$= -2\lambda_i v^{-1-\lambda_i}\int_{E^1} v^{-\lambda_i}\breve{\Omega}_i((y_i - t)v^{-\lambda_i})\,D_{tv^{-\lambda_i}}[tv^{-\lambda_i}\breve{\Omega}_i(tv^{-\lambda_i})]\,dt$$

$$= -2\lambda_i v^{-1-2\lambda_i}\int_{E^1} tv^{-\lambda_i}\breve{\Omega}_i(tv^{-\lambda_i})\,D_i\breve{\Omega}_i((y_i - t)v^{-\lambda_i})\,dt,$$

D_i denoting the derivative with respect to the ith argument.

If we substitute this derivative into formula (93), make the change of variable $y_i = y_i' + t$, and then replace y_i' with y_i, we obtain

$$\frac{\partial}{\partial v}\bar{f}_{v^\lambda}(x) =$$

$$-\sum_{i=1}^{n} v^{-1-|\lambda|-\lambda_i}\int_{E^n}\chi_i(y:v^\lambda)\,dy\int_{E^1} f(x + y + te_i)\,tv^{-\lambda_i}\breve{\Omega}_i(tv^{-\lambda_i})\,dt,$$

where

$$\chi_i(y) = 2\lambda_i D_i\breve{\Omega}_i(y_i)\prod_j{}^{(i)}\bar{\Omega}_j(y_j). \qquad (94)$$

By virtue of (88), we can transform the integral with respect to t to the form

$$\int_{E^1} f(x + y + te_i)\,tv^{-\lambda_i}\breve{\Omega}_i(tv^{-\lambda_i})\,dt$$

$$=\frac{1}{A_i}\int_{E^1}\sum_{j=0}^{k_i}(-1)^{k_i-j}C_{k_i}^j f(x+y+te_i)\frac{tv^{-\lambda_i}}{1+j\delta}K_i\left(\frac{tv^{-\lambda_i}}{1+j\delta}\right)\frac{dt}{1+j\delta}$$

$$=\int_{E^1}\Delta_i^{k_i}(\delta t)\,f(x+y+te_i)\,M_i(tv^{-\lambda_i})\,dt,$$

where

$$M_i(t)=\frac{1}{A_i}tK_i(t). \tag{95}$$

By using the expression obtained for $\frac{\partial}{\partial v}\bar{f}_{v\lambda}(x)$, we can, by virtue of formula (1), write

$$\bar{f}_{\varepsilon\lambda}(x)=\bar{f}_{h\lambda}(x)+\int_{\varepsilon}^{h}\sum_{i=1}^{n}v^{-1-|\lambda|-\lambda_i}\,dv$$
$$\times\int_{E^1}\int_{E^n}\Delta_i^{k_i}(\delta t)\,f(x+y+te_i)\,\chi_i(y:v^{\lambda})\,M_i(tv^{-\lambda_i})\,dy\,dt. \tag{96}$$

If $f\in L^{\mathrm{loc}}(G)$, by letting ε approach zero in (96), we obtain on the basis of Lemma 5.3 the identity

$$f(x)=\bar{f}_{h\lambda}(x)+\int_{0}^{h}\sum_{i=1}^{n}v^{-1-|\lambda|-\lambda_i}\,dv$$
$$\times\int_{E^1}\int_{E^n}\Delta_i^{k_i}(\delta t)\,f(x+y+te_i)\,\chi_i(y:v^{\lambda})\,M_i(tv^{-\lambda_i})\,dy\,dt. \tag{97}$$

To ascertain the set U of points x in the domain of definition G of the function $f(x)$ at which the identity (97) is valid, we first characterize the support of the representation (97).

Let us again assume that $b>0$, $a_i\neq 0$, and

$$S(K_i) = \operatorname{supp} K_i \subset \left\{ t: \frac{t}{a_i} > 0,\ 1 < \left(\frac{t}{a_i}\right)^{1/\lambda_i} < 1 + b \right\}$$
$$(i = 1, \ldots, n).$$

Then, by virtue of (88) and (90),

$$S(\breve{\Omega}_i) \subset \left\{ t: \frac{t}{a_i} > 0,\ 1 < \left(\frac{t}{a_i}\right)^{1/\lambda_i} < (1 + b)(1 + k_i\delta)^{1/\lambda_i} \right\},$$
$$S(\overline{\Omega}_i) \subset \left\{ t: \frac{t}{a_i} > 0,\ 2^{1/\lambda_i} < \left(\frac{t}{a_i}\right)^{1/\lambda_i} < 2^{1/\lambda_i}(1 + b)(1 + k_i\delta)^{1/\lambda_i} \right\}$$
$$(i = 1, \ldots, n).$$

It then follows from these formulas and formulas (94) and (95) that the following estimates are valid for the supports of the kernels $\chi_i(y:\ v^{\lambda})$ and $M_i(tv^{-\lambda_i})$:

$$2^{1/\lambda_j} v < \left(\frac{y_j}{a_j}\right)^{1/\lambda_j} < 2^{1/\lambda_j}(1 + b)(1 + k_j\delta)^{1/\lambda_j} v \quad (1 \leqslant j \leqslant n;\ j \neq i),$$
$$v < \left(\frac{y_i}{a_i}\right)^{1/\lambda_j} < (1 + b)(1 + k_i\delta)^{1/\lambda_i} v,$$
$$v < \left(\frac{t}{a_i}\right)^{1/\lambda_i} < (1 + b) v.$$

We see from these estimates that the support of the representation (97) is the translated horn $x + V\left(\frac{1}{\mu}\right)$, where

$$V\left(\frac{1}{\lambda}\right) = \bigcup_{0 < v \leqslant h} \left\{ y: \frac{y_j}{a_j} > 0,\ 2^{1/\lambda_j} v < \left(\frac{y_j}{a_j}\right)^{1/\lambda_j} \right.$$
$$\left. < 2^{1/\lambda_j}(1 + b)(1 + k_j\delta)^{1/\lambda_j} v,\ j = 1, \ldots, n \right\}. \tag{98}$$

Obviously, for suitable choice of the parameters $a_i \neq 0$, $b > 0$, and $\delta > 0$, we can make the opening of the horn $V\left(\frac{1}{\lambda}\right)$ arbitrarily small.

It is now clear that, under the assumptions made above regarding f, the identity (97) is valid for almost all x in that subset U of G for which $U + V\left(\frac{1}{\lambda}\right) \subset G$.

Let us apply the operator D_x^α to both sides of the identity (96). In the second term of the right-hand member, we apply it to the corresponding kernel. We then obtain

$$D_x^\alpha \bar{f}_{\varepsilon^\lambda}(x) = D_x^\alpha \bar{f}_{h^\lambda}(x) + (-1)^{|\alpha|} \int_\varepsilon^h \sum_{i=1}^n v^{-1-\varkappa_i} dv$$
$$\times \int_{E^1} \int_{E^n} \Delta_i^{k_i}(\delta t) f(x + y + te_i) D^\alpha \chi_i(y : v^\lambda) M_i(tv^{-\lambda_i}) dy \, dt, \tag{99}$$

where

$$\varkappa_i = |\lambda| + \lambda_i + (\alpha, \lambda). \tag{100}$$

Suppose that there exists $D^\alpha f \in L^{loc}(G)$. Then formula 6(2) is valid on the set U defined above. Consequently, $D^\alpha \bar{f}_{\varepsilon^\lambda}(x) = (D^\alpha \bar{f})_{\varepsilon^\lambda}(x)$. It follows on the basis of Lemma 5.3 that $D^\alpha f_{\varepsilon^\lambda}(x) \to D^\alpha f(x)$ almost everywhere on U as $\varepsilon \to 0$. Therefore, by letting ε approach 0 in (99), we obtain the following equation, which is valid for almost all $x \in U$:

$$D^\alpha f(x) = D_x^\alpha \bar{f}_{h^\lambda}(x) + (-1)^{|\alpha|} \int_0^h \sum_{i=1}^n v^{-1-\varkappa_i} dv$$
$$\times \int_{E^1} \int_{E^n} \Delta_i^{k_i}(\delta t) f(x + y + te_i) D^\alpha \chi(y : v^\lambda) M_i(tv^{-\lambda_i}) dy \, dt. \tag{101}$$

Since in $D_x^\alpha \bar{f}_{h^\lambda}(x)$ the differentiation can also be applied to a kernel, the identity (101) gives a representation of the mixed derivative of the function f in terms of the integrals of f and its finite differences with respect to the coordinate directions. The support of that representation is also the set $x + V\left(\frac{1}{\lambda}\right)$.

7.8.1. REMARK. Formula (101) will be used primarily to investigate the properties of functions in the classes $B^{l}_{p,\theta}(G)$ (see Chapter IV). Since we do not assume in advance that functions in these classes have mixed derivatives and formula (101) was obtained under the assumption that f has the derivative $D^{\alpha}f \in L^{\mathrm{loc}}(G)$, let us now give certain conditions ensuring that functions in the classes $B^{l}_{p,\theta}(G)$ (and somewhat more general classes) have a generalized derivative $D^{\alpha}f$.

For an arbitrary set $T \subset G$, let us define

$$\Delta_i^{k_i}(t;\ T)\,f(y) = \begin{cases} \Delta_i^{k_i}(t)\,f(y) & \text{if the difference is constructed from points in } T \text{ such that the segment between them is contained in } T, \\ 0 & \text{otherwise.} \end{cases}$$

Let f denote a member of $L^{\mathrm{loc}}(G)$. Suppose that there exist $\boldsymbol{p} = (p_1, \ldots, p_n)$ and $\boldsymbol{\theta} = (\theta_1, \ldots, \theta_n)$ with $1 \leqslant p_i, \theta_i \leqslant \infty$, and $l = (l_1, \ldots, l_n)$ with $0 < l_i < \infty$ such that for every compact subset T of G there exist constants $A_i(T)$ depending in general on T such that

$$\left(\int_0^\infty t^{-1-l_i\theta_i} \left\|\Delta_i^{k_i}(t;\ T)\,f\right\|_{p,T}^{\theta_i} dt\right)^{1/\theta_i} \leqslant A_i(T) \qquad (i = 1, \ldots, n). \tag{102}$$

Let us show that, if the function f satisfies conditions (102) and the vector $\alpha = (\alpha_1, \ldots, \alpha_n)$ satisfies the inequalities

$$\mu_i = l_i\lambda_i - (\alpha, \lambda) > 0 \qquad (i = 1, \ldots, n), \tag{103}$$

then the derivative exists, $D^{\alpha}f \in L_p^{\mathrm{loc}}(G)$, and hence the identity

(101) holds. First of all, let us show that $D^{\alpha}\bar{f}_{\varepsilon\lambda} - D^{\alpha}\bar{f}_{h\lambda} \to 0$ in the sense of convergence in $L_p^{\mathrm{loc}}(G)$ as ε and h approach 0 through values such that $0 < \varepsilon < h$. Let F denote a compact subset of G. Then, for sufficiently small $h > 0$, the set $F + V\left(\frac{1}{\lambda}\right)$ is contained in some compact subset T of G. If we use successively equation (99), Minkowski's generalized inequality 2(12), Young's inequality 2(18), and Hölder's inequality, we get

$$\left\| D^{\alpha}\bar{f}_{\varepsilon\lambda} - D^{\alpha}\bar{f}_{h\lambda} \right\|_{p,F}$$

$$\leqslant \sum_{i=1}^{n} \| D^{\alpha}\chi_i \|_1 \int_{\varepsilon}^{h} v^{-1-\varkappa_i+|\lambda|}\, dv \int_{E^1} \left| M_i\left(tv^{-\lambda_i}\right)\right| \left\| \Delta_i^{k_i}(\delta t;\ T) f \right\|_{p,T} dt$$

$$\leqslant \sum_{i=1}^{n} \tilde{C}_i \| D^{\alpha}\chi_i \|_1 \int_{\varepsilon}^{h} v^{-1-\varkappa_i+|\lambda|}\, dv \left(\int_{E^1} \left| M_i\left(tv^{-\lambda_i}\right)\right|^{\theta_i'} |t|^{\frac{\theta_i'}{\theta_i}+l_i\theta_i'} dt \right)^{\frac{1}{\theta_i'}}$$

$$\times \left(\int_{0}^{\infty} t^{-1-l_i\theta_i} \left\| \Delta_i^{k_i}(t;\ T) f \right\|_{p,T}^{\theta_i} dt \right)^{\frac{1}{\theta_i}}$$

$$\leqslant \sum_{i=1}^{n} C_i \| D^{\alpha}\chi_i \|_1 \| M_i \|_{\theta_i'} A_i(T)\, h^{\mu_i},$$

where the $\varkappa_i$ are defined in (100) and the μ_i in (103).

It follows that $\left\| D^{\alpha}\bar{f}_{\varepsilon\lambda} - D^{\alpha}\bar{f}_{h\lambda} \right\|_{p,F} \to 0$ as ε and h approach zero through values such that $0 < \varepsilon < h$. Since in addition $\bar{f}_{\varepsilon\lambda}(x) \to f(x)$ in the sense of convergence in $L^{\mathrm{loc}}(G)$ as $\varepsilon \to 0$, we conclude on the basis of Lemma 6.2 that there exists $D^{\alpha}f \in L_p^{\mathrm{loc}}(G)$, as we needed to show.

We note also that inequalities (102) are satisfied for functions in $B_{p,\theta}^{l}(G)$ by virtue of the definition of that class (see Chapter IV).

7.8.2. Remark. Let K denote a member of $C_0^{\infty}(E^n)$ such that

$$\int_{E^n} K(x)\, dx = 1.$$

Let

$$\breve{\Omega}(x) = \frac{1}{A} \sum_{j=0}^{k} \frac{(-1)^{k-j}}{(1+j\delta)^{n+1}} C_k^j K\left(\frac{x_1}{1+j\delta}, \ldots, \frac{x_n}{1+j\delta}\right),$$

where k is a natural number, $0 < \delta < 1$, and the number

$$A = \sum_{j=0}^{k} \frac{(-1)^{k+j}}{1+j\delta} C_k^j \qquad (\neq 0).$$

Let us define the functions

$$\bar{\Omega}(y) = \int_{E^n} \breve{\Omega}(y-z)\, \breve{\Omega}(z)\, dz$$

and

$$\bar{f}_{v^\lambda}(x) = v^{-|\lambda|} \int_{E^n} f(x+y)\, \bar{\Omega}(y : v^\lambda)\, dy. \tag{104}$$

Then, repeating the reasoning followed above and defining

$$N(y, z) = \sum_{i=1}^{n} \frac{\lambda_i}{A} z_i K(z)\, D_i \breve{\Omega}(y),$$

we obtain the identity

$$f(x) = \bar{f}_{h^\lambda}(x)$$
$$+ \int_0^h v^{-1-2|\lambda|}\, dv \int_{E^n} \int_{E^n} \Delta^k(\delta z) f(x+y+z) N(y : v^\lambda, z : v^\lambda)\, dy\, dz, \tag{105}$$

which holds for almost all $x \in G$ for which the support of the representation is contained in G.

For a suitable choice of the support of the function $K(x)$, the support of the representation (105) is a horn.

7.9. A general remark. In all of the above-obtained integral representations of functions and their derivatives, especially in the representations (15), (25), (43), (81), (82), (97), and (101), the right-hand members of the formulas coincide with the left-hand members almost everywhere on the set

$$U = \left\{x\colon x \in G,\ x + V\left(\frac{1}{\lambda}\right) \subset G\right\}.$$

One can easily see that, for any point $x \in U$ and any pair of functions $f(x)$ that are equivalent on G, the right-hand member of any one of these formulas has the same value for one function as for the other. Therefore, we treat such a function as a representative of the entire equivalence class of functions.

In what follows, we shall frequently assume that these integral representations are valid for every point $x \in U$ (and even for points $x \in \partial G$ if $x + V\left(\frac{1}{\lambda}\right) \subset G$). In making this assumption, we may merely replace the function f with the function defined on G equivalent to it that coincides with the right-hand member of the corresponding integral representation everywhere on U.

7.10. Multiparameter averaging and integral representations. As we saw, the method for obtaining the integral representations given above is based on examination of averaging of the type f_{v^λ}, where v^λ is the averaging parameter. We shall refer to such averaging as single-parameter averaging.

The chief purpose in obtaining these representations is to use them to establish inequalities between various norms of the partial derivatives of functions of several variables. However, many inequalities, for example, inequalities of the form

$$\left\|\frac{\partial^2 f}{\partial x_1\, \partial x_2}\right\|_p \leqslant C\left(\left\|\frac{\partial f}{\partial x_1}\right\|_p + \left\|\frac{\partial f}{\partial x_2}\right\|_p + \left\|\frac{\partial^4 f}{\partial x_1^2\, \partial x_2^2}\right\|_p\right),$$

cannot be obtained by use of these representations. The reason is that, if we begin with single-parameter averaging, we obtain a representation of some mixed derivative only in terms of derivatives of a certain type. More precisely, the derivative $D^{\alpha}f$ can be represented in terms of the derivatives $D^{l^i}f$ (for $i = 0, 1, \ldots, n$) if the vectors $\alpha = (\alpha_1, \ldots, \alpha_n)$ and $l^i = (l_1^i, \ldots. l_i^n)$ characterizing them satisfy conditions (20):

$$l_j^0 \leqslant \alpha_j \quad (j = 1, \ldots, n); \quad l_j^i \leqslant \alpha_j \quad (j = 1, \ldots, n;\ j \neq i), \quad l_i^i > \alpha_i \quad (i = 1, \ldots, n).$$

Consideration of representations based on examination of multi-parameter averaging, that is, averaging with different parameters with respect to different variables or groups of variables, considerably broadens the possibilities in the sense indicated. Here, we shall give only a very simple integral representation of this kind, one that we shall use later.

To do this, we introduce some notation. Let $e^n = \{1, \ldots, n\}$ denote a set of natural numbers. Let e denote an arbitrary (possibly empty) subset of e^n. Define $e' = e^n \setminus e$ (so that $e \cup e' = e^n$) and $\mathbf{1}^e = (\delta_1^e, \ldots, \delta_n^e)$, where δ_j^e is equal to 1 if $j \in e$ but equal to 0 if $j \in e'$, and

$$|\mathbf{1}^e| = \sum_{j=1}^{n} \delta_j^e.$$

If $l = (l_1, \ldots, l_n)$, we let l^e denote either the n-dimensional vector $l^e = (l_1^e, \ldots, l_n^e)$ such that $l_j^e = l_j$ for $j \in e$ but $l_j^e = 0$ for $j \in e'$ or the vector of dimension $|\mathbf{1}^e|$ with components l_j, where j ranges over the set e. We shall also write $l = (l^e, l^{e'})$.

The symbols

$$x^e, \ v^e, \ h^e, \ \varepsilon^e, \quad x = (x^e, x^{e'}), \quad v = (v^e, v^{e'}), \quad h = (h^e, h^{e'}),$$

where

$$x=(x_1,\ldots,x_n),\ v=(v_1,\ldots,v_n),\ h=(h_1,\ldots,h_n),\ \varepsilon=(\varepsilon_1,\ldots,\varepsilon_n),$$

are defined analogously.

We also write

$$D^{1^e}_{v^e}F=\frac{\partial}{\partial v_{j_1}}\cdots\frac{\partial}{\partial v_{j_s}}F,$$

$$\int_{\varepsilon^e}^{h^e}F\,dv^e=\int_{\varepsilon_{j_1}}^{h_{j_1}}dv_{j_1}\ldots\int_{\varepsilon_{j_s}}^{h_{j_s}}F\,dv_{j_s}=\left(\prod_{k=1}^{s}\int_{\varepsilon_{j_k}}^{h_{j_k}}dv_{j_k}\right)F,$$

where $0<\varepsilon_i<h_i$ (for $i=1,\ldots,n$) and $\{j_1,\ldots,j_s\}=e$.

Let f denote a locally summable function defined on a region G of the space E^n for which all the preceding transformations are meaningful.

Consider the average

$$F(x;v)=F(x;v_1,\ldots,v_n)=\int_{E^n}f(x+y)\prod_{j=1}^{n}v_j^{-1}\Omega(y_jv_j^{-1})\,dy, \tag{106}$$

where $v_j>0$ (for $j=1,\ldots,n$) and the $\Omega_j(t)$ are defined by (46). Obviously, $F(x;v_1,\ldots,v_n)$ is a continuously differentiable function of v_j for $v_j>0$, where $j=1,\ldots,n$.

Suppose that $\varepsilon=(\varepsilon_1,\ldots,\varepsilon_n)$, $h=(h_1,\ldots,h_n)$ such that $0<\varepsilon_j<h_j$ for $j=1,\ldots,n$. By virtue of the Newton-Leibnitz formula, we have

$$F(x;\varepsilon_1,\ldots,\varepsilon_n)$$
$$=F(x;h_1,\varepsilon_2,\ldots,\varepsilon_n)-\int_{\varepsilon_1}^{h_1}F'_{v_1}(x;v_1,\varepsilon_2,\ldots,\varepsilon_n)\,dv_1,$$

If we again apply the same equation to the right-hand member but this time with respect to the variable v_2, we obtain

$$F(x;\ \varepsilon_1, \ldots, \varepsilon_n)$$
$$= F(x;\ h_1, h_2, \ldots, \varepsilon_n) - \int_{\varepsilon_1}^{h_1} F'_{v_1}(x;\ v_1, h_2, \ldots, \varepsilon_n)\, dv_1 -$$
$$- \int_{\varepsilon_2}^{h_2} F'_{v}(x;\ h_1, v_2, \ldots, \varepsilon_n)\, dv_2$$
$$+ \int_{\varepsilon_1}^{h_1}\int_{\varepsilon_2}^{h_2} F''_{v_1v_2}(x;\ v_1, v_2, \ldots, \varepsilon_n)\, dv_1\, dv_2.$$

Continuing these operations, we obtain at the nth step

$$F(x;\ \varepsilon) = \sum_{e \subseteq e^n} (-1)^{|1^e|} \int_{\varepsilon^e}^{h^e} D^{1^e}_{v^e} F\left(x;\ v^e,\ h^{e'}\right) dv^e. \tag{107}$$

In the present case, formula (107) plays the same role as formula (1) did in the derivation of the representations examined earlier.

Futhermore, it follows from (106) and (46) that

$$D^{1^e}_{v^e} F\left(x;\ v^e,\ h^{e'}\right)$$
$$= \int_{E^n} f(x+x)\left(\prod_{i \in e} \frac{\partial}{\partial v_j}\left[v_j^{-1}\Omega_j\left(y_j v_j^{-1}\right)\right]\right)\left(\prod_{j \in e'} h_j^{-1}\Omega_j\left(y_j h_j^{-1}\right)\right) dy$$
$$= (-1)^{|1^e|} v^{-1^e} \int_{E^n} f(x+y)\left(\prod_{j \in e} v_j^{-1} D^{k_j}\mathscr{L}_j\left(y_j v_j^{-1}\right)\right)$$
$$\times\left(\prod_{j \in e'} h_j^{-1}\Omega_j\left(y_j h_j^{-1}\right)\right) dy, \tag{108}$$

where

$$\begin{aligned} v^{-\alpha} &= v_1^{-\alpha_1} \ldots v_n^{-\alpha_n} \qquad (\alpha = (\alpha_1, \ldots, \alpha_n)), \\ \mathscr{L}_j(t) &= \frac{t^{k_j}}{(k_j-1)!} K_j(t). \end{aligned} \tag{109}$$

Substituting (108) into (107), we obtain

$$F(x;\varepsilon)=\sum_{e\subseteq e^n}\int_{\varepsilon^e}^{h^e} v^{-1^e}\Phi(x;v^e,h^{e'})\,dv^e, \tag{110}$$

where

$$\Phi(x;v^e,h^{e'}) = \int_{E^n} f(x+y)\left(\prod_{j\in e} v_j^{-1}D^{k_j}\mathscr{L}_j(y_jv_j^{-1})\right)\left(\prod_{j\in e'} h_j^{-1}\Omega_j(y_jh_j^{-1})\right)dy. \tag{111}$$

It is important that the remark to Lemma 5.2 holds for averages of the form (106). It follows from this remark that, if $f\in L_p^{\text{loc}}(G)$, where $1\leqslant p<\infty$, then $F(x;\varepsilon)\to f(x)$ in the sense of convergence in $L_p^{\text{loc}}(G)$ as $\varepsilon_j\to 0$ for $j=1,\ldots,n$. Furthermore, for $p>1$, the result mentioned in the remark to Theorem 1.7 (the Jessen-Marcinkiewicz-Zygmund theorem) implies that $F(x;\varepsilon)\to f(x)$ for almost all $x\in G$. (For $\varepsilon_1=\ldots=\varepsilon_n=\varepsilon$, this last assertion holds also for $p=1$). Consequently, if $f\in L_p^{\text{loc}}(G)$, letting ε_j approach zero (for $j=1,\ldots,n$) in (110) yields the corresponding integral representation for the function $f(x)$.

Let us apply D_x^α to both sides of the identity (110). In the right-hand member, we apply the operator directly to the kernels. We obtain

$$D_x^\alpha F(x;\varepsilon)=\sum_{e\subseteq e^n}(-1)^{|\alpha|}\int_{\varepsilon^e}^{h^e} v^{-1^e}\Phi^{(\alpha)}(x;v^e,h^{e'})\,dv^e, \tag{112}$$

where

$$\Phi^{(\alpha)}(x; v^e, h^{e'}) = \int_{E^n} f(x+y)\left(\prod_{j\in e} v_j^{-1} D_{y_j}^{\alpha_j} D^{k_j}\mathscr{L}_j(y_j v_j^{-1})\right) \times\left(\prod_{j\in e'} h_j^{-1} D_{y_j}^{\alpha_j}\Omega_j(y_j h_j^{-1})\right) dy. \quad (113)$$

Let us suppose that f has generalized derivatives of the form $D^{l_e}f$ for all e contained in e^n, where the components of the vector $l_e = (l_{e,1}, \ldots, l_{e,n})$ satisfy the conditions

$$l_{e,j} \leqslant \alpha_j + k_j \quad (j\in e), \qquad l_{e,j} \leqslant \alpha_j \quad (j\in e').$$

Then, the function $\Phi^{(\alpha)}(x; v^e, h^{e'})$ can be represented in the form

$$\Phi^{(\alpha)}(x; v^e, h^{e'}) = (-1)^{|l_e|} \int_{E^n} D^{l_e} f(x+y)\left(\prod_{j\in e} v_j^{-1-\alpha_j+l_{e,j}} D^{k_j+\alpha_j-l_{e,j}}\mathscr{L}_j(y_j v_j^{-1})\right) \times\left(\prod_{j\in e'} h_j^{-1-\alpha_j+l_{e,j}} D^{\alpha_j-l_{e,j}}\Omega_j(y_j h_j^{-1})\right) dy. \quad (114)$$

The identity characterized by formulas (112) and (114) and the integral representation for $D^\alpha f$ following from it can be used, for example; to study the properties of functions in the classes $S_p^l W$, which were examined in the works of Nikol'skiĭ. What we shall need are identities following from (112) and (113) in the case of certain special transformations to be considered in §13.

Suppose that $0 < b < 1$ and that θ_j is equal to either 1 or -1. If in the construction of the kernels $\Omega_j(t)$ (see (46)) we assume that

$$\operatorname{supp} K_j \subset \left\{t\colon 1-b < \frac{t}{\theta_j} < 1\right\} \qquad (j = 1, \ldots, n),$$

then we easily conclude that, for all ε_j such that $0 < \varepsilon_j < h_j$ for $j = 1, \ldots, n$, the right-hand members of formulas (110) and (112) involve the values of the function f only at points of the rectangular parallelepiped with vertex at the point x and sides parallel to the coordinate axes, more precisely, at points of the set $x + \square(h)$, where

$$\square(h) = \left\{ y;\ 0 < \frac{y_j}{\theta_j} < h_j \quad (j = 1, \ldots, n) \right\}. \tag{115}$$

REMARK. Let us note one consequence of formula (110). Suppose that $f \in L_p(E^n)$ (where $1 \leqslant p < \infty$), that $\varepsilon_1 = \ldots = \varepsilon_n = \varepsilon$, and that $h_1 = \ldots = h_n = h$. For the general term $\mathcal{J}^e$ in the right-hand member of (110), we obtain the following estimate on the basis of Hölder's inequality

$$|\mathcal{J}^e| \leqslant \left(\prod_{j \in e} \int_{\varepsilon}^{h} v_j^{-1-\frac{1}{p}}\, dv_j \right) \| f \|_p \left(\prod_{j \in e} \| D^{k_j} \mathcal{L}_j \|_{p'} \right) \times \left(\prod_{j \in e'} h_j^{-\frac{1}{p}} \| \Omega_j \|_{p'} \right) \leqslant C \| f \|_p\, \varepsilon^{-\frac{|1e|}{p}} h^{-\frac{|1e'|}{p}},$$

where C is a constant independent of f, ε, and h.

If $e' \neq \varnothing$ (so $e \neq e^n$), then $\mathcal{J}^e \to 0$ as $h \to \infty$. Therefore, it follows from (110) that

$$F(x; \varepsilon) = \lim_{h \to \infty} \int_{\varepsilon}^{h} \cdots \int_{\varepsilon}^{h} \frac{dv_1 \ldots dv_n}{v_1^2 \ldots v_n^2} \int_{E^n} f(x + y) \prod_{j=1}^{n} D^{k_j} \mathcal{L}_j \left(y_j v_j^{-1} \right) dy$$

$$= \lim_{h \to \infty} \int_{1/h}^{1/\varepsilon} \cdots \int_{1/h}^{1/\varepsilon} du_1 \ldots du_n \int_{E^n} f(x + y) \prod_{j=1}^{n} D^{k_j} \mathcal{L}_j (y_j u_j)\, dy.$$

Noting that $F(x; \varepsilon) \to f(x)$ almost everywhere on E^n as $\varepsilon \to 0$ and replacing $1/h$ with ε and $1/\varepsilon$ with h, we then obtain

$$f(x)=$$

$$\lim_{h\to\infty}\lim_{\varepsilon\to 0}\int_{\varepsilon}^{h}\cdots\int_{\varepsilon}^{h} du_1\ldots du_n\int_{E^n} f(x+y)\prod_{j=1}^{n} D^{k_j}\mathscr{L}_j(y_j u_j)\,dy.$$

This formula is analogous to the familiar Fourier formula.

§8. The domains of definition of the functions

8.1. Suppose that $l=(l_1,\ldots,l_n)$ is a vector with positive components. Suppose that $0<h\leqslant\infty$, that $\varepsilon>0$, and that $a_i\neq 0$ for $i=1,\ldots,n$. We shall call the set

$$V(l)=V(l,\ h)$$
$$=\bigcup_{0<v<h}\left\{x\colon \frac{x_i}{a_i}>0,\ v<\left(\frac{x_i}{a_i}\right)^{l_i}<(1+\varepsilon)\,v\ \ (i=1,\ \ldots,\ n)\right\}. \tag{1}$$

an l-horn of radius h and opening ε.

Let G denote an open subset of E^n. Let K denote a positive integer. Suppose that, for $k=1,\ldots,K$, there exist open sets G_k and l-horns $V_k(l)=V_k(l,h)$ of the form (1) (with coefficients a_i depending on k) such that

$$G=\bigcup_{k=1}^{K} G_k=\bigcup_{k=1}^{K}(G_k+V_k(l,\ h)). \tag{2}$$

In such a case, we shall say that the open set G satisfies *a weak l-horn condition* and we shall write $G\in\underline{A}(l,h)$.

We shall say that the open set G satisfies *an l-horn condition* and we shall write $G\in A(l,h)$ if condition (2) holds for G and in addition

$$G = \bigcup_{k=1}^{K} G_k^{(\delta)} \qquad \text{for some} \qquad \delta > 0, \tag{3}$$

where

$$G_k^{(\delta)} = \{x\colon\ x \in G_k,\ \rho(x,\ \partial G_k \setminus \partial G) > \delta\}.$$

We shall say that the open set G satisfies *a strong l-horn condition* and shall write $G \in \bar{A}(l, h)$ if condition (2) is satisfied for G and

$$G = \bigcup_{k=1}^{K} G_k^{[\delta]} \qquad \text{for some} \qquad \delta > 0, \tag{4}$$

where

$$G_k^{[\delta]} = \{x\colon\ x \in G_k,\ \rho(x,\ G \setminus G_k) > \delta\}.$$

Let us define the classes

$$\underline{A}(l) = \bigcup_{0<h<\infty} \underline{A}(l, h), \qquad A(l) = \bigcup_{0<h<\infty} A(l, h),$$
$$\bar{A}(l) = \bigcup_{0<h<\infty} \bar{A}(l, h).$$

Obviously,

$$\underline{A}(l, h) \supset A(l, h) \supset \bar{A}(l, h), \qquad \underline{A}(l) \supset A(l) \supset \bar{A}(l).$$

It is also obvious that each of the classes $\underline{A}(l, h)$, $A(l, h)$, $\bar{A}(l, h)$ is narrowed when we increase h.

If $l_1 = \ldots = l_n$, the l-horn $V(l)$ defined by (1) is a cone and we shall say that an open set G belonging to $A(l)$ (resp. $\underline{A}(l)$, resp. $\bar{A}(l)$) satisfies *a cone condition* (resp. *a weak cone condition*, resp. *a strong cone condition*).*

*Regions satisfying a weak cone condition and similar regions were first examined in connection with integral representations of functions and imbedding theorems by Sobolev [1], [2]. Regions satisfying some horn condition

The definition of an open set G satisfying a weak cone condition can be given in a different but equivalent form as follows: an open subset G of E^n satisfies a weak cone condition if

$$x + V(e(x), H) \subset G, \qquad \forall\, x \in G,$$

where $V(e(x), H)$ is a right circular cone with vertex at the coordinate origin of fixed opening and altitude H (where $0 < H \leqslant \infty$) and with vector $e(x)$ of the direction of the axis depending on x.

To prove the equivalence of these two definitions, we need only note that the condition given implies the analogous condition that $e(x)$ assume only a finite number of distinct values though the opening of the cone may be decreased.

Let us give some examples for $n = 2$.

Euclidean space E^2 itself belongs to $\bar{A}(l)$ for every l.

The rectangular parallelepiped

$$\square = \{(x_1, x_2): |x_1| < a, |x_2| < b\, (a > 0, b > 0)\},$$

for arbitrary l, also belongs to $\bar{A}(l)$.

The disk $Q = \{(x_1, x_2): x_1^2 + x_2^2 < 1\}$ belongs to $\underline{A}(l)$ only if $\frac{1}{2} l_1 \leqslant l_2 \leqslant 2l_1$; $Q \in \bar{A}(l)$ for $l_1 = l_2$; the disk Q does not satisfy an l-horn condition if $l_1 \neq l_2$.

The annulus $R = \{(x_1, x_2): 1 < x_1^2 + x_2^2 < 4\}$ satisfies a weak cone condition. The cut annulus

$$R^- = \{(x_1, x_2): 1 < x_1^2 + x_2^2 < 4, \ x_1 < 0 \ \text{for} \ x_2 = 0\} \tag{5}$$

satisfies a cone condition but does not satisfy a strong cone condition.

Let us also look at the classes of open sets

or other were first introduced in the works of Besov and Il'in (see, for example, their joint paper [1]).

$$\underline{A}(\square) = \bigcup_{0<h<\infty} \underline{A}(\square, h), \qquad A(\square) = \bigcup_{0<h<\infty} A(\square, h),$$
$$\bar{A}(\square) = \bigcup_{0<h<\infty} \bar{A}(\square, h),$$

constructed in the same way as $\underline{A}(l)$, $A(l)$, and $\bar{A}(l)$ but with the l-horn $V(l, h)$ (defined by (1)) replaced with the cube

$$\square(h) = \{x\colon\ 0 < x_i < h \ \ (i = 1, \ldots, n)\}.$$

We shall say that an open set G belonging to $A(\square)$ (resp. $\underline{A}(\square)$, resp. $\bar{A}(\square)$) satisfies *a cube condition* (resp. *a weak cube condition*, resp. *a strong cube condition*).

Obviously, for arbitrary l,

$$\underline{A}(l) \subset \underline{A}(\square), \qquad A(l) \subset A(\square), \qquad \bar{A}(l) \subset \bar{A}(\square).$$

The disk $Q = \{(x_1, x_2)\colon\ x_1^2 + x_2^2 < 1\}$ does not satisfy a weak cube condition. The "truncated" disk

$$Q' = \{(x_1, x_2)\colon\ x_1^2 + x_2^2 < 1,\ |x_1| < 1 - \varepsilon,\ |x_2| < 1 - \varepsilon\ \ (\varepsilon > 0)\}$$

satisfies a strong cube condition.

In §13, we shall also look at open sets satisfying the weak condition of a rectangle. This condition will be formulated in subsection 13.2 in analogy with a weak cube condition but with the cube replaced with a rectangular parallelepiped having edges parallel to the coordinate axes. We shall examine the lengths of the edges (which may be infinite).

In what follows, we shall find useful the following simple relationship involving these classes of open sets: for arbitrary $c > 0$,

$$\underline{A}(l, h) = \underline{A}(cl, h^c), \quad A(l, h) = A(cl, h^c), \quad \bar{A}(l, h) = \bar{A}(cl, h^c). \tag{6}$$

To show this, we need only note that, as we see from (1), for any given $c > 0$ an arbitrary horn $V(l, h)$ contains some horn $V(cl, h^c)$.

We note, finally, that, in essence, we obtain the same classes of open sets $\underline{A}(l, h)$, $A(l, h)$, and $\bar{A}(l, h)$ if in constructing them we replace the horn $V(l, h)$ defined by (1) with the following figure (which it is also natural to call an l-horn):

$$\mathscr{V}(l, h) = \left\{ x\colon \frac{x_i}{b_i} > 0,\ \left(\frac{x_i}{b_i}\right)^{l_i} < (1+\delta)\left(\frac{x_j}{b_j}\right)^{l_j} < h; \right.$$

$$\left. i = 1, \ldots, n;\ j = 1, \ldots, n \right\}, \quad (7)$$

where $b_i \neq 0$ (for $i = 1, \ldots, n$) and $\delta > 0$.

Specifically, for every l-horn $\mathscr{V}(l, h)$ with parameters b_i and δ, there exists an l-horn $V(l, h)$ with parameters a_i and ε and positive constants c_1 and c_2 such that, for all $h > 0$,

$$V(l, c_1 h) \subset \mathscr{V}(l, h) \subset V(l, c_2 h).$$

We should mention that, with the classes $\underline{A}(l, h)$, $A(l, h)$, and $\bar{A}(l, h)$, the exact value of the constant h will not be of great significance to us.

Geometrically, the l-horn (7) is defined in n-dimensional space by surfaces described by the simple equations

$$\left(\frac{x_i}{b_i}\right)^{l_i} = (1+\delta)\left(\frac{x_j}{b_j}\right)^{l_j} \qquad (i \neq j;\ i = 1, \ldots, n;\ j = 1, \ldots, n).$$

8.2. Extendible partitions of unity. Let K denote a positive integer, and let $G, G_1, \ldots, G_K$ denote $K+1$ open subsets of E^n such that $G = \bigcup_{k=1}^{K} G_k$ and equation (4) is satisfied for some $\delta > 0$. Then, we shall call the family $\{e_k\}_1^K$ of functions *an extendible partition* of unity for the open set G corresponding to the covering $\{G_k\}_1^K$ of the set G if the following four conditions are satisfied:

a) $0 \leqslant e_k(x) \leqslant 1$ on E^n;
b) $e_k(x) = 0$ on $G \setminus G_k$;
c) $\sum_{k=1}^{K} e_k(x) = 1$ on G;
d) $|D^\alpha e_k(x)| \leqslant C_\alpha < \infty$ on E^n ($|\alpha| \geqslant 0$).

Let us indicate a method of constructing a partition of unity. Suppose that $\frac{\delta}{2} < \delta_1 < \delta_2 < \delta$ and

$$G_0 = \{x:\ x \in E^n,\ \rho(x, G) > \delta_1\}.$$

Let us define the function

$$\varphi(x) = \begin{cases} e^{-\frac{1}{1-|x|^2}} & \text{for} \quad |x| < 1, \\ 0 & \text{for} \quad |x| \geqslant 1. \end{cases}$$

Let us denote by $\chi_k(x)$ the characteristic functions of the sets G_0 (for $k = 0$) and $G_k^{[\delta]}$ (for $k \geqslant 1$). Let us define the functions

$$\eta_0(x) = \delta_1^{-n} \int \varphi((y - x) : \delta_1)\, \chi_0(y)\, dy,$$

$$\eta_k(x) = \delta_2^{-n} \int \varphi((y - x) : \delta_2)\, \chi_k(y)\, dy, \qquad k = 1, \ldots, K.$$

Obviously, for $k = 0, 1, \ldots, K$, we have $\eta_k \in C^\infty(E^n)$ and

$$|D^\alpha \eta_k(x)| \leqslant 2^{|\alpha|} \delta^{-|\alpha|} \int |D^\alpha \varphi(x)|\, dx \leqslant \mu_\alpha.$$

We note also that $\eta_0(x) = 0$ on G, that $\eta_k(x) = 0$ on $G \setminus G_k$ (for $k = 1, \ldots, K$), and that $\sum_{k=0}^{K} \eta_k(x) > 0$ on E^n.

Now, we can easily show that the system of functions

$$e_k(x) = \eta_k(x) : \sum_{i=0}^{K} \eta_i(x) \qquad (k = 1, \ldots, K)$$

constitutes an extendible partition of unity. It can be extended so as to be a partition of unity for E^n by adjoining $e_0(x)$.

In particular, for an open set G satisfying a strong l-horn condition a partition of unity corresponding to the covering $\{G_k\}$ of (2) and (4) has been constructed.

We note also that, by virtue of the construction, $e_k(x) = 0$ outside the $(\delta - \delta_2)$-neighborhood of G_k (for $k = 1, \ldots, K$).

8.3. Partitions of unity on an open set. Consider now an n-dimensional open set $G = \bigcup_{k=1}^{K} G_k$, where the G_k are open sets satisfying condition (3) for $\delta > 0$.

We shall call the family $\{e_k\}_1^K$ of functions a partition of unity on the open set G corresponding to the covering $\{G_k\}_1^K$ if the following four conditions are satisfied:

a) $0 \leqslant e_k(x) \leqslant 1$ on G;
b) $e_k(x) = 0$ on $G \setminus G_k$;
c) $\sum_{k=1}^{K} e_k(x) = 1$ on G;
d) $|D^\alpha e_k(x)| \leqslant C_\alpha < \infty$ on G $(|\alpha| \geqslant 0)$.

Obviously, the partition of unity of subsection 8.2 is also a partition of unity on an open set. The converse, however, does not always hold, as we can see with the example (with $n = 2$) of the annulus (5) cut along the radius. Furthermore, it is impossible to construct a partition of unity in the sense of subsection 8.2 for an annulus with such a cut. On the other hand, a partition of unity on an annulus with a cut along the radius does exist, as we shall now show.

Let us show that there exists a partition of unity on an open set G satisfying an l-horn condition.

Thus, suppose that conditions (2) and (3) are satisfied for $G = \bigcup_{k=1}^{K} G_k$ for some $h > 0$ and $\delta > 0$. Let us suppose that h is small, so that

$$V_k(l, h) \subset \left\{ x : |x| < \frac{\delta}{2} \right\}.$$

Then, by virtue of (2) and (3),

$$G_k^{(\delta)} + V_k(l, h) \subset G_k^{(\delta/2)}. \tag{8}$$

Let us denote by $\chi_k(x)$ the characteristic function of the set $G_k^{(\delta)}$. Let us take, for $x \in G_k$,

$$\eta_k(x) = \int L_k(y - x)\, \chi_k(y)\, dy$$
$$= \int L_k(y)\, \chi_k(x + y)\, dy = \int\limits_{V_k(l, h)} L_k(y)\, \chi_k(x + y)\, dy,$$

where

$$L_k(x) \geqslant 0, \quad L_k(x) \in C_0^\infty, \quad 0 \in \operatorname{supp} L_k \subset \overline{V}_k(l, h).$$

It follows that $\eta_k(x) \in C^\infty(G_k)$ and $\eta_k(x) > 0$ for $x \in G_k^{(\delta)}$.

By virtue of (8), the function $\eta_k(x)$ vanishes on $G_k \setminus G_k^{(\delta/2)}$. If we now define $\eta_k(x)$ to be equal to zero on $G \setminus G_k$ and keep our former notation, we obtain

$$\eta_k(x) \in C^\infty(G), \quad \eta_k(x) \geqslant 0,$$

$$\eta_k(x) > 0 \quad \text{for} \quad x \in G_k^{(\delta)}, \qquad \sum_{k=1}^{K} \eta_k(x) > 0 \quad \text{for} \quad x \in G.$$

We now construct a partition of unity on the open set G in the form

$$e_k(x) = \eta_k(x) : \sum_{i=1}^{K} \eta_i(x) \qquad (i = 1, \ldots, K).$$

Chapter 3

ANISOTROPIC SOBOLEV SPACES AND IMBEDDING THEOREMS

The theory of imbedding of function spaces of differentiable functions arose in the investigations of S. L. Sobolev [1], [2] in connection with the solution of a number of problems in mathematical physics. Its development in subsequent years was determined both by the theory of boundary-value problems and by problems associated with the theory itself. This theory is a division of the theory of functions of several real variables. In scope and terminology, it is closely related to functional analysis. A typical situation is as follows: on a sufficiently broad set of functions, we define a family of norms depending on one or several parameters characterizing in some way or other the smoothness and summability properties of the functions. The problem consists in showing that membership of a function in one of the function spaces generated by these norms implies its membership in the others. Thus, a function is regarded as an element of at least a pair of function spaces, and (from the point of view of functional analysis) it is a matter of studying the imbedding operator of one normed space in another. Usually, we are interested in the boundedness of that operator or in its complete continuity. The corresponding assertions are called imbedding or compactness theorems.

Let E denote a normed function space whose elements are defined up to equivalence with respect to Lebesgue measure (in other words, the elements of E are classes of equivalent functions, that is, functions that coincide almost everywhere). When we write $f \in E$, we mean by f either a class of equivalent functions or some function (representative) in such a class.

Let f denote a function in E defined on a subset G of E^n of positive measure. Let G^* denote a subset of G. A function $f^* = f|_{G^*}$ defined on G^* by $f^*(x) = f(x)$ for all $x \in G^*$ is called the restriction of f to G^*. If mes $G^* > 0$, we shall apply the term restriction of a class f of equivalent functions in E to the class of restrictions (obviously, equivalent on G^*) of the different functions in f to G^*. Some of the subsets G^* of $\overline{G}$ are of measure zero (for example, a section of $\overline{G}$ by a plane). In such a case, by the restriction to G^* of an element (function) $f \subset E$, we mean the trace on G^* of that element. (This trace coincides with the trace on G^* of an arbitrary function in the given class of equivalent functions [see 10.8].)

Suppose that E and F are two normed function spaces. We shall say that E is *imbedded* in F and shall write $E \hookrightarrow F$ if the following two conditions are satisfied:

(i) all elements of E (or their restrictions to the domain of definition of the elements of F) are elements of F,

(ii) there exists a constant C independent of f such that

$$\| f \|_E \leqslant C \| f \|_F, \qquad \forall f \in E.$$

This scheme includes the case in which the role of the space F is played by the space C of continuous functions (or, in general, a normed space whose elements are functions and not classes of equivalent functions). For this, we need only understand by C the space of classes of functions equivalent to continuous functions with norm equal to the supremum of the absolute value of a continuous representative of the class. In view of what was said, we can understand the imbedding $E \hookrightarrow C$ to mean that any function $f \in E$ can be modified on a set of Lebesgue measure zero in such a way that it will become continuous and there exists a constant A independent of $f \in E$ such that the inequality $\|f\|_C \leqslant A \|f\|_E$ will hold for the modified function.

This definition shows that the imbedding $E \hookrightarrow F$ is equivalent to boundedness of the identity operator or of the restriction operator defined from E into F. For simplicity, we shall refer to an assertion that a generalized differentiation operator D^α from E into F is bounded, that is, an assertion of the form

$$\| D^\alpha f \|_F \leqslant C \| f \|_E, \qquad \forall f \in E,$$

as an imbedding theorem and we shall write $D^\alpha E \hookrightarrow F$.

If G and G^* are respectively the domains of definition of elements of E and F such that $G \subset \bar{G}^*$ and $G^* \setminus G \neq \varnothing$, then the notation $E \hookrightarrow F$ will mean that there exists a bounded operator of extension of the functions from E into F.

We shall say that two distinct norms $\| f \|_E^{(1)}$ and $\| f \|_E^{(2)}$ defined on the same function space E are equivalent if there exist two positive constants c_1 and c_2 such that $0 < c_1 \leqslant c_2$ and

$$c_1 \| f \|_E^{(1)} \leqslant \| f \|_E^{(2)} \leqslant c_2 \| f \|_E^{(1)}, \qquad \forall f \in E.$$

In his investigations [1], [2], Sobolev studied the properties of the spaces $W_p^{(l)}(G)$ (for regions G with a cone condition) with norm

$$\| f \|_{W_p^{(l)}(G)} = \sum_{|\alpha| \leqslant l} \| D^\alpha f \|_{p,\, G},$$

where l is a natural number* (as well as for a more general norm), and he obtained theorems on the imbedding of these spaces in the spaces $C(G)$ and $L_q(G)$ (where $q \geqslant p$) and also in spaces L_q on

*The parameter l in the generalized spaces $W_p^{(l)}(G)$ and $W_p^l(G)$ (etc.) is a numerical parameter when it appears in parentheses and is a vector parameter, that is, $l = (l_1, \ldots, l_n)$, when it appears without them.

manifolds of different dimensions in $\overline{G}$. These investigations were supplemented by the results of Kodrašov [1] and Il'in [1].

The present chapter is devoted to generalizations of the results of Sobolev and to certain other allied questions. The generalizations consist first of all in examination, from different standpoints, of the anisotropic spaces $W_p^l(G)$ the functions in which have various differentiability properties. In connection with this, it is natural to look at regions G of more general form (satisfying some horn condition or other). In addition to its obvious generality, the use of a mixed L_p-norm enables us to obtain an estimate of the L_q-norm of a function (more accurately, of its trace) on a manifold as a special case of an estimate of a norm in $L_q(G)$.

The imbedding theorems of this chapter admit no improvement in terms of the spaces considered in it.*

In the present chapter, we shall obtain estimates of the norms of the derivatives in terms of the norms of a given set of derivatives and also in terms of the norms of differential operators (coercive estimates), we shall establish multiplicative inequalities for the norms of the derivatives, and we shall investigate the behavior of functions in $W_p^l(G)$ at infinity and the possibility of approximating them with smooth functions of compact support.

The anisotropic spaces of differentiable functions (the spaces H_p^l, where $l = (l_1, \ldots, l_n)$) were first studied in connection with imbedding theorems by Nikol'skiĭ [1]. He showed that they form a closed system with respect to the imbedding theorems and he obtained a converse to those theorems dealing with the restriction to manifolds of lower dimension without change in the metric p.

Analogous study of the anisotropic function spaces W_p^l, where $l = (l_1, \ldots, l_n)$, which generalize Sobolev spaces, was initiated by Slobodeckiĭ [1] and was continued in the works of various authors, O. V. Besov, V. P. Il'in, P. I. Lizorkin, S. V. Uspenskiĭ, and others.

*In the following chapter, we shall show, in particular, how these theorems can be strengthened by considering spaces of functions with fractional exponents for the differentiability properties. These spaces will be constructed on the basis of difference characteristics. We shall obtain inverse theorems on the traces of functions in $W_p^l(G)$ in terms of them.

More recently, imbedding theorems havé been carried over to the case of a mixed L_p-norm. The first results in this direction were obtained by Nikol'skiĭ [6], Gudiyev [1], and Džafarov [2].

§9. Properties of the anisotropic spaces $W_p^l(G)$

9.1. Let G denote an open subset of n-dimensional Euclidean space E^n, let $l = (l_1, \ldots, l_n)$ denote a vector whose components are natural numbers, and suppose that $1 \leqslant p \leqslant \infty$. We denote by $W_p^l(G)$ the space of functions f that are locally summable on G and that have (on G) continuous derivatives $D_i^{l_i} f(x)$ (for $i = 1, \ldots, n$) and a finite norm

$$\|f\|_{W_p^l(G)} = \|f\|_{p,G} + \sum_{i=1}^{n} \left\| D_i^{l_i} f \right\|_{p,G} = \|f\|_{p,G} + \|f\|_{L_p^l(G)}. \quad (1)$$

Theorem. *The space* $W_p^l(G)$, *where* $1 \leqslant p \leqslant \infty$, *is a complete normed space, that is, a Banach space.*

PROOF. Suppose that a sequence of functions $\{f_j(x)\}_1^\infty$ satisfies the Cauchy condition in $W_p^l(G)$, that is,

$$\|f_j - f_k\|_{W_p^l(G)} \to 0 \quad \text{as} \quad j, k \to \infty.$$

The completeness of $L_p(G)$ implies that there exists a function $f(x)$ for which

$$\|f_k - f\|_{p,G} \to 0 \quad (k \to \infty).$$

Since

$$\left\| D_i^{l_i} f_j - D_i^{l_i} f_k \right\|_{p,G} \to 0 \qquad (j, k \to \infty;\ i = 1, \ldots, n),$$

we conclude on the basis of Lemma 6.2 that there exists $D_i^{l_i} f \in L_p(G)$ and that

$$\left\| D_i^{l_i} f_k - D_i^{l_i} f \right\|_{p, G} \to 0 \qquad (k \to \infty;\ i = 1, \ldots, n).$$

Thus, $\| f_k - f \|_{W_p^l(G)} \to 0$ as $k \to \infty$; that is, the space $W_p^l(G)$ is complete.

Let G denote an open subset of n-dimensional Euclidean space E^n. Let $l = (l_1, \ldots, l_n)$ denote a vector whose components are natural numbers. Suppose that $1 \leqslant p^i \leqslant \infty$ (for $i = 0, 1, \ldots, n$). We denote by $W^l_{p^0;\, p^1, \ldots,\, p^n}(G)$ the space of functions f that are locally summable on G and have on G generalized derivatives $D_j^{l_i} f(x)$ (for $i = 1, \ldots, n$) and a finite norm

$$\| f \|_{W^l_{p^0;\, p^1, \ldots,\, p^n}(G)} = \| f \|_{p^0, G} + \sum_{i=1}^{n} \left\| D_i^{l_i} f \right\|_{p^i, G}.$$

The completeness of the space $W^l_{p^0;\, p^1, \ldots,\, p^n}$ is established in the same way as the completeness of $W_p^l(G)$.

9.2. Theorem. *The space $W_p^l(G)$, where $1 \leqslant p < \infty$, is separable.*

PROOF. We introduce the space of vector-valued functions $\mathbf{f}(x) = (f_0(x), f_1(x), \ldots, f_n(x))$ defined on G with norm

$$\| \mathbf{f} \|_{p, G} = \sum_{i=0}^{n} \| f_i \|_{p, G}. \tag{2}$$

The subspace of this space consisting of vector-valued functions of the form $\left(f(x), D_1^{l_1} f(x), \ldots, D_n^{l_n} f(x)\right)$ is isomorphic and isometric to the space $W_p^l(G)$. Thus, we need only establish the separability of the newly introduced space of vector-valued functions with norm (2). And for this in turn it will obviously be sufficient to establish the separability of $L_p(G)$.

Thus, let us prove the separability of the space $L_p(G)$ for $1 \leqslant p < \infty$. Since the space $L_p(G)$ can be considered the subspace of $L_p(E^n)$ consisting of functions that are equal to zero on $E^n \setminus G$, we need only prove the separability of $L_p(E^n)$ for $1 \leqslant p < \infty$. Let us exhibit a countable set that is dense in the latter space.

Since by Theorem 1.6 the set $C_0^\infty(E^n)$ is dense in $L_p(E^n)$, we need only show that any function $\varphi \in C_0^\infty(E^n)$ can be approximated with arbitrary accuracy in $L_p(E^n)$ by functions of some countable set. For this countable set we take the set of functions $\psi_{(N)}(x)$ of the form

$$\psi_{(N)}(x) = \chi_N(x) P(x),$$

where $P(x)$ is a polynomial with rational coefficients, χ_N is the characteristic function of the cube $\square_N = \{x: |x_i| \leqslant N, i = 1, \ldots, n\}$, and N is a natural number.

Obviously, there are only countably many functions of the form $\psi_{(N)}$. On the other hand, any function $\varphi_{(N)} \in C_0^\infty$ concentrated in the cube $\square_N$ can be approximated with an arbitrary degree of accuracy by a polynomial in the norm of $C(\square_N)$ (Weierstrass's theorem), hence by a polynomial with rational coefficients. The inequality

$$\| \varphi_{(N)} - \chi_N P \|_{p, E^n} = \| \varphi_{(N)} - P \|_{p, \square_N} \leqslant \| 1 \|_{p, \square_N} \max_{\square_N} | \varphi_{(N)}(x) - P(x) |$$

now leads to completion of the proof.

9.3. The space $W_p^l(G, \square)$. Suppose that $l = (l_1, \ldots, l_n)$ is a vector whose components are natural numbers, that $1 \leqslant p \leqslant \infty$, that G is a region contained in Euclidean space E^n, that $\square$ is an open n-dimensional cube with edges parallel to the coordinate axes, and that $\bar{\square} \subset G$. *Let us denote by $W_p^l(G, \square)$ the space of functions that are locally summable on G and that have (on G) generalized derivatives $D_i^{l_i} f$ (for $i = 1, \ldots, n$) and a finite norm*

$$\| f \|_{W_p^l(G, \square)} = \| f \|_{p, \square} + \sum_{i=1}^n \| D_i^{l_i} f \|_{p, G} = \| f \|_{p, \square} + \| f \|_{L_p^l(G)}. \tag{3}$$

We note that, if G^* is an open bounded set such that $\bar{G}^* \subset G$, then

$$\| f \|_{p, G^*} \leqslant C(G^*) \| f \|_{W_p^l(G, \square)}. \tag{4}$$

To see this, let us consider, in the one-dimensional case, the representation 7(15). Suppose that the support of the average kernel lies on one side of the coordinate origin and is at a positive distance away from it. Then 7(15) gives a representation of the function $f(x)$ on some interval $[a, b]$ in terms of the values of the derivative $D_1^{l_1}f$ on $[a, b]$ and the values of the function f (which take part in the construction of the average) in some smaller interval $[c, d] \subset [a, b]$. Minkowski's generalized inequality now leads to the estimate

$$\| f \|_{p,\,[a,\,b]} \leqslant C \{\| f \|_{p,\,[c,\,d]} + \| D_1^{l_1} f \|_{p,\,[a,\,b]}\}.$$

Analogous reasoning can be applied in the multidimensional case with the intervals replaced with rectangular parallelepipeds. Here, we use the one-dimensional representation 7(15) with respect to some variable x_i. Repeated use of this device obviously leads to (4).

Let us now show that *the space* $W_p^l(G, \square)$ *is complete.* The sequence $\{f_j(x)\}$, which is fundamental in $W_p^l(G, \square)$, is by (4) also fundamental in $W_p^l(G^*)$ for every bounded region G^* whose closure $\overline{G}^*$ is contained in G. The completeness of $W_p^l(G^*)$ implies the existence of a function $f(x)$ defined on $\bigcup G^* = G$ such that

$$\| f_j - f \|_{p,\,G^*} \to 0, \quad \left\| D_i^{l_i} f_j - D_i^{l_i} f \right\|_{p,\,G^*} \to 0$$

for every G^* as $j \to \infty$.

Since the sequence $\{D_i^{l_i} f_j\}_{j=1}^{\infty}$ converges also in $L_p(G)$, it follows that $\left\| D_i^{l_i} f_j - D_i^{l_i} f \right\|_{p,\,G} \to 0$ (for $i = 1, \ldots, n$) as $j \to \infty$, which completes the proof.

The space $W_p^l(G, \square)$, *where* $1 \leqslant p < \infty$, *is separable.* The proof differs from the proof of Theorem 9.2 only in that for the component $f_0(x)$ in the subspace of the auxiliary space of vector-valued functions, we need to take $f_0(x) = \chi(\square; x) f(x)$, where $\chi(\square; x)$ is the characteristic function of the set $\square$.

9.4. We denote by $\chi(G) = \chi(G; x)$ the characteristic function of the set G. Suppose that U is an open subset of E^n and that $f(x)$ is a function defined on the set $U + V(l)$, where $V(l) = V(l, h) = V$ is an l-horn (see 8.1), and possessing on that set generalized derivatives

$D_i^{l_i} f$ for $i = 1, \ldots, n$. Here and until the end of the section, we shall assume that

$$\lambda = \frac{1}{l} = \left(\frac{1}{l_1}, \ldots, \frac{1}{l_n}\right).$$

By virtue of 7(15), we have everywhere on U

$$f(x) = f_{h^\lambda}(x) + \int_0^h \sum_{i=1}^n v^{-|\lambda|} dv \int D_i^{l_i} f(x+y) \mathscr{L}_i(y : v^\lambda) dy, \quad (5)$$

where the averaging kernel and $\mathscr{L}_i$ belong to $C_0^\infty(E^n)$ and their supports are such that the horn $x + V(l)$ is the support of the representation (5).

Consider the function

$$\tilde{f}(x) = (\chi(U+V) f)_{h^\lambda}(x) +$$
$$\int_0^h \sum_{i=1}^n v^{-|\lambda|} dv \int \chi(U+V; x+y) D_i^{l_i} f(x+y) \mathscr{L}_i(y : v^\lambda) dy. \quad (6)$$

This function is defined by the right-hand member of (6) for all $x \in E^n$. With the aid of Minkowski's generalized inequality 2(12) and Young's inequality, one can easily show that $\tilde{f} \in L^{\text{loc}}(E^n)$ (and even that $\tilde{f} \in L_p^{\text{loc}}(E^n)$ for $p > 1$ and $\frac{1}{p} > 1 - |\lambda|^{-1}$). Since the right-hand members of (5) and (6) coincide for $x \in U$, we have $\tilde{f}(x) = f(x)$ almost everywhere on U. Thus, the function $\tilde{f}$ is an extension of the function f from just U to all E^n. Of course, $\tilde{f}(x)$ does not necessarily coincide with $f(x)$ outside U (that is, on $(U+V) \setminus U$).*

*In case f and $\tilde{f}$ do not coincide, it is natural to assume that the functions $f(x)$ and $\tilde{f}(x)$ are defined on different copies (sheets) of E^n glued along the set U (see subsection 9.8 and also Besov and Il'in [1] or Besov [9]).

Lemma. *Suppose either that* $|\alpha : l| < 1$ *and* $1 \leqslant p \leqslant \infty$ *or that* $|\alpha : l| = 1$ *and* $1 < p < \infty$. *Then there exists a constant C independent of f and h such that*

$$\| D^{\alpha}\tilde{f} \|_{p,\, E^n} \leqslant C h^{1-|\alpha : l|} \sum_{i=1}^{n} \| D_i^{l_i} f \|_{p,\, U+V} + C h^{-|\alpha : l|} \| f \|_{p,\, U+V}. \tag{7}$$

PROOF. Let us suppose first that $|\alpha : l| < 1$. Then, let us define, for $\varepsilon > 0$,

$$\tilde{f}(x; \varepsilon) = (\chi(U+V) f)_{h^{\lambda}}(x) + \int_{\varepsilon}^{h} \sum_{i=1}^{n} v^{-|\lambda|} dv \int \chi(U+V; x+y) D_i^{l_i} f(x+y) \mathscr{L}_i(y : v^{\lambda})\, dy. \tag{8}$$

By Minkowski's generalized inequality 2(12) and Young's inequality 2(19), we obtain from (4)

$$\| \tilde{f}(\cdot\,; \varepsilon) - \tilde{f}(\cdot) \|_p \leqslant \varepsilon \sum_{i=1}^{n} \| \mathscr{L}_i \|_1 \| D_i^{l_i} f \|_{p,\, U+V}.$$

In the same way, we obtain from (8), for $0 < \varepsilon < \eta < h$,

$$\| D^{\alpha}\tilde{f}(\cdot\,; \varepsilon) - D^{\alpha}\tilde{f}(\cdot\,; \eta) \|_p \leqslant$$

$$\sum_{i=1}^{n} \int_{\varepsilon}^{\eta} v^{-|\alpha : l|} \| D^{\alpha}\mathscr{L}_i \|_1 \| D_i^{l_i} f \|_{p,\, U+V} \leqslant C \eta^{1-|\alpha : l|} \sum_{i=1}^{n} \| D_i^{l_i} f \|_{p,\, U+V}. \tag{9}$$

By virtue of Lemma 6.2, it follows from (9) that there exists $D^{\alpha}\tilde{f} \in L_p(E^n)$ and that

$$\| D^{\alpha}\tilde{f} \|_p = \lim_{\varepsilon \to 0} \| D^{\alpha}\tilde{f}(\cdot\,; \varepsilon) \|_p.$$

Differentiating (8) and applying Young's inequality 2(19), we see that

$$\| D^\alpha \tilde{f}(\cdot;\varepsilon)\|_p \leqslant C h^{-|\alpha:l|} \| f\|_{p,U+V}$$

$$+\sum_{i=1}^{n}\int_0^h v^{-|\alpha:l|}\,dv \| D^\alpha \mathscr{L}_i\|_1 \| D_i^{l_i} f\|_{p,U+V},$$

from which we get (7) in the case $|\alpha : l| < 1$.

Our proof will follow the plan of the proof of the estimate (7). For $|\alpha : l| = 1$, it will obviously be sufficient to prove that the integrals

$$\mathscr{I}_i(x;\varepsilon)$$

$$=\int_\varepsilon^h v^{-|\lambda|-1}\,dv \int \chi(U+V;x+y) D_i^{l_i} f(x+y) D^\alpha \mathscr{L}_i(y:v^\lambda)\,dy$$

converge in L_p as $\varepsilon \to 0$ and to obtain for them estimates, independent of $\varepsilon > 0$ and $h > \varepsilon$, of the form

$$\|\mathscr{I}_i(\cdot;\varepsilon)\|_p \leqslant C_i \| D_i^{l_i} f\|_{p,U+V} \qquad (i = 1, \ldots, n).$$

However, the convergence of these integrals and this last estimate are justified by citing Theorem 4.5, condition 4(5) in which is satisfied by virtue of 7(26). This completes the proof of the lemma.

9.5. Theorem.* *Suppose that an open set G satisfies a weak l-horn condition. Suppose either that* $1 \leqslant p \leqslant \infty$ *and* $|\alpha : l| < 1$ *or that* $1 < p < \infty$ *and* $|\alpha : l| = 1$. *Then* $D^\alpha W_p^l(G) \hookrightarrow L_p(G)$ *and, for* $f \in W_p^l(G)$, *there exists a positive number h_0 (depending on G) and a constant C independent of f and h such that, for* $0 < h < h_0$,

*The isotropic case of this theorem is due to Smith [1], the general case to Besov [5] and Il'in [8] (see also their joint paper [1]). Various special cases (especially for $G = E^n$) had been obtained earlier in the works of various mathematicians (see Nikol'skiĭ [9]).

$$\|D^{\alpha}f\|_{p, G} \leqslant Ch^{1-|\alpha:l|} \sum_{i=1}^{n} \|D_i^{l_i}f\|_{p, G} + Ch^{-|\alpha:l|}\|f\|_{p, G}$$

$$\leqslant C(h_0)\|f\|_{W_p^l(G)}. \quad (10)$$

PROOF. Condition 8(2) is satisfied for the set G, so that it will be sufficient to estimate $\|D^{\alpha}f\|_{p, G_k}$. For the corresponding l-horn $V_k(l, h)$, we have $G_k + V_k(l, h) \subset G$, so that the needed estimate follows from Lemma 9.4.

We note also that we can take (for $h_0 > 0$) the radius of l-horns $V_k(l, h)$ that satisfy condition 8(2). In particular, for certain open sets G (e.g., the entire space E^n, a half-space, etc.), we may take $h_0 = +\infty$.

In the case $|\alpha : l| = 1$, strict inequality $1 < p < \infty$ is needed for validity of the assertion. Thus, with $p = 1$, Ornstein [1] has shown that, for $f \in C_0^{\infty}(E^2)$, the norm

$$\int |D^{(1, 1)}f(x_1, x_2)|\, dx_1\, dx_2$$

cannot be estimated in terms of

$$C \int (|D^{(2, 0)}f| + |D^{(0, 2)}f|)\, dx_1\, dx_2.$$

For $p = \infty$, continuity of the second unmixed derivatives $D^{(2, 0)}f(x_1, x_2)$ and $D^{(0, 2)}f(x_1, x_2)$ and the function $f(x_1, x_2)$ does not imply essential boundedness of the generalized mixed derivative $D^{(1, 1)}f(x_1, x_2)$ (see Mityagin [1]). A simpler example of a function of this kind (exhibited by V. I. Yudovich) is the function

$$f(x_1, x_2) = x_1 x_2 \ln \ln \frac{1}{x_1^2 + x_2^2},$$

considered in a neighborhood of the point (0, 0).

We have the following estimate for the growth of the mixed

derivative. Suppose that a function $f(x_1, x_2)$ is of compact support, that it is concentrated in the square

$$\square = \{(x_1, x_2): |x_1| < 1, |x_2| < 1\},$$

and that

$$\text{ess sup}\,\{|D^{(2,0)}f| + |D^{(0,2)}f|\} \leqslant 1.$$

Then, there exist constants $\mu > 0$ and M, both independent of f, such that

$$\int_{\square} \exp\{\mu|D^{(1,1)}f(x_1, x_2)|\}\,dx_1\,dx_2 \leqslant M.$$

If $D^{(2,0)}f$ and $D^{(0,2)}f$ are continuous, then, for every $\mu > 0$,

$$\int_{\square} \exp\{\mu|D^{(1,1)}f(x_1, x_2)|\}\,dx_1\,dx_2 < \infty.$$

(See Yudovič [1], which also contains generalizations to the case of an elliptic operator. A generalization to the anisotropic case and other cases is included in the article by Besov [15].)

9.6. Theorem.* *Let G denote an open set satisfying a strong l-horn condition and suppose that* $1 < p < \infty$. *Then, the space* $W_p^l(G)$ *coincides with the restriction of the space* $W_p^l(E^n)$ *to G. Also, there exists a linear bounded operator extending functions in* $W_p^l(G)$ *to* $W_p^l(E^n)$, *that is,*

$$W_p^l(G) \ni f \to \tilde{f} \in W_p^l(E^n), \qquad \tilde{f}|_G = f. \tag{11}$$

*The possibility of extending functions in the Sobolev space $W_p^l(G)$ outside a region G with a nonsmooth boundary satisfying only a strong cone condition was established by Calderón [1]. The case $l_1 = \ldots = l_n$ of the present theorem is confined in the results of Calderón [1] and Smith [1]. For the general case, see Besov [5], Il'in [8], and the paper written jointly by those two authors [1].

PROOF. Let us use Calderón's method for constructing the required extension operator. This method can be characterized as a method of extending a representation. Suppose that the system of open sets $\{G_k\}_1^K$ forms a covering of G that satisfies, together with the l-horns $V_k(l, h)$, conditions 8(2) and 8(4). Suppose that $\{e_k\}_1^K$ is the partition of unity corresponding to the covering $\{G_k\}_1^K$ (see 8.2).

Let f denote a function belonging to $W_p^l(G)$. We denote by $\tilde{f}_k(x)$ the function constructed in accordance with formula (6) from the function f, the set $U = G_k$, and the l-horn $V = V_k(l, h)$. The function $\tilde{f}(x)$ is defined on E^n and is thus an extension of $f(x)$ beyond the boundary of G_k.

Let us show now that the function

$$\tilde{f}(x) = \sum_{k=1}^{K} e_k(x) \tilde{f}_k(x), \qquad x \in E^n, \tag{12}$$

is an extension of $f(x)$ outside G and that the mapping $f \to \tilde{f}$ is linear and bounded in the sense of (11). By virtue of properties b) and c) of a partition of unity, we have for $x \in G$

$$\tilde{f}(x) = \sum_{k=1}^{K} e_k(x) \tilde{f}_k(x) = \sum_{k=1}^{K} e_k(x) f(x) = f(x).$$

By virtue of properties a) and d) of a partition of unity and Lemma 9.4, we have

$$\| \tilde{f} \|_{p,\, E^n} \leqslant \sum_{k=1}^{K} \| \tilde{f}_k \|_{p,\, E^n} \leqslant C \| f \|_{W_p^l(G)},$$

$$\left\| D_i^{l_i} \tilde{f} \right\|_{p,\, E^n} \leqslant C_1 \sum_{k=1}^{K} \sum_{j=0}^{l_i} \left\| D_i^{l_i - j} e_k D_i^j \tilde{f}_k \right\|_{p,\, E^n} \leqslant$$

$$\leqslant C_2 \sum_{k=1}^{K} \sum_{j=0}^{l_i} \left\| D_i^j \tilde{f}_k \right\|_{p,\, E^n} \leqslant C_3 \| f \|_{W_p^l(G)},$$

as we wished to show.

We note also that all the functions $e_k(x)$ of the partition of unity constructed in subsection 8.2 are concentrated in a δ-neighborhood of G. Therefore, $\tilde{f}(x) = 0$ outside a δ-neighborhood of G.

The significance of this theorem consists primarily in the fact that it enables us to reduce examination of a number of properties of functions in $W_p^l(G)$ to examination of the corresponding properties of functions in the space $W_p^l(E^n)$, where the investigative tool is simpler and better developed.

9.7. In Theorem 9.6, we cannot replace a strong l-horn condition for an open set G with a more general l-horn condition. One can see this by taking for G the two-dimensional annulus cut along the radius 8(5)

$$R^- = \{(x_1, x_2)\colon\ 1 < x_1^2 + x_2^2 < 4,\ x_1 < 0 \ \text{ for } \ x_2 = 0\},$$

which satisfies a cone condition but not a strong cone condition. Suppose that the function $\psi(x)$ is infinitely differentiable outside the interval $\{1 < x_1 < 2,\ x_2 = 0\}$ and is equal to 0 in a lower half-neighborhood of that interval but equal to unity in an upper. Let us show that for no extension to that interval does the function $\psi(x)$ have a derivative $\frac{\partial}{\partial x_2}\psi(x)$ in a neighborhood of any point of the form $(a, 0)$ in it. Let us suppose, to the contrary, that there exists a function $\chi(x) \in L^{\text{loc}}(1 < x_1^2 + x_2^2 < 4)$ for which, in accordance with the definition of a generalized derivative 6(1), we have

$$\int \chi(x)\,\varphi(x)\,dx = -\int \psi(x)\frac{\partial}{\partial x_2}\varphi(x)\,dx$$

for all functions $\varphi \in C_0^\infty(1 < x_1^2 + x_2^2 < 4)$ concentrated in a sufficiently small neighborhood of the point $(a, 0)$. Integrating by parts, we see that

$$\int\int \chi(x_1, x_2)\,\varphi(x_1, x_2)\,dx_1\,dx_2 = \int \varphi(x_1, 0)\,dx_1. \tag{13}$$

Let us now take for $\varphi(x) = \varphi(x_1, x_2)$ the function

$$\varphi(x) = e^{\frac{(x_1-a)^2}{(x_1-a)^2-\delta^2}} e^{\frac{x_2^2}{x_2^2-\varepsilon^2}} \qquad (|x_1-a|<\delta,\ |x_2|<\varepsilon),$$

extended to E^2 by assigning it the value zero otherwise. For fixed δ, by letting ε approach 0, we see that the right-hand member of (13) is positive and does not change, whereas the left-hand member of (13) does not exceed in absolute value the integral

$$\int_{-\varepsilon}^{\varepsilon} \int_{a-\delta}^{a+\delta} |\chi(x_1, x_2)|\, dx_1\, dx_2,$$

which approaches 0 as ε approaches 0. Thus, we have arrived at a contradiction.

In Theorem 9.5, a weak l-horn condition for G cannot be dropped or replaced with a weak s-horn condition with vector $s = (s_1, \ldots, s_n)$ that is not collinear with l. This question will be taken up in detail in subsections 12.2 and 12.3.

9.8. Theorem 9.5 on the estimation of derivatives on a region and Theorem 9.6 on the extension of functions in $W_p^l(G)$ were proven for open sets G satisfying respectively a weak and a strong l-horn condition. These classes of domains of definition of functions for which Theorems 9.5 and 9.6 are valid can be broadened somewhat by preliminarily applying Lemma 9.4 one or more times. We illustrate this device with the example of the disk $G = \{x = (x_1, x_2)\colon x_1^2 + x_2^2 < 1\}$ and the space $W_p^l(G)$, where $l = (l_1, 2l_1)$ and $1 < p < \infty$. The disk G does not satisfy an l-horn condition. Therefore, we cannot use Theorem 9.6 to extend the functions to E^2. Let us apply Lemma 9.4, taking for the set U the quadrant of the disk lying in the open first quadrant $\mathcal{I}_1$. When we do this, we obtain a partial extension of the functions to $G \cup \mathcal{I}_1$. We again apply Lemma 9.4, this time to the region $G \cup \mathcal{I}_1$ and the set U, which is the quadrant of the disk G contained in the second quadrant $\mathcal{I}_2$. This gives us an extension of the functions to $G \cup \mathcal{I}_1 \cup \mathcal{I}_2$. After a third application of Lemma 9.4, we obtain an extension of the functions from $G \cup \mathcal{I}_1 \cup \mathcal{I}_2$ to $G \cup \mathcal{I}_1 \cup \mathcal{I}_2 \cup \mathcal{I}_3$, where $\mathcal{J}_3$ is the open third quadrant. Now, the region $G \cup \mathcal{J}_1 \cup \mathcal{J}_2 \cup \mathcal{J}_3$ satisfies an l-horn condition and we can use Theorem 9.5 to extend the functions to E^2.

As a second example, consider the two-dimensional (1, 2)-horn

$$G = \{(x_1, x_2): ax_1^2 < x_2 < bx_1^2, 0 < x_1 < \infty\}, \qquad b > a > 0.$$

The region G does not satisfy (in a neighborhood of the point $x = 0$) a weak (1, 2)-horn condition. However, for $b > 2a$, it is possible to extend the functions of $W_p^l(G)$, where $l = (l_1, l_2) = (l_1, 2l_1)$, in accordance with Lemma 9.4 and equation 9(6) to the region

$$\tilde{G} = \left\{(x_1, x_2): x_2 > 0, \sqrt{\frac{x_2}{a}} > x_1\right\}$$

with a weak l-horn condition. The estimates of Lemma 9.4 and Theorem 9.5 for the region $\tilde{G}$ lead to the validity of Theorem 9.5 for G.

This device of auxiliary extension of functions proves useful not only for broadening the class of regions in Theorems 9.5 and 9.6 but also in a number of other questions (in imbedding theorems, estimates of coerciveness, etc.) which we shall study later. One should also keep in mind that, with such an auxiliary extension, it is sometimes convenient to consider the multiple-valued functions obtained and the corresponding "many-sheeted Euclidean spaces" (see footnote, p. 169).

9.9. The methods of Hestenes, Calderón, and Stein for extending functions. Let us look at the question of extending functions belonging to the function space $W_p^l(E_+^n)$ and defined on

$$E_+^n = \{x: x = (x_1, \ldots, x_n) \in E^n, x_n > 0\},$$

to the entire space E^n in such a way that their properties are maintained. The bounded linear extension operator

$$W_p^l(E_+^n) \ni f \to \tilde{f} \in W_p^l(E^n), \qquad \tilde{f}\,|_{E_+^n} = f$$

can be constructed in the present case by Hestenes' method

$$\tilde{f}(x_1, \ldots, x_{n-1}, x_n) = \sum_{k=1}^{m} \lambda_k f\left(x_1, \ldots, x_{n-1}, -\frac{x_n}{k}\right), \qquad x_n < 0.$$

Here, we need to take $m \geqslant l_n$ and to find the coefficients λ_k from the system

$$\sum_{k=1}^{m} \lambda_k \left(-\frac{1}{k}\right)^s = 1 \qquad (s = 0, 1, \ldots, m-1).$$

For functions $f(x)$ that are continuous and have continuous derivatives $\frac{\partial^j}{\partial x_n^j} f(x)$ (for $j = 0, 1, \ldots, m-1$) on $\overline{E_+^n}$, such a choice of the coefficients λ_k guarantees maintenance of continuity of those derivatives for the extended function $\tilde{f}$. Hestenes' method is applicable in the case of extension across a smooth boundary (or cross section of a smooth boundary) and in those cases in which the boundary (or a portion of it) can be "rectified" by a sufficiently smooth transformation of the domain of definition of the functions that is invariant with respect to the properties of the functions in the given function space.

Calderón's extension method (the method of extending the representation), which was used in subsection 9.6, is considerably more general with regard to the structure of the domain of definition of the functions but it does not take care of all cases where Hestenes' method is applicable. Thus, for example, when we extend functions in $W_p^l(E_+^n)$ by Calderón's method, we must make the restrictions $1 < p < \infty$ whereas Hestenes' method can be used for $1 \leqslant p \leqslant \infty$.

Both Calderón's and Hestenes' methods can be applied in the extension of other function spaces characterized by integro-differential-difference properties of functions in them.

Stein [1] proposed yet another method of extending functions in an isotropic space $\widetilde{W}_p^{(l)}(G)$ with norm $\sum_{|\alpha| \leqslant l} \| D^\alpha f \|_{p,G}$, where l is a natural number, $1 \leqslant p \leqslant \infty$, and the region G satisfies a strong cone condition. With the aid of a partition of unity, the question can be reduced to the case of a very simple region G of the form

$$G = \{x = (x_1, \ldots, x_n): x_n > \varphi(x_1, \ldots, x_{n-1})\},$$

where the function $\varphi(x_1, \ldots, x_{n-1}) = \varphi(x')$ is defined on E^{n-1} and satisfies a Lipschitz condition. In this case, the extension operator is defined by the formulas

$$\tilde{f}(x', x_n) = f(x', x_n), \quad x_n > \varphi(x'),$$

$$\tilde{f}(x', x_n) = \int_1^\infty f(x', x_n + t\delta(x', x_n))\,\psi(t)\,dt, \quad x_n \leqslant \varphi(x'), \tag{14}$$

where the infinitely differentiable function (equivalent to a distance) $\delta(x', x_n) > 2(\varphi(x') - x_n)$, $|\delta^\alpha(x', x_n)| \leqslant C_\alpha[\delta(x', x_n)]^{-|\alpha|}$ and the function $\psi(t)$ (defined and continuous on $[1, \infty)) = O(t^{-m})$ as $t \to \infty$ for arbitrary m and satisfies the condition

$$\int_1^\infty \psi(t)\,dt = 1, \quad \int_1^\infty t^k \psi(t)\,dt = 0 \qquad (k = 1, 2, \ldots).$$

Let $G^\infty(\bar{G})$denote the set of funcitons that, together with all their derivatives, are bounded and continuously extendible to $\bar{G}$. The extension operator defined by equations (14) is defined on $G^\infty(\bar{G})$, and we have

$$\|\tilde{f}\|_{\widetilde{W}_p^{(l)}(E^n)} \leqslant A_l \|f\|_{\widetilde{W}_p^{(l)}(G)}. \tag{15}$$

The definition of the extension operator and this last estimate can be carried over to arbitrary functions $f \in \widetilde{W}_p^{(l)}(G)$ by taking the limit with respect to a sequence of suitable averages $f_j(x)$ (for $j = 1, 2, \ldots$) (see, for example, 19(2)). Since $f_j \to f$ (as $j \to \infty$) in $\widetilde{W}_p^{(l)}(G)$ for $1 \leqslant p < \infty$ and in $W_p^{(l-1)}(G)$ for $l \geqslant 1$ and $1 \leqslant p \leqslant \infty$, it follows that the $\tilde{f}_j$ constitute a fundamental sequence in $\widetilde{W}_p^{(l)}(E^n)$ (for $1 \leqslant p < \infty$) and in $\widetilde{W}_p^{(l-1)}(E^n)$ (for $1 \leqslant p \leqslant \infty$ and $l \geqslant 1$)

that the estimate (15) remains valid for the limit function $\widetilde{f}$.

It is interesting to note that the form of the extension operator (14) does not depend either on p (for $1 \leqslant p \leqslant \infty$) or on the index of smoothness l of the functions.

§10. The imbedding of $W_p^l(G)$ and $L_q(G)$ in $C(G)$ and in an Orlicz class. Estimates for the trace of a function

In this section, we shall obtain theorems on the imbedding of anisotropic spaces $W_p^l(G)$, for $l = (l_1, \ldots, l_n)$ in $L_q(G)$ and in $C(G)$. These theorems generalize the corresponding classical imbedding theorems of Sobolev [2]. We shall look at certain consequences of these theorems, in particular, the behavior of functions in $W_p^l(G)$ on cross sections of $\bar{G}$ with manifolds of different dimensions. Throughout the section, we shall assume that

$$\lambda = \frac{1}{l} = \left(\frac{1}{l_1}, \ldots, \frac{1}{l_n}\right).$$

The basic index in the formulation of the imbedding theorems will be the quantity

$$\varkappa = \left|\left(\alpha + \frac{1}{p} - \frac{1}{q}\right) : l\right|,$$

which has a simple geometrical interpretation. The condition $\varkappa = 1$ means that the point $\alpha + \frac{1}{p} - \frac{1}{q}$ in n-dimensional space (the space of indices) lies on the hyperplane passing through the n points $(l_1, 0, \ldots, 0), \ldots, (0, \ldots, 0, l_n)$. For $\varkappa \leqslant 1$, the quantity $1 - \varkappa$ is proportional to the distance from the point $\alpha + \frac{1}{p} - \frac{1}{q}$ to the hyperplane mentioned.

10.1. Let U denote an open subset of E^n. Let $f(x)$ denote a function defined on the set $U + V(l)$, where $V(l) = V(l, h) = V$

is an l-horn (see subsection 8.1), and possessing on that set generalized derivatives $D_i^{l_i} f$ for $i = 1, \dots, n$.

Lemma. *Suppose that* $1 \leqslant p \leqslant q \leqslant \infty$ *and*

$$\varkappa = \left| \left(\alpha + \frac{1}{p} - \frac{1}{q} \right) : l \right| \leqslant 1.$$

For $\varkappa = 1$, *suppose that one of the three conditions*

$$1 < p = q < \infty,$$
$$1 < p_n < q_n < \infty,$$
$$1 = p_n < q_n = \infty$$

holds. Then, $D^\alpha W_p^l (U + V) \hookrightarrow L_q (U)$ *and, for* $f \in W_p^l (U + V)$, *there exist a constant* C_1 *independent of* f *and* h *and a constant* C_2 *independent of* f, h, *and* q *such that*

$$\| D^\alpha f \|_{q,\, U} \leqslant C_1 h^{1-\varkappa} \sum_{i=1}^{n} \| D_i^{l_i} f \|_{p,\, U+V} + C_2 h^{-\varkappa} \| f \|_{p,\, U+V}. \quad (1)$$

In the left-hand member of (1), $D^\alpha f$ *can be replaced with* $D^\alpha f_\varepsilon$ *for any* ε *in* $(0, h]$. *For* $\varkappa < 1$, *there exists a constant* C_1' *independent of* q *such that*

$$C_1 = C_1' \left(\frac{1}{1-\varkappa} \right)^{1 - \frac{1}{p_n} + \frac{1}{q_n}} \left(\frac{1}{q_n} \right)^{\frac{1}{q_n}}.$$

PROOF. Let us look at equation 7(22):

$$D^\alpha f_{\varepsilon\lambda} (x) = D^\alpha f_{h\lambda} (x) + \int_\varepsilon^h \sum_{i=1}^{n} v^{-|\lambda| - (\alpha,\, \lambda)}\, dv \int D_i^{l_i} f (x + y) M_i (y : v^\lambda)\, dy, \quad (2)$$

where $0 < \varepsilon < h$. The averaging kernel and M_i belong to $C_0^\infty (E^n)$ and their supports are such that the horn $V(l)$ will serve as support of the representation (2).

Let us consider first the case $\varkappa < 1$. Suppose that $0 < \varepsilon < \eta < h$ and $1 - \frac{1}{r} = \frac{1}{p} - \frac{1}{q}$. By virtue of Minkowski's generalized inequality 2(12) and Young's inequality 2(18), we obtain from (2)

$$\| D^{\alpha} f_{\varepsilon\lambda} - D^{\alpha} f_{\eta\lambda} \|_{q,\, U}$$

$$\leqslant \sum_{i=1}^{n} \int_{\varepsilon}^{\eta} v^{-\varkappa}\, dv \, \| M_i \|_r \, \| D_i^{l_i} f \|_{p,\, U+V} \leqslant C \eta^{1-\varkappa} \sum_{i=1}^{n} \| D_i^{l_i} f \|_{p,\, U+V}. \qquad (3)$$

Thus, $D^{\alpha} f_{\varepsilon\lambda}$ is Cauchy-convergent in $L_q(U)$ as $\varepsilon \to 0$. On the other hand, $f_{\varepsilon\lambda} \to f$ in $L_p(U)$ as $\varepsilon \to 0$, as was shown in the proof of Lemma 9.3. We conclude on the basis of Lemma 6.2 that there exists, on U, the generalized derivative

$$D^{\alpha} f \in L_q(U), \quad \| D^{\alpha} f - D^{\alpha} f_{\varepsilon\lambda} \|_{q,\, U} \to 0 \qquad (\varepsilon \to 0).$$

From this fact and the relations (2) and (3), we obtain

$$\| D^{\alpha} f \|_{q,\, U} \leqslant \| D^{\alpha} f_{h\lambda} \|_{q,\, U} + C_1 h^{1-\varkappa} \sum_{i=1}^{n} \| D_i^{l_i} f \|_{p,\, U+V}.$$

Using 2(18) to estimate the first term in the right-hand member:

$$\| D^{\alpha} f_{h\lambda} \|_{q,\, U} \leqslant C_2 h^{-\left(\lambda,\, \alpha + \frac{1}{p} - \frac{1}{q}\right)} \| f \|_{p,\, U+V}, \qquad (4)$$

we complete the proof of inequality (1) for $\varkappa < 1$. One can easily see that the constant C_2 can be taken proportional to the maximum absolute value of the αth-order derivative of the averaging kernel with coefficient independent of q.

Suppose now that $\varkappa = 1$. The case $1 < p = q < \infty$ was examined in Lemma 9.4, so that we shall assume that $1 < p_n < q_n < \infty$ or that $1 = p_n < q_n = \infty$.

Assuming that

$$x=(\bar{x},\ x_n),\quad y=(\bar{y},\ y_n),\quad p=(\bar{p},\ p_n),\quad q=(\bar{q},\ q_n),$$

$$1-\frac{1}{r}=\frac{1}{p}-\frac{1}{q},\quad r=(\bar{r},\ r_n),\quad U|_{x_n}=\{\bar{x}:\ (\bar{x},\ x_n)\in U\},$$

we use Minkowski's generalized inequality 2(12) and Young's inequality 2(18) to obtain from (2), for $0<\varepsilon<\eta<h$,

$$\|D^\alpha f_{\varepsilon^\lambda}(\cdot,\ x_n)-D^\alpha f_{\eta^\lambda}(\cdot,\ x_n)\|_{\bar{q},\ U|_{x_n}}$$

$$\leqslant\sum_{i=1}^{n}\int_{\varepsilon}^{\eta} v^{-1-\left(\frac{1}{r},\ \lambda\right)}dv\left\|\int\left|M_i(y:v^\lambda)\,D_i^{l_i}f(\cdot+\bar{y},\ x_n+y_n)\right|dy\right\|_{\bar{q},\ U|_{x_n}}$$

$$\leqslant\sum_{i=1}^{n}\int_{\varepsilon}^{\eta} v^{- -\frac{\lambda_n}{r_n}}\,dv\int M(y_n:v^{\lambda_n})\,\|f_i(\cdot,\ x_n+y_n)\|_p\,dy_n,\qquad(5)$$

where

$$M(y_n)=\max_i\|M_i(\cdot,\ y_n)\|_{\bar{r}},$$

$$f_i(x)=D_i^{l_i}f(x)\quad(x\in U+V),\quad f_i(x)=0\quad(\text{outside } U+V).$$

Since the support of $M(y_n)$ is compact and at a positive distance from the coordinate origin (this last is significant only for $r_n=\infty$), we have

$$N_{\varepsilon\eta}(y_n)$$

$$=\int_{\varepsilon}^{\eta} v^{-1-\frac{\lambda_n}{r_n}}M(y_n:v^{\lambda_n})\,dv\leqslant N_{0\eta}(y_n)\leqslant C_1|y_n|^{-\frac{1}{r_n}}=N(y_n).$$

Furthermore, if supp $M\subset[a,\ b]$, where $0<a<b<\infty$, then $N_{0\eta}(y_n)=0$ for $y_n>b\eta^{\lambda_n}$. By using the Hardy-Littlewood inequality 2(32) for $1<p_n<q_n<\infty$ or Young's inequality 2(18) for $1=p_n<q_n=\infty$, we obtain from (5)

$$\| D^{\alpha} f_{\varepsilon\lambda} - D^{\alpha} f_{\eta\lambda} \|_{q,\,U}$$

$$\leqslant \sum_{i=1}^{n} \left\{ \int_{E^1} \left[\int_{E^1} N_{0\eta}(y_n) \| f_i(\cdot\,, x_n + y_n) \|_{\bar{p}} \, dy_n \right]^{q_n} dx_n \right\}^{1/q_n}$$

$$\leqslant \sum_{i=1}^{n} \left\{ \int_{E^1} \left[\int_{E^1} N(y_n) \| f_i(\cdot\,, x_n + y_n) \|_{\bar{p}} \, dy_n \right]^{q_n} dx \right\}^{1/q_n}$$

$$\leqslant C \sum_{i=1}^{n} \| f_i \|_p = C \sum_{i=1}^{n} \| D_i^{l_i} f \|_{p,\,U+V}. \qquad (6)$$

Since $N_{0\eta}(y_n) \to 0$ as $\eta \to 0$ for $r_n < \infty$ and every y_n and since $N_{0\eta}(y_n) \leqslant N(y_n)$, it follows from the Lebesgue dominated-convergence theorem that we obtain from (6)

$$\| D^{\alpha} f_{\varepsilon\lambda} - D^{\alpha} f_{\eta\lambda} \|_{q,\,U} \to 0 \qquad (0 < \varepsilon < \eta,\ \eta \to 0). \qquad (7)$$

The relation (7) is also valid for $r_n = \infty$. In this case,

$$\| D^{\alpha} f_{\varepsilon\lambda} - D^{\alpha} f_{\eta\lambda} \|_{q,\,U}$$

$$\leqslant \sum_{i=1}^{n} \operatorname*{ess\,sup}_{x_n} \int_{0}^{b\eta^{\lambda_n}} N_{0\eta}(y_n) \| f_i(\cdot\,, x_n + y_n) \|_{\bar{p}} \, dy_n,$$

and it only remains to use the absolute continuity of the Lebesgue integral.

Continuing as in the proof of inequality (1) for $\varkappa < 1$, we prove (1) also for the case $\varkappa = 1$.

We note that the proof of the last case of inequality (1) for $q < \infty$ can also be obtained in a different way (see Sobolev [2], §8). Specifically, instead of (7), we use the boundedness of the set $D^{\alpha} f_{\varepsilon\lambda}(x)$ in $L_q(U)$ for all sufficiently small ε (this follows from (2), (4), and (6)). We then use the weak compactness of a bounded set for the existence of a generalized derivative $D^{\alpha} f \in L_q(U)$ and the estimate for the weak limit

$$\| D^{\alpha} f \|_{q,\,U} \leqslant \varliminf_{\varepsilon \to 0} \| D^{\alpha} f_{\varepsilon\lambda} \|_{q,\,U}.$$

It remains to show that, *under the conditions of the lemma, we can, for* $\varkappa < 1$, *represent the constant* C_1 *in inequality* (1) *in the form*

$$C_1 = C_1' \left(\frac{1}{1-\varkappa}\right)^{1-\frac{1}{p_n}+\frac{1}{q_n}},$$

where C_1' *is independent of* f, h, *and* q. This fact will be used in studying the imbedding of W_p^l in an Orlicz class. It follows from the proof of the preceding part of the lemma that we need only obtain an estimate for $\|\mathcal{J}_i(\cdot, h)\|_{q,U}$,

$$\mathcal{J}_i(x, h) = \int_0^h v^{-|\lambda|-(\alpha,\lambda)}\,dv \int_{E^n} \left| D_i^{l_i} f(x+y)\, M_i(y : v^\lambda)\right| dy,$$

in which the constant has the required form. Just as in the proof of the lemma for the case $\varkappa = 1$, we obtain

$$\|\mathcal{J}_i(\cdot, x_n, h)\|_{\bar{q}, U|x_n} \leqslant \int_0^h v^{-\varkappa-\frac{\lambda_n}{r_n}}\,dv \int_{E^1} \varphi_i(x_n+y_n)\, M(y_n v^{-\lambda_n})\,dy \leqslant \int_{E^1} \varphi_i(x_n+y_n)\, N_h(y_n)\,dy_n,$$

where

$$\varphi_i(x_n) = \|f_i(\cdot, x_n)\|_{\bar{p}},$$

$$f_i(x) = D_i^{l_i} f(x) \text{ for } x \in U+V, \qquad f_i(x) = 0 \text{ for } x \notin U+V,$$

$$M(y_n) = \max_i \|M_i(\cdot, y_n)\|_{\bar{r}},$$

$$N_h(y_n) = \int_0^h v^{-\varkappa-\frac{\lambda_n}{r_n}} M(y_n v^{-\lambda_n})\,dv.$$

Since the support of the function $M(y_n)$ is compact and at a positive distance from the coordinate origin, that is, supp $M \subset [a, b]$, where $0 < a < b < \infty$, and since

$$\frac{\varkappa - 1}{\lambda_n} + \frac{1}{r_n} - 1 < 0,$$

we have the following estimate for $N_h(y_n)$:

$$N_h(y_n) \leqslant |y_n|^{\frac{1-\varkappa}{\lambda_n} - \frac{1}{r_n}} \frac{1}{\lambda_n} \int_a^b M(u)\, u^{\frac{\varkappa-1}{\lambda_n} + \frac{1}{r_n} - 1}\, du$$

$$\leqslant C |y_n|^{\frac{1-\varkappa}{\lambda_n} - \frac{1}{r_n}} a^{\frac{\varkappa-1}{\lambda_n} + \frac{1}{r_n} - 1} b \leqslant C' |y_n|^{\frac{1-\varkappa}{\lambda_n} - \frac{1}{r_n}},$$

where C' is a constant independent of q. The function $N_h(y_n)$ is also of compact support since $N_h(y_n) = 0$ for $|y_n| > bh^{\lambda_n}$. Using the estimate obtained for $N_h(y_n)$ and setting

$$\frac{1-\varkappa}{\lambda_n} = \varepsilon, \qquad \frac{1}{r_n} = 1 - \frac{1}{p_n} + \frac{1}{q_n} = \frac{1}{p_n'} + \frac{1}{q_n},$$

we obtain

$$\| \mathcal{I}_i(\cdot, x_n, h) \|_{\bar{q}, U|x_n}$$

$$\leqslant C' \int_{|y_n| \leqslant bh^{\lambda_n}} \varphi_i(x_n + y_n) |y_n|^{\varepsilon - \left(\frac{1}{p_n'} + \frac{1}{q_n}\right)} dy_n.$$

Let us suppose first that $1 < p_n < q_n < \infty$ and let us estimate the integral on the right by using Hölder's inequality for three factors with exponents

$$q_n, \quad \frac{p_n q_n}{q_n - p_n}, \quad p_n' \left(\frac{1}{q_n} + \frac{q_n - p_n}{q_n p_n} + \frac{1}{p_n'} = 1 \right).$$

We get

$$\|\mathcal{J}_i(\cdot, x_n, h)\|_{\overline{q}, U|x_n} \leqslant C' \int\limits_{|y_n| \leqslant bh^{\lambda_n}} \left(|\varphi_i(x_n+y_n)|^{\frac{p_n}{q_n}} |y_n|^{-\frac{1}{q_n}+\frac{\varepsilon}{2}} \right)$$

$$\times \left(|\varphi_i(x_n+y_n)|^{1-\frac{p_n}{q_n}} \right) \left(|y_n|^{-\frac{1}{p_n'}+\frac{\varepsilon}{2}} \right) dy_n$$

$$\leqslant C'' \left(\int\limits_{|y_n| \leqslant bh^{\lambda_n}} |\varphi_i(x_n+y_n)|^{p_n} |y_n|^{-1+\frac{\varepsilon q_n}{2}} dy_n \right)^{\frac{1}{q_n}}$$

$$\times \|\varphi_i\|_{p_n}^{1-\frac{p_n}{q_n}} \left(\frac{1}{\varepsilon p_n'} \right)^{\frac{1}{p_n'}} h^{\frac{\varepsilon \lambda_n}{2}}.$$

It follows that

$$\|\mathcal{J}_i(\cdot, h)\|_{q, U} = \Big\| \|\mathcal{J}_i(\cdot, x_n, h)\|_{\overline{q}, U|x_n} \Big\|_{q_n, \{x_n : U|x_n \neq \varnothing\}}$$

$$\leqslant C''' \left(\frac{1}{\varepsilon p_n'} \right)^{1/p_n} \left(\frac{1}{\varepsilon q_n} \right)^{1/q_n} h^{\varepsilon \lambda_n} \|\varphi_i\|_{p_n}$$

$$= C_1' \left(\frac{1}{1-\varkappa} \right)^{1/p_n'} \left(\frac{1}{(1-\varkappa) q_n} \right)^{1/q_n} h^{1-\varkappa} \left\| D_i^{l_i} f \right\|_{p, U+V},$$

where C_1' is independent of f, h, and $\boldsymbol{q}$.

An analogous estimate holds for other relations between the parameters p_n and q_n (where $1 \leqslant p_n \leqslant q_n \leqslant \infty$). We mention that, in these cases, the estimate is based on Hölder's inequality for two factors. The inequality obtained proves our assertion.

10.2. Theorem. *Suppose that the open set G satisfies a weak l-horn condition, that $\boldsymbol{1} \leqslant \boldsymbol{p} \leqslant \boldsymbol{q} \leqslant \infty$, and that* $\varkappa = \left| \left(\alpha + \frac{1}{\boldsymbol{p}} - \frac{1}{\boldsymbol{q}} \right) : l \right| \leqslant$ 1. *For* $\varkappa = 1$, *suppose that one of the following is true*:

$$1 < p = q < \infty,$$
$$1 < p_n < q_n < \infty,$$
$$1 = p_n < q_n = \infty.$$

Then, $D^\alpha W_p^l(G) \hookrightarrow L_q$; *or, more precisely, for* $f \in W_p^l(G)$ *there exists on* G *a generalized derivative* $D^\alpha f \in L_q(G)$ *and there exist numbers* $h_0 > 0$ *and* $C > 0$ *such that*

$$\| D^\alpha f \|_{q,G} \leqslant C h^{1-\varkappa} \sum_{i=1}^{n} \| D_i^{l_i} f \|_{p,G} + C h^{-\varkappa} \| f \|_{p,G}, \tag{8}$$

where the constant C *is independent of* f *and* $h \in (0, h_0)$. *In particular, for* $\alpha = 0$, *we have*

$$W_p^l(G) \hookrightarrow L_q(G).$$

PROOF. Condition 8(2) holds for G, so that it will be sufficient to estimate $\| D^\alpha f \|_{q, G_k}$ for $k = 1, \ldots, K$. Since $G_k + V_k(l, h) \subset G$ for the corresponding l-horn $V_k(l, h)$, the needed estimate follows from Lemma 10.1.

We note that the basic case of the theorem ($p > 1$, $\varkappa < 1$) is due to Sobolev [1], [2]. The case $1 < p < q < \infty$, $\varkappa = 1$ is due to Hardy and Littlewood for $n = 1$ and to Sobolev for $n > 1$. The case $p = 1$, $q = \infty$, $\varkappa = 1$ is due to Sobolev [1]. The case $1 = p < q < \infty$, $\varkappa = 1$ (which we are not proving) and the case $p = 1$, $q = (1, \ldots, 1, \infty, \ldots, \infty)$, $\varkappa = 1$ (see subsection 18.14) are due to Gagliardo [2], [1]. The case $1 < p < q < \infty$, $q = (q, \ldots, q, \infty, \ldots, \infty)$, $\varkappa = 1$ is due to Il'in [1].

10.3. REMARK. Suppose that the conditions of Theorem 10.2 are satisfied and that some of the components of q are infinite. In the left-hand member of inequality (8), we can replace the L_q-norm $\| D^\alpha f \|_{q,G}$ with the modified $\tilde{L}_q$-norm, which differs from it only in that the essential supremum with respect to the variables corresponding to the infinite components q is replaced with the usual upper bound with respect to those variables.

The reason we can do this is that, in Lemma 10.1, we obtain $D^\alpha f$ as the limit of continuous (in fact infinitely differentiable) functions $D^\alpha f_{e\lambda}$ and the L_q-norm and the $\tilde{L}_q$ norm coincide for continuous functions. Thus, Cauchy convergence in one of these norms implies Cauchy convergence in the other, and it only remains for us to use the completeness of the space with the $\tilde{L}_q$-norm (see remark in subsection 1.1).

We note in connection with what has been said that, under the conditions of the theorem, if $p = (p, \ldots, p)$, $q = (q, \ldots, q, \infty, \ldots, \infty)$, $1 < p < q < \infty$, and, for example, $G = E^n$ instead of the familiar estimate for

$$\varkappa = \sum_1^n \frac{\alpha_i}{l_i} + \frac{1}{p} \sum_{i=1}^n \frac{1}{l_i} - \frac{1}{q} \sum_{i=1}^m \frac{1}{l_i} \leqslant 1,$$

namely,

$$\sup_{x'' \in E^{n-m}} \| D^\alpha f(\cdot, x'') \|_{L_q(E^m)} \leqslant C \| f \|_{W_p^l(E^n)},$$

we obtain the stronger estimate,

$$\Big\| \sup_{x'' \in E^{n-m}} | D^\alpha f(\cdot, x'') | \Big\|_{L_q(E^m)} \leqslant C \| f \|_{W_p^l(E^n)}.$$

We can obtain an analogous estimate for $f \in W_p^l(G)$.

10.4. Theorem. *Suppose that an open set G satisfies a weak l-horn condition, $1 \leqslant p \leqslant \infty$, and either $\varkappa = \left| \left(\alpha + \frac{1}{p} \right) : l \right| < 1$ or $\varkappa = 1$ and $p_n = 1$. Then, $D^\alpha W_p^l(G) \hookrightarrow C(G)$; more precisely, for $f \in W_p^l(G)$, the derivative $D^\alpha f$ is continuous on G and there exists a number h such that $0 < h < h_0$ and a constant C independent of f and h such that*

$$\sup_G | D^\alpha f | \leqslant C h^{1-\varkappa} \sum_{i=1}^n \| D_i^{l_i} f \|_{p, G} + C h^{-\varkappa} \| f \|_{p, G}. \tag{9}$$

PROOF. Since $D^\alpha f$ is defined only up to equivalence, what the theorem asserts is that $D^\alpha f$ is equivalent to a continuous function. We must keep this in mind. It will be sufficient to establish the continuity of $D^\alpha f(x)$ on G since the estimate (9) is then a special case of the estimate (8) (for $q = \infty$).

Suppose that $G = \bigcup_{1}^{K} G_k$ and that condition 8(2) holds for the open sets G_k. As was shown in the proof of Lemma 10.1,

$$\left\| D^\alpha f - D^\alpha f_{\varepsilon\lambda} \right\|_{\infty, G_k} \to 0 \quad \text{as} \quad \varepsilon \to 0.$$

Since $D^\alpha f_{\varepsilon\lambda}$ is continuous on G_k, convergence in $L_\infty(G_k)$ coincides in the present case with uniform convergence, so that the limit function $D^\alpha f$ is continuous on every G_k and hence on $G = \bigcup_{k=1}^{K} G_k$. This completes the proof of the theorem.

We mention that this theorem was proven by Sobolev [1, 2] for the isotropic case with $p_1 = \ldots = p_n$.

10.5. The imbedding of W_p^l in an Orlicz class. Let us look in greater detail at the properties of the derivative $D^\alpha f$ of the function $f \in W_p^l(G)$, where $\mathbf{1} \leqslant \boldsymbol{p} \leqslant \boldsymbol{\infty}$, for the case in which

$$\varkappa = \left| \left(\alpha + \frac{1}{p} \right) : l \right| = 1. \tag{10}$$

It follows from Theorem 10.2 that, if (10) holds and if $p_n = 1$, then $D^\alpha f \in L_\infty(G)$ (more precisely, by virtue of Theorem 10.4, we have $D^\alpha f \in C(G)$) but if $p_n > 1$, then $D^\alpha f \in L_q(G)$, where $\boldsymbol{p} \leqslant \boldsymbol{q} \leqslant \boldsymbol{\infty}$ and at least one component q_j is finite (and this case, the number $\varkappa$ is less than 1). Let us show that this result can be strengthened somewhat if $1 < p_n < \infty$.

Let us introduce some necessary notation and definitions. We define $\bar{\boldsymbol{p}} = (p_1, \ldots, p_{n-1})$, $\bar{\boldsymbol{q}} = (q_1, \ldots, q_{n-1})$, $\bar{x} = (x_1, \ldots, x_{n-1})$, $\boldsymbol{p} = (\bar{\boldsymbol{p}}, p_n)$, $\boldsymbol{q} = (\bar{\boldsymbol{q}}, q_n)$, $x = (\bar{x}, x_n)$, $G\,|_{x_n} = \{\bar{x}: (\bar{x}, x_n) \in G\}$, $\Pi_n G =$

$\{x_n\colon G|_{x_n}\neq\varnothing\}$, where $\Pi_n G$ is the projection of G onto the x_n-axis.

Suppose that $\Phi(t)$ is a continuous real-valued convex even function of a real variable t and that

$$\lim_{t\to 0}\frac{\Phi(t)}{t}=0, \qquad \lim_{t\to\infty}\frac{\Phi(t)}{t}=\infty.$$

We shall call a function $\Phi(t)$ with these properties an N-function.

We shall say that a function f defined on G belongs to the class $L_{(\bar{p},\Phi)}(G)$ if $\|f(\cdot, x_n)\|_{\bar{p}, G|_{x_n}}$, as a function of x_n, belongs to the Orlicz class with N-function Φ, that is, if

$$\int_{\Pi_n G}\Phi\left(\|f(\cdot, x_n)\|_{\bar{p}, G|_{x_n}}\right)dx_n<\infty.$$

Theorem. *Suppose that G satisfies a weak l-horn conditon, that $f\in W_{\bar{p}}^l(G)$, where $\mathbf{1}\leqslant\mathbf{p}\leqslant\infty$ and $1<p_n<\infty$, that $\alpha=(\alpha_1, \ldots, \alpha_n)$ (where the α_i are nonnegative integers), and that*

$$\left|\left(\alpha+\frac{1}{\mathbf{p}}\right):l\right|=1. \tag{10}$$

Then, f has a generalized derivative $D^\alpha f$ on G. For arbitrary $\bar{\mathbf{q}}$ such that $\bar{\mathbf{p}}\leqslant\bar{\mathbf{q}}\leqslant\overline{\infty}$, this derivative belongs to $L_{(\bar{q},\Phi)}(G)$ (and in particular, to $L_{(\overline{\infty},\Phi)}(G)$). That is, there is a constant C^* that does not depend on f such that if

$$\mu<\left[e\left(C^*\|f\|_{W_p^l(G)}\right)^{p'_n}\right]^{-1}$$

and Φ is defined by

$$\Phi(t)=e^{\mu|t|^{p'_n}}-\sum_{0\leqslant k<p_n-1}\frac{1}{k!}\left(\mu|t|^{p'_n}\right)^k \qquad \left(p'_n=\frac{p_n}{p_n-1}\right),$$

then

$$\int_{\Pi_n G} \Phi\left(\| D^\alpha f(\cdot\,,x_n)\|_{\bar{q},\,G|x_n}\right) dx_n < A, \tag{11}$$

where A is a constant that depends on $\| f \|_{W^l_p(G)}$. (See inequality (12) below.)

PROOF. The existence of the derivative $D^\alpha f$ follows from Theorem 10.2. Therefore, it is sufficient to prove (11). We have

$$\mathcal{J} = \int_{\Pi_n G} \Phi\left(\| D^\alpha f(\cdot\,,x_n)\|_{\bar{q},\,G|x_n}\right) dx_n = \int_{\Pi_n G} \Bigg[e^{\mu \| D^\alpha f(\cdot,\,x_n)\|^{p'_n}_{\bar{q},\,G|x_n}} - \sum_{0 \leqslant k < p_n - 1} \frac{1}{k!}\left(\mu \| D^\alpha f(\cdot\,,x_n)\|^{p'_n}_{\bar{q},\,G|x_n}\right)^k \Bigg] dx_n$$

$$\leqslant \sum_{k \geqslant p_n - 1} \frac{\mu^k}{k!} \int_{\Pi_n G} \| D^\alpha f(\cdot\,,x_n)\|^{kp'_n}_{\bar{q},\,G|x_n} dx_n = \sum_{k \geqslant p_n - 1} \frac{\mu^k}{k!} \| D^\alpha f \|^{q_n^{(k)}}_{q^{(k)},\,G},$$

where $q^{(k)} = (\bar{q},\, q_n^{(k)})$, $q_n^{(k)} = kp'_n$, and $1 < p_n \leqslant q_n^{(k)} < \infty$.

We estimate the norm under the summation sign with the aid of inequality (1) (by Lemma 10.1). Before applying that inequality, we note that, by virtue of (10),

$$\varkappa^{(k)} = \left|\left(\alpha + \frac{1}{p} - \frac{1}{q^{(k)}}\right) : l\right| = 1 - \left|\frac{1}{q^{(k)}} : l\right| \leqslant 1 - \frac{1}{q_n^{(k)} l_n} < 1.$$

It follows that we can take the constant C_1 in inequality (1) in the form

$$C_1 = C'_1 \left(\frac{1}{1 - \varkappa^{(k)}}\right)^{1/p'_n} \left(\frac{1}{(1 - \varkappa^{(k)})\, q_n^{(k)}}\right)^{1/q_n^{(k)}}.$$

Therefore, since

$$1 - \varkappa^{(k)} \geqslant \frac{1}{q_n^{(k)} l_n},$$

we have

$$C_1 \leqslant C_1' \left(q_n^{(k)} l_n\right)^{1/p_n'} l_n^{1/q_n^{(k)}} \leqslant C_1'' k^{1/p_n'},$$

where C_1'' is independent of k.

By virtue of inequality (1), we now obtain

$$\| D^\alpha f \|_{q^{(k)}, G} \leqslant C^* k^{1/p_n'} \| f \|_{W_p^l(G)},$$

where

$$C^* = C^*(h) = \sup_k \left(C_1'' h^{1-\varkappa^{(k)}} + C_2 h^{-\varkappa^{(k)}} \right)$$

is independent of k.

Thus

$$\mathcal{I} \leqslant \sum_{k \geqslant p_n - 1} \frac{1}{k!} \left(\mu k \left[C^* \| f \|_{W_p^l(G)} \right]^{p_n'} \right)^k = A. \qquad (12)$$

One can easily see that, under the condition mentioned in the theorem with regard to μ, the series converges. Denoting its limit by A, we obtain (11). This completes the proof of the theorem.

10.6. REMARK. Suppose that, under the conditions of the theorem, with $p = (p, \ldots, p)$, where $1 < p < \infty$, the function $\Phi(t)$ is defined by the same formula as in the theorem. Then,

$$\int_G \Phi(| D^\alpha f(x) |)\, dx < \infty.$$

The proof is analogous to the proof of inequality (11). In applications, one often uses just this result.

We mention that the theorems on the imbedding of the isotropic classes $W_p^{(l)}(G)$ in Orlicz classes were considered in the works of Kalugina [1], Krasnosel'skiĭ and Rutickiĭ [1], Yudovič [1], Pohožayev [1], and Trudinger [1]. Analogous results have been obtained by Kalugina [2], Ahmetžanov [1], and Brudnyĭ [1] for the anisotropic classes $H_p^l(G)$, $W_p^l(G)$, and $B_{p,\theta}^l(G)$.

10.7. Theorem 10.2 does not take care of the case of imbedding when $1 = p_n < q_n < \infty$. For this case, we have the following results:

Gagliardo [2] obtained the estimate

$$\|f\|_{q,G^{(m)}} \leqslant C \sum_{|\alpha| \leqslant l} \|D^\alpha f\|_{1,G},$$

where $1 \leqslant q \leqslant \infty$, $l - n + \frac{m}{q} = 0$, G is an open subset of E^n that satisfies a cone condition, $G^{(m)}$ is the cross section of the set $\overline{G}$ by any m-dimensional hyperplane parallel to the coordinate plane, and $0 \leqslant m \leqslant n$ (the case $m = n$ was examined earlier by Sobolev [1].)

Solonnikov established [4] the imbedding

$$D^\alpha W_1^l(E^n) \hookrightarrow L_q(E^n),$$

where $l = (l_1, \ldots, l_n)$, $1 \leqslant q < \infty$, and

$$\left|\left(1 - \frac{1}{q} + \alpha\right) : l\right| = 1, \quad \text{where} \quad \alpha_k = 0 \quad \text{for} \quad q_k = 1.$$

10.8. The trace of a function. Suppose that $f(x)$ is a function defined on an open subset G of E^n. We shall now introduce the concept of the trace of this function on the cross sections of $\overline{G}$ with a plane of dimension $m < n$ and shall establish some of its properties. We shall use this concept in a number of places later. A more general concept of a trace on a Lipschitzian manifold and its properties will be studied in §20, and the concept of a trace and its properties on a sufficiently smooth manifold will be studied in §§24 and 25.

A function $f(x)$ that belongs to the function space $W_p^l(E^n)$ or to any of the other spaces studied in this book is defined on E^n only up to a set of n-dimensional measure zero. Therefore, the trace of the function f

$$f|_{E^m} = \varphi(x') = \varphi(x_1, \ldots, x_m), \qquad 0 \leqslant m < n,$$

is meaningless on any subspace E^m of E^n if we understand it literally. Below, we shall give, in particular, a definition of the trace of the function f on E^m that leads to a unique function on E^m up to a subset of E^m of m-dimensional measure zero. The trace of the function f is connected naturally and "stably" (with respect to change in f on a set of n-dimensional measure zero) with f. For a continuous function, it coincides with the restriction of f to that set. The definition that we shall give differs slightly from that given in the book by Nikol'skiĭ [9].

The following definitions and notations will be used for the remainder of the subsection.

Definition of a trace. *Let G denote an open subset of E^n. Let m denote an integer in* $[1, n)$. *Suppose that* $x = (x', x'')$, *where* $x' = (x_1, \ldots, x_m)$ *and* $x'' = (x_{m+1}, \ldots, x_n)$. *Let D_k denote an open bounded subset of E^m. For* $k = 1, 2, \ldots$, *suppose that*

$$\Gamma_k = \{x = (x', x''): \ x' \in D_k, \ x'' = \mathring{x}'' = \text{const}\} \subset \overline{G}.$$

We shall say that the function

$$f(x', \mathring{x}'') = f(x') \in L^{\text{loc}}\left(\bigcup_{k=1}^{\infty} D_k\right)$$

is the trace on $\Gamma = \bigcup_{k=1}^{\infty} \Gamma_k$ *of a function* $F(x) \in L^{\text{loc}}(G)$ *if, for* $k = 1, 2, \ldots$, *there exists a function* $F_k(x)$ *for* $x \in G \cup \Gamma_k$ *such that*

a) $F_k(x) = F(x)$ *for almost every* $x \in G$;

b) $F_k(x', \mathring{x}'') = f(x', \mathring{x}'')$ *for all* $(x', \mathring{x}'') \in \Gamma_k$;

c) $\int\limits_{D_k} |f(x')|\, dx' < \infty$;

d) $\int\limits_{D_k} |\tilde{\Delta}(z; G) F_k(x', \overset{\circ}{x}{}'')|\, dx' \to 0$ *as* $z = (0, z'') \to 0$, *where*

$$\tilde{\Delta}(z; E)\, \Phi = \begin{cases} \Phi(x+z) - \Phi(x) & \text{if} \quad x + z \in E, \\ 0 & \text{if} \quad x + z \notin E. \end{cases}$$

This definition of a trace is formulated for a cross section of $\overline{G}$ by a plane $x_i = \overset{\circ}{x}_i$ (for $i = m+1, \ldots, n$). It is defined analogously for sections by a plane parallel to any other coordinate plane:

$$x_{j_i} = \overset{\circ}{x}_{j_i} \qquad (i = m+1, \ldots, n;\ 1 \leqslant j_{m+1} < \ldots < j_n \leqslant n).$$

Its formulation then differs from that given only in that the notation $x = (x', x'')$ has a different meaning. Specifically, $x'' = (x_{j_{m+1}}, \ldots, x_{j_n})$ and $x' = (x_{j_1}, \ldots, x_{j_m})$, where $1 \leqslant j_1 < \ldots < j_m \leqslant n$ for $j_k \neq j_s$ for $k \neq s$.

We note that the set Γ, so defined, is open in the cross section of $\overline{G}$ with the plane $x'' = \overset{\circ}{x}{}''$. Furthermore, every open set in that section can be represented in the form $\Gamma = \bigcup\limits_{k=1}^{\infty} \Gamma_k$. For given Γ, let us suppose that we have its representation in that form. For example, if G coincides with the space E^n or with the subspace of it $E_+^n = \{x\colon\ x \in E^n,\ x_n > 0\}$ and Γ is an m-dimensional plane, it is natural to understand the D_k to mean the m-dimensional cubes with edges of length 2 parallel to the coordinate axes and centers at the nodes of the integer lattice.

If the function $F(x)$ is continuous (continuously extendible) on $G \cup \Gamma$, its values on Γ form a function $f(x', \overset{\circ}{x}{}'')$, which is obviously the trace of $F(x)$ on Γ.

Consider the case in which the function $F(x) = F(x', x'')$ is defined on $E^n = E^m \times E^{n-m} = \Gamma \times E^{n-m}$ and belongs to $L_q(E^n)$, where

$$\boldsymbol{q}=(\boldsymbol{q}',\infty)=(q_1,\ldots,q_m,\ \infty,\ldots,\infty).$$

This condition alone is not sufficient for the existence of a trace. Let us require in addition that the norm

$$\sup_{x''\in E^{n-m}} \|F(\cdot,x'')\|_{q'}$$

be finite and that $F(x)$ be continuous in that norm with respect to translation by the vector $z=(0,z'')$. It is natural to denote the space of such functions with this norm by $L_{q'}C(E^m\times E^{n-m})$. Obviously, for $F(x',x'')\in L_{q'}C(E\ \ \times E^{n-m})$, its values $F(x',0)$ form a function which is the trace of $F(x)$ on $\Gamma=E^m$.

An analogous situation arises when $F(x',x'')$ belongs to $L_{q'}C$ only locally, that is, in a neighborhood of each point of $\boldsymbol{\Gamma}$ (we shall not give here a precise definition of local membership of F in $L_{q'}C$). We note that, in the simplest geometrical situation, the existence of a trace for $F(x)$ is equivalent to local membership of F (or a function equivalent to it) in L_1C.

This approach defines the trace of a function only up to equivalence with respect to m-dimensional measure on $D=\bigcup_{k=1}^{\infty}D_k$. If a function $f(x',\overset{\circ}{x}'')$ is the trace of a function $F(x)$ on Γ and if, for a function $f^*(x',\overset{\circ}{x}'')$,

$$\int_{D_k}|f(x',\overset{\circ}{x}'')-f^*(x',\overset{\circ}{x}'')|\,dx'=0 \qquad (k=1,2,\ldots),$$

the function $f^*(x',\overset{\circ}{x}'')$ is obviously also a trace of $F(x)$ on Γ. On the other hand, under rather general assumptions regarding the structure of G in a neighborhood of Γ, any two traces of a function $F(x)$ on Γ differ only on a set of m-dimensional measure zero, as asserted in the following theorem:

10.9. A theorem on the uniqueness of a trace. *Suppose that each of the functions* $f(x', \overset{\circ}{x}'') = f(x')$ *and* $f^*(x', \overset{\circ}{x}'') = f^*(x')$ *is the trace of a function* $F(x)$ *for* $x \in G \subset E^n$ *on an open m-dimensional plane set* $\Gamma = \bigcup_{k=1}^{\infty} \Gamma_k$ *contained in the cross section of* $\overline{G}$ *with the plane* $x'' = \overset{\circ}{x}''$. *Suppose that for every positive integer k and every point* $(\overset{\circ}{x}', \overset{\circ}{x}'') \in \Gamma_k$, *there exist an open set* $\Gamma_k^{(0)} \ni (\overset{\circ}{x}', \overset{\circ}{x}'')$ *contained in* Γ_k *and an open subset* $V_k^{(0)}$ *of* E^{n-m} *whose closure includes zero such that* $\Gamma_k^{(0)} \times V_k^{(0)} \subset G$. *Then,*

$$\int_{D_k} | f(x', \overset{\circ}{x}'') - f^*(x', \overset{\circ}{x}'' \, | dx' = 0 \qquad (k = 1, 2, \ldots).$$

PROOF. Let $F_k(x)$ and $F_k^*(x)$ denote modifications of $F(x)$ on a set of n-dimensional measure zero corresponding to the functions $f(x', \overset{\circ}{x}'')$ and $f^*(x', \overset{\circ}{x}'')$, each of which is a trace of $F(x)$. We denote by $D_k^{(0)} \subset D_k$ the projection of $\Gamma_k^{(0)}$ onto the coordinate plane $x'' = 0$. Then, for $z'' \in V_k^{(0)}$,

$$\int_{D_k^{(0)}} | f(x', \overset{\circ}{x}'') - f^*(x', \overset{\circ}{x}'') \, | dx'$$

$$\leqslant \int_{D_k^{(0)}} | F_k(x', \overset{\circ}{x}'' + z'') - f(x', \overset{\circ}{x}'') \, | dx'$$

$$+ \int_{D_k^{(0)}} | F_k^*(x', \overset{\circ}{x}'' + z'') - f^*(x', \overset{\circ}{x}'') \, | dx' + \mathcal{I}_k(z''),$$

where

$$\mathcal{I}_k(z'') = \int_{D_k^{(0)}} | F_k(x', \overset{\circ}{x}'' + z'') - F_k^*(x', \overset{\circ}{x}'' + z'') \, | dx'.$$

By virtue of the definition of a trace and the conditions of the theorem, the first and second integrals in the right-hand member of this inequality approach zero as z'' approaches zero through values in $V_k^{(0)}$. Let us show that the last integral $\mathcal{J}_k(z'')$ is equal to 0 on some sequence of points $\overset{(i)}{z''}$ that approaches 0 through values in $V_k^{(0)}$ as $i \to \infty$. This will imply vanishing of the integral in the left-hand member of the inequality and hence the assertion of the theorem.

Let us define $\Phi(x) = \Phi(x', x'') = | F_k(x) - F_k^*(x) |$, so that $\Phi(x) = 0$ almost everywhere on $\Gamma_k^{(0)} \times V_k^{(0)}$. Consequently,

$$\int_{V_k^{(0)}} \mathcal{J}_k(z'')\, dz'' = \int_{V_k^{(0)}} \int_{D_k^{(0)}} \Phi(x', \overset{\circ}{x}'' + z'')\, dx'\, dz''$$

$$= \int_{\Gamma_k^{(0)} \times V_k^{(0)}} \Phi(x)\, dx = 0.$$

Therefore, $\mathcal{J}_k(z'') = 0$ for almost all $z'' \in V_k^{(0)}$, which implies the existence of a sequence of points $\overset{(i)}{z''}$ that approaches zero through values in $V_k^{(0)}$. This completes the proof of the theorem.

10.10. In the study of the trace of functions in various function spaces on a plane section, the following theorem (similar to Lemma 6.10.3 in Nikol'skiĭ [10]) often proves useful:

Theorem on the existence of a trace. *Let G denote an open subset of E^n. Let m denote a number in $[1, n)$, and*

$$\Gamma = \bigcup_{k=1}^{\infty} \Gamma_k \subset \overline{G} \cap \{x = (x', \overset{\circ}{x}'')\}.$$

Suppose that, for $k = 1, 2, \ldots$, there exists an open bounded subset G_k of G such that $\overline{G}_k \supset \Gamma_k$ and $\overline{G \setminus G_k} \cap \overline{\Gamma}_k = 0$. Let $F(x)$ denote a function defined on G. For every k and s, define $\Gamma_{ks} = \Gamma_k \cap \Gamma_s$, $D_{ks} = D_k \cap D_s$, and $G_{ks} = G_k \cap G_s$. Suppose that the following conditions are satisfied for all values of k and s:

α) *There exist sequences* $\{F_{kj}\}_{j=1}^{\infty}$ *of continuous functions on* $\bar{G}_k$ *such that*

$$\operatorname*{ess\,sup}_{z''} \int\limits_{\substack{x' \in D_k \\ (x',\, \mathring{x}''+z'') \in G_k}} | F(x', \mathring{x}'' + z'') - F_{kj}(x', \mathring{x}''+z'') | \, dx' \to 0$$

$$\text{as } j \to \infty;$$

β) *for every function* $\Phi(x)$ *that is continuous on* $\bar{G}_{ks}$,

$$\int\limits_{x' \in D_{ks}} | \Phi(x', \mathring{x}'') | \, dx' = \overline{\lim_{z'' \to 0}} \int\limits_{\substack{x' \in D_{ks} \\ (x',\, \mathring{x}''+z'') \in G_{ks}}} | \Phi(x', \mathring{x}'' + z'') | \, dx'.$$

Then, the function $F(x)$ *has on* Γ *a trace* $f(x', \mathring{x}'') = f(x') \in L(D_k)$, *for* $k = 1, 2, \ldots$, *that coincides with the limit in* $L(D_k)$ *of the traces on* Γ_k *of the continuous functions* F_{kj}, *for* $k = 1, 2, \ldots$, *as* $j \to \infty$,

In many cases, it is not difficult to test whether condition β) is satisfied or not. For example, condition β) is obviously satisfied if, for

$$G_{ks}(z'') = \{x = (x', \mathring{x}'' + z''): \ x \in G_{ks}, \ x' \in D_{ks}\}$$

and some sequence of points $\overset{(i)}{z''}$ that converges to zero as $i \to \infty$, the sequence of measures $\operatorname{mes} G_{ks}(\overset{(i)}{z''})$ converges to $\operatorname{mes} D_{ks}$ as $i \to \infty$. This last condition is satisfied in a situation sufficiently general for us:

β′) *for* $k = 1, 2, \ldots$, *there exists an open l-horn* V_k *such that* $\Gamma_k + V_k \subset G$.

PROOF. Conditions α) and β) imply that the sequence $\{F_{kj}\}_{j=1}^{\infty}$ is a fundamental sequence and hence converges in the metric

$$\sup_{z''} \int\limits_{\substack{x' \in D_k \\ (x', \overset{\circ}{x}''+z'') \in G_k \cup \Gamma_k}} | F_k(x', \overset{\circ}{x}'' + z'') - F_{kj}(x', \overset{\circ}{x}'' + z'') | \, dx' \to 0 \quad \text{as } j \to \infty \tag{13}$$

to some function $F_k(x)$ (see Remark 1.1 on the completeness of the corresponding space). Here,

$$\sup_{z''} \int\limits_{\substack{x' \in D_k \\ (x', \overset{\circ}{x}''+z'') \in G_k \cup \Gamma_k}} | F_k(x', \overset{\circ}{x}'' + z'') | \, dx' < \infty, \tag{14}$$

$$F_k(x) = F(x) \qquad \text{for almost all} \qquad x \in G_k. \tag{15}$$

To establish equation (15), we set, for fixed k,

$$\Phi(x) = \Phi(x', x'') = | F_k(x) - F(x) | \quad \text{for} \quad x \in G_k$$

and we define $\Phi(x)$ to be equal to zero on $E^n \setminus G_k$. By virtue of condition α) of the theorem and the limit relation (13),

$$\operatorname*{ess\,sup}_{z''} \int\limits_{\substack{x' \in D_k \\ (x', \overset{\circ}{x}''+z'') \in G_k}} \Phi(x', \overset{\circ}{x}'' + z'') \, dx' = 0.$$

Then, on the basis of Fubini's theorem, we obtain

$$0 = \int\limits_{E^{n-m}} \int\limits_{D_k} \Phi(x', x'') \, dx' \, dx'' = \int\limits_{D_k} \int\limits_{E^{n-m}} \Phi(x', x'') \, dx'' \, dx',$$

so that $\Phi(x) = 0$ almost everywhere on $D_k \times E^{n-m}$, and this leads to (15).

Let us define, for $x' \in D_k$,

$$f_{kj}(x') = f_{kj}(x', x'') = F_{kj}(x', \overset{\circ}{x}''), \; f_k(x') = f_k(x', \overset{\circ}{x}'') = F_k(x', \overset{\circ}{x}'').$$

It follows in particular from (13) and (14) that

$$\int_{D_k} | f_k(x') - f_{kj}(x') | \, dx' \to 0 \qquad (j \to \infty), \tag{16}$$

$$\int_{D_k} | f_k(x') | \, dx' < \infty. \tag{17}$$

If D_k and D_s are not disjoint, the functions $f_k(x')$ and $f_s(x')$ are compatible in the sense that, for arbitrary k and s,

$$\int_{D_{ks}} | f_k(x') - f_s(x') | \, dx' = 0, \tag{18}$$

that is,

$$f_k(x') = f_s(x') \qquad \text{for almost all} \qquad x' \in D_{ks}. \tag{19}$$

Let us prove this. By virtue of condition α) and the continuity of F_{kj} and F_{sj}, we have

$$\sup_{z''} \int\limits_{\substack{x' \in D_{ks} \\ (x', \mathring{x}'' + z'') \in G_{ks}}} | F_{kj}(x', \mathring{x}'' + z'') - F_{sj}(x', \mathring{x}'' + z'') | \, dx' \to 0$$

$$(j \to \infty),$$

so that, by virtue of condition β),

$$\int_{D_{ks}} | f_{kj}(x') - f_{sj}(x') | \, dx' \to 0 \qquad (j \to \infty),$$

which, by virtue of (16), leads to (18).

Let us now define a function $f(x') = f(x', \mathring{x}'')$ for $(x', \mathring{x}'') \in \Gamma$ by

$$f(x') = f_k(x'), \text{ where } \quad k = k(x') = \min\{s\colon\ \Gamma_s \ni (x', \overset{\circ}{x}{}'')\}, \tag{20}$$

and let us show that $f(x', \overset{\circ}{x}{}'')$ is the trace of $F(x)$ on Γ. Since the functions $f_k(x')$ are defined to begin with only up to equivalence, we now take for $f_k(x')$ the function $f(x')$ that we have just defined. Thus,

$$f_k(x') = f(x') \qquad \text{for all} \qquad x' \in D_k \qquad \text{and all } k = 1,\ 2,\ \ldots$$

Let us now show that $f(x)$ satisfies all the conditions of subsection 10.8 for the trace of the function $F(x)$.

By virtue of (15), condition a) is satisfied for the functions $F_k(x)$. Condition b) is satisfied because of the way in which these functions were constructed. Inequality (17) establishes condition c). Therefore, let us see whether condition d) is satisfied or not. If $z = (0, z'')$ and $|z| < \rho(\Gamma_k, G \setminus G_k)$, we have

$$\int_{D_k} |\tilde{\Delta}(z; G_k) F_k(x', \overset{\circ}{x}{}'')|\,dx'$$
$$\leqslant \int_{D_k} |\tilde{\Delta}(z; G_k) F_{kj}(x', \overset{\circ}{x}{}'')|\,dx' + \int_{D_k} |f_k(x') - f_{kj}(x')|\,dx'$$
$$+ \sup_{z''} \int_{\substack{x' \in D_k \\ (x', \overset{\circ}{x}{}''+z'') \in G_k \cup \Gamma_k}} |F_k(x', \overset{\circ}{x}{}'' + z'') - F_{kj}(x', \overset{\circ}{x}{}'' + z'')|\,dx',$$

By virtue of (16) and (13), the second and third terms in the right-hand member of this inequality are arbitrarily small for sufficiently large values of j. Since $F_{kj}(x)$ is continuous on the closed bounded set $\overline{G}_k \subset \Gamma_k$, for fixed j the first term in the right-hand member approaches 0 as $z = (0, z'') \to 0$. Consequently, the left-hand member of the last inequality approaches 0 as $z = (0, z'') \to 0$, as we needed to show. This completes the proof of the theorem.

This theorem shows that, if sequences $\{F_{kj}(x)\}_{j=1}^{\infty}$ (for $k = 1, 2, \ldots$) of functions that are continuous on $\overline{G}_k$ converge (in the sense of condition α)) to a function $F(x)$ for $x \in G$, the traces of these continuous functions converge on Γ (in the sense of (16)) to some function $f(x)$ (see (20)) that is the trace of $F(x)$. For the function $F_k(x)$, it is often convenient to take some integral representation or other of the function $F(x)$ that coincides with $F(x)$ almost everywhere on $G_k \subset G$. Then, it is natural to take for $F_{kj}(x)$ the averages of $F(x)$ that generate the corresponding representations with averaging parameters that approach zero. The functions $F_{kj}(x)$ are thus represented in the same form as $F(x)$ but with zero replaced with $\varepsilon_j > 0$ in the lower limit of the outer integral of the basic part of the representation.

We note also that the significance of the condition $\overline{G \setminus G_k} \cap \overline{\Gamma}_k = 0$ in the theorem can be seen from the example of a two-dimensional region G in the form of the annulus cut along the horizontal radius 8(5). Obviously, for a function to have a trace on this cut, it is necessary that the traces of the function considered separately on the upper and lower halves of the annulus coincide (up to equivalence) on it.

10.11. Let us look at a typical geometrical situation in which the properties of traces of functions are studied. Suppose that the open set G satisfies a weak l-horn condition 8(2), that is,

$$G = \bigcup_{k=1}^{K} G_k = \bigcup_{k=1}^{K} (G_k + V_k(l)),$$

where $V_k(l)$ is an open l-horn.

Suppose again that

$$\Gamma = \bigcup_{s=1}^{\infty} \Gamma_s \subset \overline{G} \cap \{x\colon\ x_{m+1} = \mathring{x}_{m+1}, \ldots, x_n = \mathring{x}_n\},$$

just as in the definition of a trace given in subsection 10.8 and suppose that the following condition is satisfied:

γ) *for* $s = 1, 2, \ldots,$ *there exists a natural number* $k = k(s) \leqslant K$ *such that*

$$\Gamma_s \subset \overline{G}_{k(s)}, \qquad \overline{G \setminus G_{k(s)}} \cap \overline{\Gamma}_s = 0.$$

We recall that the sets Γ_s are bounded though the sets $G_{k(s)}$ may not be. Therefore, let us define the bounded open sets

$$G'_s = G_{k(s)} \cap \{x:\ \rho(x, \Gamma_s) < \rho_s\}, \qquad 0 < \rho_s < \infty,$$

which, for $s = 1, 2, \ldots$, satisfy the condition

$$\gamma') \qquad \Gamma_s \subset \overline{G'_s}, \quad \overline{G \setminus G'_s} \cap \overline{\Gamma}_s = 0, \quad \overline{G'_s} + V_{k(s)}(l) \subset G.$$

(The inclusion relation at the end holds because the arithmetic sum of two open sets remains the same if we replace one of them with its closure.)

Let us show that the conditions of the imbedding Theorem 10.2 established in this section ensure, when we shift to a $\boldsymbol{q}$ with one or more infinite components, the existence of the trace $D^\alpha f$ and provide an estimate of it in the corresponding norm.

Suppose that the conditions of Theorem 10.2 are satisfied for $\boldsymbol{q} = (q_1, \ldots, q_m, \infty, \ldots, \infty)$, where $1 \leqslant m < n$. Suppose that Γ satisfies condition γ) and hence condition γ'). Then, condition β') and hence condition β) of Theorem 10.10 on the existence of a trace must hold.

Let us show that condition α) of Theorem 10.10 must also hold for the function $D^\alpha f \in L_{\boldsymbol{q}}(G)$. Let us represent $\Gamma = \bigcup_{s=1}^{\infty} \Gamma_s$ in the form $\Gamma = \bigcup_{k=1}^{K} \Gamma^{(k)}$, where $\Gamma^{(k)} = \bigcup_{k(s)=k} \Gamma_s$, where in turn $\Gamma_s = \{x = (x', \overset{\circ}{x}''):\ x' \in D_s\}$ (D_s being an open bounded subset of E^m). Define $D^{(k)} = \bigcup_{k(s)=k} D_s$. It is shown in Lemma 10.1 that, for a function f with finite right-hand member (1),

$$D^\alpha f_{\varepsilon\lambda} \to D^\alpha f \quad \text{in} \quad L_{\boldsymbol{q}}(G_k) \quad \text{as} \quad \varepsilon \to 0. \tag{21}$$

Then, for $k(s) = k$, we have on the basis of Hölder's inequality

$$D^{\alpha}f_{\varepsilon\lambda} \to D^{\alpha}f \text{ in } L_{q^*}(G_s') \quad \text{as } \varepsilon \to 0, \; q^* = (1, \ldots, 1, \infty, \ldots, \infty). \tag{22}$$

This last property implies condition α) of Theorem 10.10 since the $D^{\alpha}f_{\varepsilon\lambda}$ are continuous on $\overline{G}_s'$ (in fact, they are extendible to $C_0^{\infty}(E^n)$ [see 9(8)]). Thus, by virtue of Theorem 10.10, $D^{\alpha}f$ has a trace on Γ, and that trace coincides on Γ_s with the limit

$$\lim_{\varepsilon\to 0} D^{\alpha}f_{\varepsilon\lambda}(x', \overset{\circ}{x}{}'') \quad \text{in} \quad L_1(D_s). \tag{23}$$

The limit relation (21) and the continuity of $D^{\alpha}f_{\varepsilon\lambda}(x)$ on $\overline{G}_k$ imply the Cauchy convergence of $D^{\alpha}f_{\varepsilon\lambda}(x', \overset{\circ}{x}{}'')$ in $L_{q'}(D^{(k)})$, where $q' = (q_1, \ldots, q_m)$. Obviously, the limit coincides on D_s with the limit (23); that is, it is the trace of $D^{\alpha}f$.

From Lemma 10.1, we also have

$$\left\| D^{\alpha}f_{\varepsilon\lambda}(\cdot, \overset{\circ}{x}{}'') \right\|_{q', D^{(k)}} \leqslant \left\| D^{\alpha}f_{\varepsilon\lambda} \right\|_{q, G_k}$$

$$\leqslant Ch^{1-\varkappa} \sum_{i=1}^{n} \left\| D_i^{l_i}f \right\|_{p, G_k+V_k(l)} + Ch^{-\varkappa} \| f \|_{p, G_k+V_k(l)}.$$

Taking the limit as $\varepsilon \to 0$ and summing the estimates with respect to k from 1 to K, we finally obtain the result that, *under the conditions of Theorem* 10.2 *with* $q_{m+1} = \ldots = q_n = \infty$,

$$\| D^{\alpha}f(\cdot, \overset{\circ}{x}{}'') \|_{q', \bigcup_{k=1}^{K} D^{(k)}} \leqslant \tilde{C}h^{1-\varkappa} \sum_{i=1}^{n} \left\| D_i^{l_i}f \right\|_{p, G} + \tilde{C}h^{-\varkappa} \| f \|_{p, G}. \tag{24}$$

One can easily show that the constant $\widetilde{G}$ is independent of the choice of the point $\overset{\circ}{x}{}''$, that is, invariant under parallel translation of the plane making the section (with condition γ) still holding). Therefore, in the left-hand member of (24), we can take an upper bound over all such sections.

It is interesting to note that, although the estimate (24) of the norm of the trace was obtained under the assumption of finiteness of the norm $\|D^\alpha f\|_{q, G}$, where $q=(q_1, \ldots, q_m, \infty, \ldots, \infty)$, mere finiteness of this norm would not be a sufficient condition even for existence of the trace. In the proof of (24), we used the stronger property that $D^\alpha f$ can be approximated by continuous functions in the norm of $L_q(G_k)$. This property follows from the fact that $f \in W_p^l(G)$ (by the condition of Theorem 10.2). The possibility of such approximation is closely connected with the continuity in the wide sense of the given function $(D^\alpha f)$ in L_q and will be taken up again in §19.

Up to now, we have examined the properties of the trace of a function on the sections of $\overline{G}$ by the m-dimensional plane $x_{m+1} = \overset{\circ}{x}_{m+1}, \ldots, x_n = \overset{\circ}{x}_n$. Let us now look at the section of $\overline{G}$ by the m-dimensional plane $x_{i_{m+1}} = \overset{\circ}{x}_{i_{m+1}}, \ldots, x_{i_n} = \overset{\circ}{x}_{i_n}$ (where $1 \leqslant m < n$, $1 \leqslant i_{m+1} < i_{m+2} < \ldots < i_n \leqslant n$). As has been mentioned, the very definition of a trace is formulated in an analogous manner (differing only in a transformation of variables). Obviously, the same is true with regard to the theorems on the existence and uniqueness of a trace. One can easily show that the estimate (24) remains valid in the corresponding transformations.

§11. Coerciveness in the space $W_p^l(G)$

The paper by Aronszajn [1] was the first of a number of papers on the problem of coerciveness. His results were later expanded by Agmon [1], Schechter [1], Hörmander [1], Smith [1], Figueiredo [1], and Nečas [1]. The present section is devoted to further study of the problem in the direction of the last three works and it considers anisotropic Sobolev spaces and the mixed L_p-norm. In the case of the unmixed L_p-norm, the corresponding results were treated in the paper by Besov [8]. Throughout the section, we shall use the following notation: $l = (l_1, \ldots, l_n)$ denotes a vector with natural numbers as components, S and m_s for $s = 1, \ldots, S$ will denote natural numbers, $l^{(s)} = m_s l = (m_s l_1, \ldots, m_s l_n)$, $\lambda^{(s)} = 1/l^{(s)}$, $f(x) = (f_1(x), \ldots, f_s(x))$, $1 < p < \infty$, $1 < \boldsymbol{p} < \infty$, and $D = (D_1, \ldots, D_n) = (\partial/\partial x_1, \ldots, \partial/\partial x_n)$,

$$P_j(D)f = P_j(x,\ D)f = \sum_{s=1}^{S} \sum_{|\beta:l^{(s)}|\leqslant 1} c_{js\beta}(x)\, D^{\beta} f_s(x),$$

$P_{js}(D)f_s$ is the principal part of the corresponding term in $P_j(D)f$, that is,

$$P_{js}(D)f_s = P_{js}(x,\ D)f_s = \sum_{|\beta:l^{(s)}|=1} c_{js\beta}(x)\, D^{\beta} f_s(x),$$

where the $c_{js\beta}$ are complex numbers.

We denote by $\prod W_p^{l^{(s)}}(G) = \prod_{s=1}^{S} W_p^{l^{(s)}}(G)$ the space of vector-valued functions $f = (f_1,\ \ldots,\ f_S)$ with norm

$$\sum_{s=1}^{S} \| f_s \|_{W_p^{l^{(s)}}(G)}.$$

11.1. Definition. A system of differential operators $\{P_j(x,\ D)\}_1^N$ is said to be *coercive* in $\prod W_p^{l^{(s)}}(G)$ if there exists a single constant C independent of f such that, for all $f \in \prod W_p^{l^{(s)}}(G)$,

$$\sum_{s=1}^{S} \sum_{|\alpha:l^{(s)}|\leqslant 1} \| D^{\alpha} f_s \|_{p,\,G} \leqslant C \sum_{j=1}^{N} \| P_j(D) f \|_{p,\,G} + C \sum_{s=1}^{S} \| f_s \|_{p,\,G}. \tag{1}$$

11.2. Theorem. *Suppose that the l-polynomials with constant coefficients*

$$P_j(\xi) = \sum_{|\beta:l|=1} c_{j\beta}\xi^{\beta} \qquad (j = 1,\ \ldots,\ N)$$

have no nonzero complex roots in common and that the region G satisfies the weak condition of an l-horn. Then, for all $f \in \prod W_p^{l^{(s)}}(G)$, *if* $|\alpha : l| \leqslant 1$,

$$\|D^\alpha f\|_{p,\,G} \leqslant Ch^{1-|\alpha:l|} \sum_{j=1}^{N} \|P_j(D)f\|_{p,\,G} + Ch^{-|\alpha:l|}\|f\|_{p,\,G}, \tag{2}$$

where $0 < h \leqslant h_0(G)$, $h_0(G)$ *is a constant that depends only on G, and C is a constant that does not depend on f or h.*

The proof is based on the representations 7(29) and 7(30) and is carried out in the same way as the proofs of Lemma 9.4 and Theorem 9.5.

We mention also that, for $|\alpha : l| < 1$, we can assume in (2) that $1 \leqslant p \leqslant \infty$.

Absence of a common nonzero complex root of the polynomials $P_j(\xi)$ is also a necessary condition for validity of the estimate (2), as we shall see from examples to be given below.

It follows from the theorem just proven that

$$\|f\|_{W_p^l(G)} \sim \|f\|_{p,\,G} + \sum_{j=1}^{N} \|P_j(D)f\|_{p,\,G}.$$

11.3. Theorem. *Suppose that the region G satisfies a weak l-horn condition. The system of differential operators with constant coefficients*

$$P_j(D)f = \sum_{s=1}^{S} \sum_{|\beta : l^{(s)}| = 1} c_{js\beta} D^\beta f_s = \sum_{s=1}^{S} P_{js}(D) f_s, \quad j = 1, \ldots, N, \tag{3}$$

is coercive in $\prod W_p^{l^{(s)}}(G)$ *if the rank of the matrix* $(P_{js}(\xi))$ *is equal to S for an arbitrary nonzero complex value of* ξ. *This last condition is also a necessary condition if the region G is bounded.*

PROOF. Sufficiency. We denote by $\Delta_r = \Delta_r(\xi)$ for $r = 1, \ldots, R$ the minors of order S of the matrix $(P_{js}(\xi))$ and we denote by $\Delta_{rjs} = \Delta_{rjs}(\xi)$ the cofactor of the element $P_{js}(\xi)$ in the square matrix corresponding to the minor Δ_r if $P_{js}(\xi)$ is an element of that matrix and $\Delta_{rjs} = 0$ otherwise.

From the properties of determinants, we obtain from (3) the result that, for $r = 1, \ldots, R$ and $s = 1, \ldots, S$,

$$\Delta_r(\xi) = \sum_{j=1}^{N} \Delta_{rjs}(\xi) P_{js}(\xi), \quad 0 = \sum_{j=1}^{N} \Delta_{rjs}(\xi) P_{jk}(\xi) \qquad (k \neq s). \tag{4}$$

Since the $|m|l$-polynomials $\Delta_r(\xi)$, for $r = 1, \ldots, R$, do not have any nonzero complex root in common, it follows from a theorem of Hilbert that the identies 7(27) hold for sufficiently large positive M:

$$\xi_i^{l_i^{(s)} M} = \sum_{r=1}^{R} b_{irs}(\xi) \Delta_r(-\xi) \qquad (i = 1, \ldots, n),$$

where the $b_{irs}(\xi)$ are $(m_s M - |m|)\, l$-polynomials.

Keeping (4) in mind, we obtain the result that

$$\begin{aligned} \xi_i^{l_i^{(s)} M} &= \sum_{j=1}^{N} a_{ijs}(\xi) P_{js}(-\xi), \\ 0 &= \sum_{j=1}^{N} a_{ijs}(\xi) P_{jk}(-\xi) \qquad (k \neq s), \end{aligned} \tag{5}$$

where the $(M - 1)\, m_s l$-polynomials $a_{ijs}(\xi)$ are defined by

$$a_{ijs}(\xi) = \sum_{r=1}^{R} b_{irs}(\xi) \Delta_{rjs}(-\xi).$$

On the basis of 7(27), 7(28), and 7(30), we obtain from the identities (5) the representation

$$D^{\alpha}\varphi_{_{\varepsilon\lambda^{(s)}}}(x) = D^{\alpha}\varphi_{_{h\lambda^{(s)}}}(x) + (-1)^{|\alpha|}$$
$$\times \sum_{j=1}^{N} \int_{\varepsilon}^{h} v^{-|\lambda^{(s)}|-(\alpha,\lambda^{(s)})}\,dv \int T_{js}^{(\alpha)}\left(y : v^{\lambda^{(s)}}\right) P_{js}(D)\,\varphi(x+y)\,dy \tag{6}$$

for

$$T_{js}(x) = \sum_{i=1}^{n} \lambda_i^{(s)} a_{ijs}(D)\,\tilde{\mathscr{L}}_i(x). \tag{7}$$

We note that, for $k \neq s$,

$$\sum_{j=1}^{N} \int T_{js}^{(\alpha)}\left(y : v^{\lambda^{(s)}}\right) P_{jk}(D)\,f_k(x+y)\,dy$$
$$= v^{-m_k} \int \sum_{j=1}^{N} \left(D^{\alpha}P_{jk}(-D)\,T_{js}\right)\left(y : v^{\lambda^{(s)}}\right) f_k(x+y)\,dy = 0 \tag{8}$$

since, by virtue of (7) and (5),

$$\sum_{j=1}^{N} P_{jk}(-D)\,T_{js}(x) = \sum_{i=1}^{n} \lambda_i^{(s)} \sum_{j=1}^{N} P_{jk}(-D)\,a_{ijs}(D)\,\tilde{\mathscr{L}}_i(x) = 0.$$

By applying the representation (6) to $\varphi(x) = f_s(x)$ and adding to the right-hand member integrals that, by virtue of (8), are equal to 0, we obtain the representation

$$D^{\alpha}(f_s)_{_{\varepsilon\lambda^{(s)}}}(x) = D^{\alpha}(f_s)_{_{h\lambda^{(s)}}}(x)$$
$$+ (-1)^{|\alpha|} \sum_{j=1}^{N} \int_{\varepsilon}^{h} v^{-|\lambda^{(s)}|-(\alpha,\lambda^{(s)})} \int T_{js}^{(\alpha)}\left(y : v^{\lambda^{(s)}}\right) P_j(D)\,f(x+y)\,dy. \tag{9}$$

From this we obtain, just as in the proof of Lemma 9.4 and Theorem

9.5, the estimate (1) and, for arbitrary δ such that $0 < \delta < \delta_0(G)$, the estimate

$$\sum_{s=1}^{S} \sum_{|\alpha : l^{(s)}| < 1} \| D^{\alpha} f_s \|_{p,\,G} \leqslant \delta \sum_{j=1}^{N} \| P_j(D) f \|_{p,\,G} + C(\delta) \sum_{s=1}^{S} \| f_s \|_{p,\,G}. \tag{10}$$

This proves the sufficiency of the conditions of the theorem.

Necessity. Suppose now that, for some $\xi \neq 0$, the rank of the matrix $(P_{js}(\xi))$ is less than S. Then, there exists a nonnegative solution of the system

$$\sum_{s=1}^{S} P_{js}(\xi)\, b_s = 0 \qquad (j = 1, \ldots, N). \tag{11}$$

For $t > 0$, we set

$$f_s(x, t) = b_s e^{\sum_{1}^{n} t^{\lambda_j^{(s)}} \xi_j x_j} \qquad (s = 1, \ldots, S). \tag{12}$$

Since the system is coercive, we obtain from (1) the result that, for a bounded region G,

$$\sum_{s=1}^{S} \sum_{j=1}^{n} \left\| D_j^{l_j^{(s)}} f_s(\cdot, t) \right\|_{p,\,G}$$
$$= t \sum_{s=1}^{S} \sum_{j=1}^{n} |\xi_j|^{l_j^{(s)}} \| f_s(\cdot, t) \|_{p,\,G} \leqslant C \sum_{s=1}^{S} \| f_s(\cdot, t) \|_{p,\,G},$$

which is impossible for sufficiently large t.

We note that we have in fact proven that the estimate (1) is valid in case (3) not only for $f(x) \in \prod W_p^{l^{(s)}}(G)$ (when we have coerciveness) but also for any vector-valued function $f(x)$ for which the right-hand member of (1) is finite.

11.4. Theorem. *Suppose that the bounded region G satisfies a weak l-horn condition. The system of differential operators*

$$P_j(D)f = \sum_{s=1}^{S} \sum_{|\beta : l^{(s)}| \leqslant 1} c_{js\beta}(x)\, D^\beta f_s \qquad (j = 1, \ldots, N),$$

where the $c_{js\beta}(x)$ are measurable and bounded and, for $|\beta : l^{(s)}| = 1$, are continuous on $\overline{G}$, is coercive in $\prod W_p^{l^{(s)}}(G)$ if the rank of the matrix $(P_{js}(x, \xi))$ is equal to S for arbitrary nonzero complex numbers ξ and arbitrary x in $\overline{G}$. This last condition is a necessary condition for coerciveness if the coefficients $c_{js\beta}$ are constant for $|\beta : l^{(s)}| = 1$.

PROOF. The necessity is proven in the same way as in Theorem 11.3. Let us prove the sufficiency. We partition the space E^n by means of hyperplanes parallel to the coordinate planes into a grid of open cubes. We partition each cube into 2^n open cubes and we denote by $Q^*_{11}, \ldots, Q^*_{1j_1}$ those cubes in this last partition that intersect $\overline{G}$. We partition each of the cubes Q^*_{1j} into 2^n open cubes and we denote by $Q^*_{21}, \ldots, Q^*_{2j_2}$ those cubes in *this* partition that intersect with $\overline{G}$. Continuing the process, we obtain a countable system of cubes $\{Q^*_{ij}\}$, where $j = 1, 2, \ldots, j_i$ for $i = 1, 2, \ldots$. For each cube Q^*_{ij} we denote by Q_{ij} the open cube with the same center and parallel edges but with diameter twice as great. Obviously, the system of cubes $\{Q_{ij}\}$ forms a covering of the closed region $\overline{G}$.

It follows from the hypothesis of the theorem that the region G has a covering $\{G_k\}_1^K$ for which condition 8(2) is satisfied. For every k such that $1 \leqslant k \leqslant K$, we choose a subsystem $\{Q_i^{(k)}\}$ (for $i = 1, 2, \ldots$) of the system $\{Q_{ij}\}$ consisting of those cubes Q_{ij} that have a nonempty intersection with $\overline{G}_k$. Let us assume that the cubes of the subsystem are numbered in increasing order of their diameters.

Let $x^{(0)}$ denote a point in $\overline{G}_k$. Let us use the following rule to assign to this point a cube $Q^{(k)}(x^{(0)}) = Q^{(k)}_{i(x^{(0)})}$ of sufficiently small diameter to which it belongs:

Suppose that $U_k(x^{(0)}) \subset G_k \cap Q^{(k)}(x^{(0)})$. By virtue of 8(2), we have $U_k(x^{(0)}) + V_k(l, h) \subset G$ for all h in $0 < h < h_0(G)$. The continuity of the "smooth" coefficients $c_{js\beta}(x)$ (that is, those coefficients for which $|\beta : l^{(s)}| = 1$) implies that, for every $\varepsilon > 0$, there exists an $h(\varepsilon) > 0$ such that

$$\left\| P_j(\cdot, D) f - \sum_{s=1}^{S} P_{js}(x^{(0)}, D) f \right\|_{p, U_k(x^{(0)}) + V_k} \leqslant \varepsilon \sum_{s=1}^{S} \sum_{|\beta : l^{(s)}| = 1} \| D^{\beta} f_s \|_{p, U_k(x^{(0)}) + V_k} + C_1 \sum_{s=1}^{S} \sum_{|\beta : l^{(s)}| < 1} \| D^{\beta} f_s \|_{p, U_k(x^{(0)}) + V_k}$$

for every cube $Q^{(h)}(x^{(0)})$ of sufficiently small diameter (depending on ε). We assign one of these cubes to the point $x^{(0)}$. The last inequality holds for every $\varepsilon > 0$. It will be used for a particular value, which we shall identify below.

By virtue of Theorem 11.3 for constant coefficients (more precisely, by virtue of the intermediate estimate [analogous to Lemma 9.3] of the right-hand member of the representation (9) obtained in the proof of that theorem), we get from the above inequality

$$\sum_{s=1}^{S} \sum_{|\alpha : l^{(s)}| = 1} \| D^{\alpha} f_s \|_{p, U_k(x^{(0)})} \leqslant C(x^{(0)}) \sum_{j=1}^{N} \left\| \sum_{s=1}^{S} P_{js}(x^{(0)}, D) f_s \right\|_{p, U_k(x^{(0)}) + V_k} + C(x^{(0)}, h) \sum_{s=1}^{S} \| f_s \|_{p, U_k(x^{(0)}) + V_k}$$

$$\leqslant C(x^{(0)}) \sum_{j=1}^{N} \| P_j(\cdot, D) f \|_{p, U_k(x^{(0)}) + V_k}$$

$$+C(x^{(0)})\varepsilon \sum_{s=1}^{S} \sum_{|\beta:l^{(s)}|=1} \| D^{\beta} f_s \|_{p,\, U_k(x^{(0)})+V_k}$$

$$+C_2(x^{(0)}, h) \sum_{s=1}^{S} \sum_{|\beta:l^{(s)}|<1} \| D^{\beta} f_s \|_{p,\, U_k(x^{(0)})+V_k}. \qquad (13)$$

We now choose ε from the condition $C(x^{(0)})\varepsilon < \theta$, where θ is a sufficiently small number independent of both $x^{(0)} \in \bar{G}_k$ and k. The exact value of θ will be indicated below. Thus, we assign to each point $x^{(0)} \in \bar{G}_k$ a cube $Q^{(k)}(x^{(0)})$ to which it belongs. For $k = 1, \ldots, K$, we choose from the covering of $\bar{G}_k$ with the cubes $Q^{(k)}(x)$ for $x \in \bar{G}_k$ a finite covering $\{Q^{(k)}(x^{(r)})\}_{r=1}^{r(k)}$. Let d denote the smallest of the diameters of the cubes $Q^{(k)}(x^{(r)})$ for $1 \leqslant r \leqslant r(k)$, where $1 \leqslant k \leqslant K$. We partition each of the cubes $Q^{(k)}(x^{(r)})$ into open cubes $Q^{(k)}(x^{(r)})$ of diameter d and we leave in the system $\{Q_i^{(k)}(x^{(r)})\}$, for $i = 1, \ldots, i(k, r)$ and $r = 1, \ldots, r(k)$, only distinct cubes that intersect with G_k. We note that the point $x^{(r)}$ does not necessarily belong to $Q_i^{(k)}$. After the transformation, we obtain for each k a system of open disjoint cubes $\{q_i^{(k)}\}_{i=1}^{i(k)}$ such that $G_k \subset \bigcup_{i=1}^{i(k)} \bar{q}_i^{(k)}$.

Suppose that $U_{ki} = G_k \cap q_i^{(k)}$. By virtue of (13), we have, for $r = r(i)$,

$$\sum_{s=1}^{S} \sum_{|\alpha:l^{(s)}|=1} \| D^{\alpha} f_s \|_{p,\, U_{ki}}^{p} \leqslant C_3(p)\, C(x^{(r)}) \sum_{j=1}^{N} \| P_j(\cdot, D) f \|_{p,\, U_{ki}+V_k}^{p}$$

$$+ C_3(p)\,\theta \sum_{s=1}^{S} \sum_{|\beta:l^{(s)}|=1} \| D^{\beta} f_s \|_{p,\, U_{ki}+V_k}^{p}$$

$$+ C_3(p)\, C_2(x^{(r)}, h) \sum_{s=1}^{S} \sum_{|\beta:l^{(s)}|<1} \| D^{\beta} f_s \|_{p,\, U_{ki}+V_k}^{p}.$$

Let us assume the radius h of the l-horn $V_k = V_k(l, h)$ to be sufficiently small that the horn $V_k(l, h)$ is contained in a cube of diameter d. Remembering that every point of the region G is covered no more than C_4 times (where C_4 is independent of d) by the sets $U_{ki} + V_k$ (where $k = 1, \ldots, K$ and $i = 1, \ldots, i(k)$), we take $\theta < \frac{1}{2C_4C_3(p)}$ and sum the last inequality with respect to i and k:

$$[1 - C_4C_3(p)\theta] \sum_{s=1}^{S} \sum_{|\alpha : l^{(s)}|=1} \| D^\alpha f_s \|_{p,G}^p$$
$$\leqslant C_5 \sum_{j=1}^{N} \| P_j(\cdot, D) f \|_{p,G}^p + C_5 \sum_{s=1}^{S} \sum_{|\beta : l^{(s)}|<1} \| D^\beta f_s \|_{p,G}^p. \quad (14)$$

If we now use Theorem 9.5 to estimate this last summation with sufficiently small h and transpose the terms containing the norms of the derivatives (with sufficiently small coefficients) from the right side to the left, we get the estimate (1) that we need.

11.5. The function $\Phi(\xi)$ is called a *multiplier from* $L_p(E^n)$ *into* $L_p(E^n)$ if the operator $F^{-1}[\Phi F(f)]$ is a bounded operator from $L_p(E^n)$ into $L_p(E^n)$, where $F(f) = \hat{f}$ is the Fourier transform of the function f.

A theorem on multipliers. *Suppose that the function* $\Phi(\xi)$ *and its derivatives* $\xi^\alpha D^\alpha \Phi(\xi)$, *where* α_i *is* 0 *or* 1 *for* $i = 1, \ldots, n$, *are continuous and bounded on the set*

$$E_*^n = \{\xi \colon \xi \in E^n, \ \xi_i \neq 0 \ \ (i = 1, \ldots, n)\}.$$

Then $\Phi(\xi)$ *is a multiplier from* $L_p(E^n)$ *into* $L_p(E^n)$ *for* $1 < p < \infty$. *The norm of the operator* $F^{-1}[\Phi F(f)]$ *does not exceed* C_pM, *where*

$$M = \sup |\xi^\alpha D^\alpha \Phi(\xi)| \qquad (\xi \in E_*^n, \ 0 \leqslant \alpha_i \leqslant 1, \ i = 1, \ldots, n). \quad (15)$$

This theorem was proven by Lizorkin [1].

11.6. We denote by $\mathring{W}_p^l(G)$ the closure, with respect to the norm of the space $W_p^l(G)$, of the set of infinitely differentiable functions $f(x)$ of compact support contained in G. We mention that every function $f(x) \in W_p^l(G)$ of compact support in G belongs to the space $\overset{\circ}{W}{}_p^l(G)$ since such a function is approximated to an arbitrary degree of accuracy in $W_p^l(G)$ by infinitely differentiable functions of compact support in G. We can take averagings, for example, as the approximating functions (see Lemma 5.2 and equation 6(6)).

We denote by $\prod \overset{\circ}{W}{}_p^{l(s)}(G)$ the closure, with respect to the norm of the space $\prod W_p^{l(s)}(G)$, of the vector-valued functions $f(x)$ of compact support in G.

Theorem. *The system of differential operators* (3) *with constant coefficients is coercive in* $\prod \overset{\circ}{W}{}_p^{l(s)}(G)$; *that is, the estimate* (1) *is valid for all* $f \in \prod \overset{\circ}{W}{}_p^{l(s)}(G)$ *if and only if the rank of the matrix* $(P_{js}(i\xi))$ *is equal to* S *for arbitrary nonzero real* ξ.

PROOF. Sufficiency. Since $\widehat{D^\alpha f_s}(\xi) = (i\xi)^\alpha \hat{f}_s(\xi)$, we obtain from (3) by taking the Fourier transforms

$$\widehat{P_j(D) f} = \sum_{s=1}^{S} P_{js}(i\xi) \hat{f}_s, \quad \hat{f} = F(f).$$

By virtue of (4),

$$\sum_{j=1}^{N} \Delta_{rjs}(i\xi) \widehat{P_j(D) f}(\xi) = \sum_{j=1}^{N} \Delta_{rjs}(i\xi) P_{js}(i\xi) \hat{f}_s(\xi) = \Delta_r(i\xi) \hat{f}_s(\xi).$$

It follows that

$$\hat{f}_s(\xi) = \frac{\sum_{r=1}^{R} \sum_{j=1}^{N} \overline{\Delta_r(i\xi)} \Delta_{rjs}(i\xi) \widehat{P_j(D) f}(\xi)}{\sum_{r=1}^{R} |\Delta_r(i\xi)|^2},$$

so that, to get an estimate for the norms of $D^\alpha f_s$ for $|\alpha : l^{(s)}| = 1$ in the left-hand member of (1), it will be sufficient to show that the ratio

$$\Phi(\xi) = \frac{\xi^\alpha \overline{\Delta_r(i\xi)}\, \Delta_{rjs}(i\xi)}{\sum_{r=1}^{R} |\Delta_r(i\xi)|^2}, \qquad |\alpha : l^{(s)}| = 1, \tag{16}$$

is a multiplier from $L_p(E^n)$ into itself. This fact can be shown with the aid of Theorem 11.5 on multipliers.

The norms of the derivatives $D^\alpha f_s$, for $|\alpha : l^{(s)}| < 1$, are estimated with the aid of Theorem 9.5 in terms of the norms of $D_i^{l_i^{(s)}} f_s$ (for $i = 1, \ldots, n$) and the norm of the function f_s. Here, we may assume that $G = E^n$ since all the functions in question can be extended to E^n without destroying their differentiability properties by setting them equal to zero on $E^n \setminus G$. This completes the proof of the sufficiency.

Necessity. Let us suppose that, for some nonzero real number ξ, the rank of the matrix $(P_{js}(i\xi))$ is less than S. Then, there exists a nontrivial solution of the system

$$\sum_{s=1}^{S} P_{js}(i\xi)\, b_s = 0 \qquad (j = 1, \ldots, N).$$

Suppose that $t > 0$ and that a function φ, not identically equal to 0, is infinitely differentiable and of compact support in G. Let us define

$$f_s(x, t) = \varphi(x)\, t^{-1} b_s e^{i \sum_1^n t^{\lambda_j^{(s)}} \xi_j x_j}. \tag{17}$$

As $t \to +\infty$, the left-hand member of (1) for $f_s(x, t)$ approaches a positive constant whereas the right-hand member approaches 0. Therefore, the system of differential operators is noncoercive.

11.7. Lemma. *Let G denote a bounded region and let G_0 denote an open set whose closure $\overline{G}_0$ is contained in G. Suppose that the coefficients in the system of differential operators*

$$P_j(x, D)f = \sum_{s=1}^{S} \sum_{|\beta : l^{(s)}|=1} c_{js\beta}(x) D^\beta f_s = \sum_{s=1}^{S} P_{js}(x, D) f_s$$
$$(j=1, \ldots, N)$$

are continuous on G. If the rank of the matrix $(P_{js}(x, i\xi))$ *is equal to S for arbitrary nonzero real* ξ *and arbitrary* $x \in G$*, then the system* $\{P_j(x, D\}$ *is coercive in* $\prod \overset{\circ}{W}{}_p^{l(s)}(G_0)$.

PROOF. Consider a cubic grid of step δ on E^n. Let $a^{(r)}$ (for $r = 1, 2, \ldots, R$) denote the nodes of that grid lying within a δ-neighborhood of G_0. Suppose that δ is sufficiently small that $\rho(E^n \setminus G, a^{(r)}) > \delta$ for $r = 1, \ldots, R$. Let us define

$$\varphi_\delta(x) = \begin{cases} e^{-\frac{1}{\delta^2 - |x|^2}} & \text{for} \quad |x| < \delta, \\ 0 & \text{for} \quad |x| \geqslant \delta, \end{cases}$$

$$\eta_r(x) = \varphi_\delta(x - a^{(r)}) \left(\sum_{s=1}^{R} \varphi_\delta(x - a^{(s)}) \right)^{-1/p}.$$

Obviously, supp η_r coincides with the closed δ-neighborhood of $a^{(r)}$, η_r belongs to $C^\infty(\bar{G}_0)$, and

$$\sum_{1}^{R} \eta_r^p(x) = 1 \quad \text{on} \quad G_0. \tag{18}$$

Suppose that $f(x) \in \prod \overset{\circ}{W}{}_p^{l(s)}(G_0)$. Obviously,

$$\left\| \eta_r P_j(\cdot, D) f - \sum_{s=1}^{S} P_{js}(a^r, D)(\eta_r f_s) \right\|_{p, G_0}$$
$$\leqslant \varepsilon \sum_{s=1}^{S} \sum_{|\beta : l^{(s)}|=1} \| D^\beta(\eta_r f_s) \|_{p, G_0} + C_1 \sum_{s=1}^{S} \sum_{|\beta : l^{(s)}| < 1} \| D^\beta f_s \|_{p, G_0},$$

where ε depends on δ and on the moduli of continuity of the coefficients $c_{js\beta}(x)$ and is small for small δ.

Then, with the aid of Theorem 11.6 (for constant coefficients), we obtain

$$\sum_{s=1}^{S}\sum_{|\alpha:l^{(s)}|=1}\|D^{\alpha}(\eta_s f_s)\|_{p,G_0}\leqslant C_2\sum_{j=1}^{N}\sum_{s=1}^{S}\|P_{js}(a^{(r)},D)(\eta_r f_s)\|_{p,G_0}$$

$$+C_2\sum_{s=1}^{S}\|\eta_r f_s\|_{p,G_0}\leqslant C_2\sum_{j=1}^{N}\|\eta_r P_j(\cdot\,,D)f\|_{p,G_0}$$

$$+C_2\varepsilon\sum_{s=1}^{S}\sum_{|\beta:l^{(s)}|=1}\|D^{\beta}(\eta_r f_s)\|_{p,G_0}+C_1C_2\sum_{s=1}^{S}\sum_{|\beta:l^{(s)}|<1}\|D^{\beta}f_s\|_{p,G_0}.$$

We emphasize that C_2 is independent of $a^{(r)}$. In fact, as we see from the proof of Theorem 11.6, the constant C_2 is estimated in terms of the constants M of (15) for the multipliers $\Phi(\xi)$ in (16). Since $\Phi(\xi)$ in (16) is a rational function of ξ with coefficients that (in the present case) depend continuously on the point $a^{(r)}$ of the region G, the corresponding M is continuous on G and hence bounded on any compact subset of G.

Let us suppose that $C_2\varepsilon < 1/2$. If in the last chain of inequalities we transpose the terms with a factor ε to the left-hand member, we obtain

$$\sum_{s=1}^{S}\sum_{|\alpha:l^{(s)}|=1}\|\eta_r D^{\alpha}f_s\|_{p,G_0}$$

$$\leqslant C_3\sum_{j=1}^{N}\|\eta_r P_j(\cdot\,,D)f\|_{p,G_0}+C_3\sum_{s=1}^{S}\sum_{|\beta:l^{(s)}|<1}\|D^{\beta}f_s\|_{p,G_0}.$$

By virtue of the finiteness of the number of terms, the same inequality (with a different constant) holds when we raise each term to the power p. If we then sum with respect to r, keeping (18) in mind, we obtain

$$\sum_{s=1}^{S}\sum_{|\alpha:l^{(s)}|=1}\|D^{\alpha}f_s\|_{p,G_0}^{p}$$

$$\leqslant C_4 \sum_{j=1}^{N} \| P_j(\cdot, D) f \|_{p,G}^p + C_4 \sum_{s=1}^{S} \sum_{|\beta : l^{(s)}| < 1} \| D^\beta f_s \|_{p,G_0}^p.$$

If we now estimate the double summation on the right in accordance with Theorem 9.5 with sufficiently small h and transpose the terms with factor h to the left-hand member of the inequality, we complete the proof of the lemma.

11.8. Theorem. *Let G denote a bounded region satisfying the weak l-horn condition. Suppose that the coefficients in the system of differential operators*

$$P_j(x, D) f = \sum_{s=1}^{S} \sum_{|\beta : l^{(s)}| \leqslant 1} c_{js\beta}(x) D^\beta f_s \qquad (j = 1, \ldots, N)$$

are measurable and bounded and that, for $|\beta : l^{(s)}| = 1$, they are continuous on $\overline{G}$.

For the system $\{P_j(x, D)\}$ to be coercive in $\prod W_p^{l^{(s)}}(G)$, it is sufficient that the rank of the matrix $(P_{js}(x, i\xi))$ be equal to S for arbitrary nonzero real ξ when $x \in G$ and for arbitrary nonzero complex ξ when $x \in \partial G = \overline{G} \setminus G$.

The same conditions for interior points of G are necessary for coerciveness of the system. For constant coefficients, they are also necessary (when $|\beta : l^{(s)}| = 1$) for points of the boundary ∂G.

Proof. The necessity in the case of constant "principal" coefficients $c_{js\beta}$, for $|\beta : l^{(s)}| = 1$, was shown already in Theorem 11.4. Suppose now that, for some real $\xi \neq 0$ and $x^{(0)} \in G$, the rank of $(P_{js}(x^{(0)}, i\xi))$ is less than S. In this case, just as in the proof of the necessity of Theorem 11.6, the question is resolved by examining the functions $f_s(x, t)$ defined by (17). We need only take, in (17), functions $\varphi(x)$ concentrated in a sufficiently small neighborhood of the point $x^{(0)}$.

Sufficiency. Suppose that $G = G_0 \cup G_*$, where G_0 and G_* are open sets such that $\overline{G}_0 \subset G$ and G_* is contained in a sufficiently small neighborhood of ∂G. Let $\psi_0(x)$ and $\psi_*(x)$ denote nonnegative infinitely differentiable functions defined on E^n such that $\psi_0(x) + \psi_*(x) \equiv 1$, $\psi_0(x)$ is concentrated in G_0, and $\psi_*(x) = 0$ on $G \setminus G_*$.

The set G_0 and the functions ψ_0 and ψ_1 are easily constructed for G from the set G_*.

For $|\alpha : l^{(s)}| = 1$,

$$\begin{aligned} \| D^\alpha f_s \|_{p,\,G} &\leqslant \| \psi_0 D^\alpha f_s \|_{p,\,G_0} + \| \psi_* D^\alpha f_s \|_{p,\,G_*} \\ &\leqslant \| D^\alpha (\psi_0 f_s) \|_{p,\,G_0} + \| D^\alpha (\psi_* f_s) \|_{p,\,G_*} + C \sum_{|\beta : l^{(s)}| < 1} \| D^\beta f_s \|_{p,\,G}. \end{aligned} \tag{19}$$

We estimate the second term in the right-hand member of (19) with the aid of the constructions used in proving Theorem 11.4. Specifically, to every point $x^{(0)}$ belonging to $\partial G \cap \partial G_k$ we assign a cube $Q^{(k)}(x^{(0)})$ covering it according to the same rule (and with the same restrictions) as in the proof of Theorem 11.4. Thus, the estimate (13) holds for $C(x^{(0)})\varepsilon < \theta$, where θ is independent of both $x^{(0)} \in \partial G$ and k. For each k, let us take a finite covering $\{Q^{(k)}(x^{(r)})\}$ and let us consider the open sets

$$G_k^* = G_k \cap \bigcup_r Q^{(k)}(x^{(r)})$$

and

$$G_* = \bigcup_{k=1}^{K} G_k^*.$$

Let d denote the smallest diameter of any of the cubes $Q^{(k)}(x^{(r)})$. Let us partition each of the cubes $Q^{(k)}(x^{(r)})$ into open cubes $Q_i^{(k)}(x^{(r)})$ of diameter d and let us leave in the system $\{Q_i^{(k)}(x^{(r)})\}$ only distinct cubes that intersect with G_k^*. After re-indexing, we obtain for each k a finite system of open disjoint cubes $\{q_i^{(k)}\}$ such that $G_k^* \subset \bigcup_i \overline{q_i^{(k)}}$. Since $\psi_*(x) = 0$ on $G \setminus G_*$, we obtain in a manner analogous to the derivation of (14)

$$\frac{1}{2}\sum_{s=1}^{S}\sum_{|\alpha:l^{(s)}|=1}\|D^\alpha(\psi_* f_s)\|_{p,G_*}^p$$

$$\leqslant C_5\sum_{j=1}^{N}\|P_j(\cdot,D)(\psi_* f)\|_{p,G_*}^p + C_5\sum_{s=1}^{S}\sum_{|\beta:l^{(s)}|<1}\|D^\beta(\psi_* f_s)\|_{p,G_*}^p.$$

Then, for $|\alpha:l^{(s)}|=1$,

$$\|D^\alpha(\psi_* f_s)\|_{p,G_*}$$

$$C_6\sum_{j=1}^{N}\|P_j(\cdot,D)f\|_{p,G_*}+C_6\sum_{s=1}^{S}\sum_{|\beta:l^{(s)}|<1}\|D^\beta f_s\|_{p,G_*}. \quad (20)$$

We estimate the first term in the right-hand member of (19) with the aid of Lemma 11.7:

$$\|D^\alpha(\psi_0 f_s)\|_{p,G_0}\leqslant C_7\sum_{j=1}^{N}\|P_j(\cdot,D)(\psi_0 f)\|_{p,G_0}+C_7\sum_{s=1}^{S}\|f_s\|_{p,G_0}$$

$$\leqslant C_7\sum_{j=1}^{N}\|P_j(\cdot,D)f\|_{p,G_0}+C_8\sum_{s=1}^{S}\sum_{|\beta:l^{(s)}|<1}\|D^\beta f_s\|_{p,G_0}.$$

Then, by virtue of (19) and (20), we have

$$\sum_{s=1}^{S}\sum_{|\alpha:l^{(s)}|=1}\|D^\alpha f_s\|_{p,G}$$

$$\leqslant C_9\sum_{j=1}^{N}\|P_j(\cdot,D)f\|_{p,G}+C_9\sum_{s=1}^{S}\sum_{|\beta:l^{(s)}|<1}\|D^\beta f_s\|_{p,G}.$$

It now remains only to estimate the last double summation on the basis of Theorem 9.5 for sufficiently small h and transfer to the left-hand member those terms containing the factor h.

§12. Imbedding of $W_p^l(G)$ when l does not correspond to the type of the region G

In §10, we obtained imbedding theorems for function spaces $W_p^l(G)$ under the assumption that the open set G satisfies a weak l-horn condition, that is, under the assumption that G is a region belonging to the class $\underline{A}(l, H)$. Coincidence of the parameter l characterizing the class of regions with the differential parameter of the function space reflected a definite agreement of the properties of the domain of definition G of functions with the properties of the class of functions defined on G that was being considered. In the case of such an agreement, the imbedding theorems for the spaces $W_p^l(G)$ are valid in the same formulation as in the case $G = E^n$. In subsection 12.3, we shall show that the conditions imposed on the open set G are in a certain sense also necessary conditions for validity of the same theorems, at least in terms of the classes of regions being considered.

The basic purpose of the present section is to obtain imbedding theorems for the spaces $W_p^l(G)$ under the assumption that the open set G satisfies a weak s-horn condition, where, in general, $s = (s_1, \ldots, s_n) \neq l$. We shall see that, when the parameter s of the class $\underline{A}(s, H)$ of domains of definition G of the functions does not coincide with the differential parameter l of the class of functions defined on G that we are considering, the results will be different from those for the case $G = E^n$. It follows in particular that the imbedding theorems that are valid for the entire space E^n do not necessarily hold for subregions, even those with fairly smooth boundaries.

We note that, in the case of isotropic $W_p^l(G)$, the question of dependence of the properties of the functions on the geometric properties of their domains of definition was studied by Maz'ya [1], [2], Globenko [1], and Campanato [2]. For example, Maz'ya obtained in a number of cases necessary and sufficient conditions that the region G must satisfy for certain imbedding theorems to hold. However, we cannot stop here to treat all these investigations in detail.

Since the procedure for proving the imbedding theorems for the spaces $W_p^{l}(G)$ in the case in which $G \in A(s, h)$, where $s \neq l$, does not differ in any significant way from the procedure used in the case $s = l$, we shall confine ourselves to proving a single theorem, analogous to Theorem 10.2, and to looking at a number of consequences of it and examples clarifying the nature of the question.

12.1. Theorem. *Suppose that* $1 \leqslant p \leqslant q \leqslant \infty$. *Let* G *denote a member of* $\underline{A}(s, H)$, *and let* f *denote a member of* $W_p^l(G)$. *Suppose that* $\alpha = (\alpha_1, \ldots, \alpha_n)$ *is a vector with nonnegative-integer components,*

$$\delta_i = \frac{l_i}{s_i} - \left(\alpha + \frac{1}{p} - \frac{1}{q}, \frac{1}{s}\right) \quad (i = 1, \ldots, n), \quad \delta = \min_i \delta_i \geqslant 0. \tag{1}$$

For $\delta = 0$, *suppose that one of the following three holds*:

$$1 < p_n < q_n < \infty,$$
$$1 = p_n < q_n = \infty,$$
$$1 < p = q < \infty.$$

Then, $D^{\alpha} W_p^l(G) \hookrightarrow L_q(G)$. *More precisely, a function* f *that belongs to* $W_p^l(G)$ *has on* G *a generalized derivative* $D^{\alpha} f \in L_q(G)$ *and for all* h *in* $(0, H]$ *there is a constant* C, *independent of* f *and* h, *such that*

$$\| D^{\alpha} f \|_{q, G} \leqslant C \left(h^{-\delta_0} \| f \|_{p, G} + \sum_{i=1}^{n} h^{\delta_i} \| D_i^{l_i} f \|_{p, G} \right), \tag{2}$$

where $\delta_0 = \left(\alpha + \frac{1}{p} - \frac{1}{q}, \frac{1}{s}\right)$,

PROOF. Let $\{G_k\}_{k=1}^{K}$ denote a collection of open sets constituting a covering of G according to the definition of the class $\underline{A}(s, H)$. Just as in the proof of Theorem 10.2, we need only prove that $D^{\alpha} f$ exists on G_k and obtain an estimate for $\| D^{\alpha} f \|$ for arbitrary fixed k in $[1, K]$. Let us fix k.

We set $\lambda = \frac{1}{s}$ (that is, $\lambda_i = \frac{1}{s_i}$ for $i = 1, \ldots, n$) and write the relationships (1) in the form

$$\delta_i = l_i\lambda_i - \left(\alpha + \frac{1}{p} - \frac{1}{q}, \lambda\right) \quad (i = 1, \ldots, n), \quad \delta = \min_i \delta_i \geqslant 0. \tag{1}$$

In what follows, we distinguish two cases:

1) $\delta > 0$ or $\delta = 0$ and $1 < p_n < q_n < \infty$ or $\delta = 0$ and $1 = p_n < q_n = \infty$;
2) $\delta = 0$ and $1 < p = q < \infty$.

Let us suppose first that case 1) holds. Then, from the relationships (1) we easily obtain the inequalities

$$(\alpha, \lambda) < l_i\lambda_i \qquad (i = 1, \ldots, n). \tag{3}$$

Setting $l^i = (0, \ldots, l_i, \ldots, 0)$ for $i = 1, \ldots, n$ and keeping (3) in mind, we see that 7(20) and 7(24) of subsection 7.3 hold for the vectors α and l^i and that $D^{l^i} f \in L_p(G)$ for $i = 1, \ldots, n$. On the basis of the results of this section, we may conclude that the generalized derivative $D^\alpha f$ exists on G_h and that equation 7(25) holds for almost all $x \in G_h$:

$$D^\alpha f(x) = D^\alpha f_{h^\lambda}(x) + \int_0^h \sum_{i=1}^n v^{-1-\varkappa_i} dv \int_{E^n} D_i^{l_i} f(x+y) M_i(y : v^\lambda)\, dy, \tag{4}$$

where

$$0 < h \leqslant H, \lambda = \frac{1}{s}, \mu = \left(1 - \frac{1}{p} + \frac{1}{q}, \lambda\right), \varkappa_i = (1 + \alpha, \lambda) - l_i\lambda_i$$

$$= \left(1 - \frac{1}{p} + \frac{1}{q}, \lambda\right) - \left[l_i\lambda_i - \left(\alpha + \frac{1}{p} - \frac{1}{q}, \lambda\right)\right]$$

$$= \mu - \delta_i \; (i = 1, \ldots, n),$$

the averaging kernel and the functions M_i are functions in the class $C_0^\infty(E^n)$ such that the support of the representation is contained in the translated horn $x + V_k(s) \subset G$.

It follows from (4) that

$$\| D^\alpha f \|_{q, G_k} \leqslant \| D^\alpha f_{h^\lambda} \|_{q, G_k} + \sum_{i=1}^{n} \mathscr{I}_i(h), \tag{5}$$

where

$$\mathscr{I}_i(h) = \left\| \int_0^h v^{-1-\mu+\delta_i} dv \int_{E^n} D_i^{l_i} f(\cdot + y) M_i(y : v^\lambda)\, dy \right\|_{q, G_k}.$$

The first term in the right-hand member of (5) is estimated with the aid of Young's inequality 2(18) as follows:

$$\| D^\alpha f_{h^\lambda} \|_{q, G_k} \leqslant h^{-(1+\alpha, \lambda)} \left\| \int_{E^n} f(\cdot + y)\, \Omega(y : h^\lambda)\, dy \right\|_{q, G_k}$$

$$\leqslant C_0 h^{-\delta_0} \| f \|_{p, G_k + V_k}, \tag{6}$$

where

$$\delta_0 = \left(\alpha + \frac{1}{p} - \frac{1}{q},\ \lambda \right)$$

and C_0 is independent of f and h.

To estimate $\mathscr{I}_i(h)$ for $\delta_i > 0$, we use successively Minkowski's generalized inequality 2(12) and Young's inequality 2(18). We obtain

$$\mathscr{I}_i(h) \leqslant \int_0^h v^{-1-\mu+\delta_i}\, dv \left\| \int_{E^n} D_i^{l_i} f(\cdot + y) M_i(y : v^\lambda)\, dy \right\|_{q, G_k}$$

$$\leqslant C_i h^{\delta_i} \| D_i^{l_i} f \|_{p, G_k + V_k}, \tag{7}$$

where C_i is a constant independent of f and h.

Suppose now that $\delta_i = 0$ and that either $1 < p_n < q_n < \infty$ or $1 = p_n < q_n = \infty$. Setting $x = (\bar{x}, x_n)$, $y = (\bar{y}, y_n)$, and $\boldsymbol{q} = (\bar{\boldsymbol{q}}, q_n)$, we first estimate for fixed x_n the norm

$$\mathcal{I}_i(h; x_n) = \left\| \int_0^h v^{-1-\mu}\, dv \int_{E^n} D_i^{l_i} f(\cdot + \bar{y}, x_n + y_n)\, M_i(y : v^\lambda)\, dy \right\|_{\bar{q}},$$

applying Minkowski's generalized inequality 2(12) with respect to v and y_n. We obtain

$$\mathcal{I}_i(h; x_n) \leqslant \int_0^h v^{-1-\frac{\lambda_n}{r_n}}\, dv \left\| \int |M_i(y : v^\lambda)\, f_i(\cdot + \bar{y}, x_n + y_n)|\, dy \right\|_{\bar{q}},$$

where $\frac{1}{\boldsymbol{r}} = \boldsymbol{1} - \frac{1}{\boldsymbol{p}} + \frac{1}{\boldsymbol{q}}$ and $f_i(x) = D_i^{l_i} f(x)$ on $G_k + V$ but $f_i(x) = 0$ outside $G_k + V$.

Setting $\boldsymbol{p} = (\bar{\boldsymbol{p}}, p_n)$ and $\boldsymbol{r} = (\bar{\boldsymbol{r}}, r_n)$, we get by using Young's inequality 2(18)

$$\mathcal{I}_i(h; x_n) \leqslant \int_0^h v^{-1-\frac{\lambda_n}{r_n}}\, dv \int_{E_1} M(y_n : v^{\lambda_n}) \| f_i(\cdot, x_n + y_n) \|_{\bar{p}}\, dy_n,$$

where $M(y_n) = \max \| M_i(\cdot, y_n) \|_{\bar{r}}$.

Other estimates are obtained with the aid of the Hardy–Littlewood inequality 2(32) for $1 < p_n < q_n < \infty$ and Young's inequality 2(18) for $1 = p_n < q_n = \infty$ (with allowance for the position of the support of M_i); these estimates are no different from the estimates of the right-hand member of 10(5). As a result, we obtain the estimate (7) for $\mathcal{I}_i(h) = \| \mathcal{I}_i(h; \cdot) \|_{q_n}$ in this case too. From (5)-(7) we get the estimate

$$\| D^\alpha f \|_{q, G_k} \leqslant C_0 h^{-\delta_0} \| f \|_{p, G_k + V_k} + \sum_{i=1}^n C_i h^{\delta_i} \| D_i^{l_i} f \|_{p, G_k + V_k}.$$

Since k is an arbitrary number in $[1, K]$, this last estimate leads to the estimate (2) and hence to the conclusion of the theorem in case 1).

Suppose now that case 2) holds; that is, suppose $\delta = 0$ and $1 < p = q < \infty$. In this case, to prove the assertion of the theorem we use the identity 7(22), which is valid at every point x in G_k. From it we get

$$\| D^\alpha f_{\varepsilon\lambda} - D^\alpha f_{h\lambda} \|_{p, G_k}$$

$$\leqslant \sum_{i=1}^{n} \left\| \int_{\varepsilon}^{h} v^{-1-\mu+\delta_i} dv \int_{E^n} D_i^{l_i} f(\cdot + y) M_i(y: v^\lambda) dy \right\|_{p, G_k} = \sum_{i=1}^{n} \mathcal{J}_i(\varepsilon, h), \tag{8}$$

where $\mu = |\lambda|$ and $\delta_i = l_i \lambda_i - (\alpha, \lambda)$ for $i = 1, \ldots, n$.

If $\delta_i > 0$, we obtain, just as in the estimate for $\mathcal{J}_i(h)$ (see (7)),

$$\mathcal{J}_i(\varepsilon, h) \leqslant C_i h^{\delta_i} \| D_i^{l_i} \|_{p, G_k + V k}. \tag{9}$$

If $\delta_i = 0$, it follows on the basis of Lemma 9.4 that

$$\mathcal{J}_i(\varepsilon, h) \leqslant C_i \| D_i^{l_i} f \|_{p, G_k + V_k} \tag{10}$$

and $\mathcal{J}_i \to 0$ as ε and h approach 0 through combinations such that $0 < \varepsilon < h$.

Consequently, $D^\alpha f_{\varepsilon\lambda}$ is Cauchy-convergent in $L_p(G_k)$ as $\varepsilon \to 0$. Furthermore, since $f_{\varepsilon\lambda} \to f$ in $L_p(G_k)$ as $\varepsilon \to 0$, it follows on the basis of the remark to Lemma 6.2 that there exists $D^\alpha f \in L_p(G_k)$ and

$$\| D^\alpha f \|_{p, G_k} = \lim_{\varepsilon \to 0} \| D^\alpha f_{\varepsilon\lambda} \|_{p, G_k} \leqslant \| D^\alpha f_{h\lambda} \|_{p, G_k} + \lim_{\varepsilon \to 0} \| D^\alpha f_{\varepsilon\lambda} - D^\alpha f_{h\lambda} \|_{p, G_k}.$$

On the basis of inequalities (6)-(10), we now obtain

$$\|D^{\alpha}f\|_{p,\,G_k} \leqslant C_0 h^{-\delta_0} \|f\|_{p,\,G_k+V_k} + \sum_{i=1}^{n} C_i h^{\delta_i} \|D_i^{l_i} f\|_{p,\,G_k+V_k}.$$

This completes the proof of the theorem.

REMARK. If the conditions of the theorem are satisfied for the vector s, they will also be satisfied for the vector cs, for any positive number c. This follows, first of all, from the fact that $G \in \underline{A}(cs, H^c)$ for arbitrary positive c whenever $G \in \underline{A}(s, H)$ (see 8(6)) and, second, from the fact that inequality (1) remains valid when we multiply s by a positive number since δ_j (for $j = 0, 1, \ldots, n$) are homogeneous functions with respect to s_i for $i = 1, \ldots, n$.

It follows that we can confine ourselves to those classes of regions that are characterized by vectors s normalized in some way or other. For example, we may assume that $\left\|\frac{1}{s}\right\| = \left(\sum_{i=1}^{n} \frac{1}{s_i^2}\right)^{1/2} = 1$ or that some component of the vectors s in question is equal to a fixed number.

12.2. The necessity of inequality (1) in Theorem 12.1. Let us show with a simple example in the two-dimensional case ($n = 2$) that satisfaction of inequality (1), which we shall find more convenient to write in the form

$$\min_{i=1,\ldots,n} \frac{l_i}{s_i} - \left(\alpha + \frac{1}{p} - \frac{1}{q},\ \frac{1}{s}\right) \geqslant 0, \qquad (1')$$

is in a certain sense necessary for the validity of inequality (2). More precisely, let us show that, for arbitrary vectors $s = (s_1, s_2)$ (for $s_i > 0$) and $l = (l_1, l_2)$ (where the l_i are positive integers), there exists a region G in the class $\underline{A}(s, H)$ such that inequality (2) for an arbitrary function in $W_p^l(G)$ can hold only when inequality $(1')$ is satisfied.

Suppose that we are given the vectors s and l. Let us suppose for definiteness that $\frac{l_1}{s_1} \leqslant \frac{l_2}{s_2}$.

Consider a plane region G bounded by the curve $x_1^{s_1} = |x_2|^{s_2}$ and the lines $x_2 = \frac{1}{2}$, $x_2 = -\frac{1}{2}$, and $x_1 = 1$ (see Fig. 1).

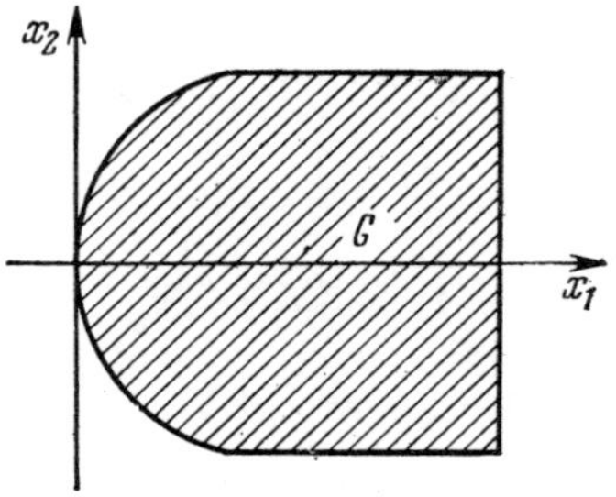

FIG. 1

One can easily see that $G \in \underline{A}(s, H)$, where $s = (s_1, s_2)$ and H is a positive number (see §8). Obviously, $G \in \underline{A}(s', H')$, where $s' = (s_1, s_2')$ if $0 < s_2' \leqslant s_2$ but $G \notin \underline{A}(s', H')$ if $s_2' > s_2$.

We define on G the sequence of functions

$$f_n(x_1, x_2) = x_2^{l_2-1} e^{-nx_1} \qquad (n = 1, \ldots)$$

belonging to the space $W_{\boldsymbol{p}}^l(G)$, where $\boldsymbol{p} = (p_1, p_2)$ (where in turn $1 \leqslant p_i \leqslant \infty$ for $i = 1, 2$). Elementary calculations show that

$$\| f_n \|_{\boldsymbol{p}, G} = O(n^{-\nu(\boldsymbol{p})}), \ \| D_1^{l_1} f_n \|_{\boldsymbol{p}, G} = O(n^{-\nu(\boldsymbol{p}) + l_1}), \ \| D_2^{l_2} f_n \|_{\boldsymbol{p}, G} = 0,$$

where

$$\nu(\boldsymbol{p}) = \frac{1}{p_1} + \frac{s_1}{s_2 p_2} + (l_2 - 1)\frac{s_1}{s_2}.$$

Consequently,

$$\| f_n \|_{W_{\boldsymbol{p}}^l(G)} = O(n^{-\nu(\boldsymbol{p}) + l_1}). \tag{11}$$

Suppose that $\alpha = (\alpha_1, \alpha_2)$ (where the α_i are nonnegative integers and $\alpha_2 < l_2$) and $\boldsymbol{q} = (q_1, q_2)$ (where $1 \leqslant q_i \leqslant \infty$ for $i = 1, 2$). Then,

$$\| D^{\alpha} f_n \|_{q,\,G} = O\left(n^{-\nu(q)+\alpha_1+\alpha_2 \frac{s_1}{s_2}}\right). \tag{12}$$

It follows from (11) and (12) that inequality (2),

$$\| D^{\alpha} f_n \|_{q,\,G} \leqslant C \| f_n \|_{W_p^l(G)},$$

with constant C independent of f_n is possible only when $-\nu(q) + \alpha_1 + \alpha_2 \frac{s_1}{s_2} \leqslant -\nu(p) + l_1$, that is, when

$$\frac{l_1}{s_1} - \sum_{i=1}^{2} \left(\alpha_i + \frac{1}{p_i} - \frac{1}{q_i}\right) \frac{1}{s_i} \geqslant 0.$$

Since $\frac{l_1}{s_1} \leqslant \frac{l_2}{s_2}$, this inequality coincides with inequality (1′), which proves our assertion.

We note that, if $\frac{l_1}{s_1} \geqslant \frac{l_2}{s_2}$, then the region G satisfies a weak l-horn condition. Suppose that $s' = (s_1, s_2')$, where $s_2' \leqslant s_2$ and $\frac{l_1}{s_1} = \frac{l_2}{s_2'}$. As was mentioned above, $G \in \underline{A}(s', H)$, so that $G \in \underline{A}(cs', H^c)$ for arbitrary $c > 0$. Setting $c = \frac{l_1}{s_1} = \frac{l_2}{s_2'}$, we obtain the result that $cs' = l$ and hence, $G \in \underline{A}(l, H^c)$. It follows that Theorem 10.2 is valid for $\frac{l_1}{s_1} \geqslant \frac{l_2}{s_2}$.

It is easy to construct an analogous example for $n > 2$.

12.3. Consequences of Theorem 12.1. Let us look at some consequences of the theorem that we have proven and the example constructed.

12.3.1. Suppose that, under the conditions of Theorem 12.1,

$$\min_{i=1,\ldots,n} \frac{l_i}{s_i} = \frac{l_1}{s_1}$$

and the vector s is normalized so that $s_1 = l_1$ (see remark to

Theorem 12.1), that is, that

$$s_1 = l_1, \quad s_i \leqslant l_i \qquad (i = 2, \ldots, n). \tag{13}$$

Then, inequality (1) takes the form

$$\left(\alpha + \frac{1}{p} - \frac{1}{q}, \frac{1}{s}\right) \leqslant 1. \tag{14}$$

Thus, for an arbitrary function in the class $W_p^l(G)$, where $G \in \underline{A}(s, H)$ (with $s = (l_1, s_2, \ldots, s_n)$) and the vector l satisfies the relations (13), inequality (2) can be guaranteed only for α and q for which inequality (14) holds.*

We see from inequality (14) that the sets of vectors α and q satisfying inequality (2) depend essentially on the vectors $s = (l_1, s_2, \ldots, s_n)$ and not on the vector l provided this vector satisfies conditions (13).

Let us suppose for simplicity that the components of the vector $s = (l_1, s_2, \ldots, s_n)$ are positive integers. Since the relations (13) are satisfied when we replace the vector l with the vector s, we may assert that, for functions in the class $W_p^s(G)$, inequality (2) with $l = s$ will hold also for those α and q for which (14) is valid.

It follows from what has been said that, if $G \in \underline{A}(s, H)$, where $s = (s_1, \ldots, s_n)$, then for arbitrary l satisfying conditions (13) *for functions in the class $W_p^l(G)$, inequality* (2) *can be guaranteed only for those vectors α and q for which it is valid for functions in the class $W_p^s(G)$.* In other words, increase in the vector index l satisfying conditions (13) does not in general improve the properties of functions in the classes $W_p^l(G)$ asserted by Theorem 12.1. This result agrees with the example of 12.2, which shows that, for arbitrary $s > 0$, there exists a region G in the class $\underline{A}(s, H)$ such that, for arbitrary l satisfying conditions (13), inequality (2) for an arbitrary function in the class $W_p^l(G)$ is valid only for vectors α and q satisfying (14).

*For $\left(\alpha + \frac{1}{p} - \frac{1}{q}, \frac{1}{s}\right) = 1$, the vectors p and q must, by Theorem 12.1, satisfy supplementary conditions.

12.3.2. Let us now compare the properties of functions in $W_p^l(G)$ (where l and p are fixed) defined on regions G belonging to the classes $\underline{A}(s, H)$ for different vectors $s < \mathbf{0}$.

For given l, we partition the set of all vectors s into n groups $S^1, \ldots, S^n$, where

$$S^i = \left\{ s: \min_{j=1,\ldots,n} \frac{l_j}{s_j} = \frac{l_i}{s_i} \right\} \qquad (i = 1, \ldots, n).$$

Let us look first at classes of regions that are defined by the vectors $s \in S^1$. Since we shall be interested only in those properties of functions $W_p^l(G)$ that are asserted in Theorem 12.1, we can, on the basis of the remark following it, confine ourselves to those vectors $s \in S^1$ for which $s_1 = l_1$. Consequently, we may assume that the components of the vectors s under consideration satisfy (13).

In what follows, we shall say that functions belonging to $W_p^l(G)$ possess better properties on open sets in the class $\underline{A}(s^1, H)$, where $s^1 = (l_1, s_2^1, \ldots, s_n^1)$, than on open sets in the class $\underline{A}(s^2, H)$, where $s^2 = (l_1, s_2^2, \ldots, s_n^2)$, if the sets of vectors α and $\boldsymbol{q}$ for which inequality (2) is valid are broader for $s = s^1$ than for $s = s^2$.

When conditions (13) are satisfied, these sets of vectors α and $\boldsymbol{q}$ are defined by inequality (14). We see from the form of the left-hand member of (14) that, the greater the components of the vector $s = (l_1, s_2, \ldots, s_n)$, the broader the sets of vectors α and $\boldsymbol{q}$ (for $\alpha \geqslant 0$ and $\boldsymbol{q} \geqslant \boldsymbol{p}$) that will satisfy that inequality. Of all the vectors s satisfying conditions (13), the best in this sense is the vector $s = l$. Obviously, the same result will be obtained from examination of other groups of vectors.

It follows that *functions in the space* $W_p^l(G)$ *possess the best properties* (in the sense of the definition given above) *on open sets in the class* $\underline{A}(l, H)$, that is, *on open sets satisfying a weak l-horn condition.* These properties will hold, for example, when $G = E^n$ since $E^n \in \underline{A}(l, H)$ for arbitrary l.

Noting also that the class $\underline{A}(s, H)$ of open sets characterized by the vector s coincides with the class $\underline{A}(l, H)$ of open sets characterized by the vector l if and only if there exists a positive number

c such that $cs = l$, that is, if and only if

$$\frac{l_1}{s_1} = \frac{l_2}{s_2} = \ldots = \frac{l_n}{s_n}, \tag{15}$$

we can rephrase the result obtained in the following somewhat more general form: *Functions in the space* $W_p^l(G)$ *possess best properties on open sets* G *of an arbitrary class* $\underline{A}(s, H)$ *characterized by the vector* s *satisfying equations* (15).

Let us suppose that conditions (15) are not satisfied, that is, that

$$\min_{j=1,\ldots,n} \frac{l_j}{s_j} < \max_{j=1,\ldots,n} \frac{l_j}{s_j}.$$

Suppose, for example, that $\min \frac{l_j}{s_i} = \frac{l_1}{s_1}$, where we may assume $s_1 = l_1$. Then, equations (13) are valid for at least one j in $1 \leqslant j \leqslant n$ with $s_j < l_j$. Then, keeping example 12.2 in mind, we conclude from this that there exists a region G in the class $\underline{A}(s, H)$ on which the properties of functions in $W_p^l(G)$ are worse than on regions in the class $\underline{A}(l, H)$. All this leads us to the following conclusion: *For the same imbedding theorem of the type* 12.1 *to be valid for an arbitrary open set* G *in the class* $\underline{A}(s, H)$ *as for* $G = E^n$, *it is necessary and sufficient that the components of the vector* s *satisfy equations* (15).

12.3.3. Let us look at some examples. Specifically, let us look at some plane regions G_i and let us ascertain on the basis of equations (15) what conditions, in the case of each region, the parameters l_1 and l_2 must satisfy for the imbedding Theorem 12.1 to coincide with the analogous theorem for $G = E^n$ for the function space $W_p^l(G_i)$, where $l = (l_1, l_2)$.

Suppose that G_1 is the triangle shown in Figure 2. Suppose that G_2 is the trapezoid (shown in Figure 3) bounded by the straight lines $x_2 = kx_1$ (where $k > 0$), $x_2 = h > 0$, $x_2 = 0$, and $x_1 = d$. Suppose that G_3 is the disk shown in Figure 4. Suppose that G_4 is the region bounded by semicircles of radius H and segments of the lines $x_2 = \pm h$ (shown in Figure 5). Suppose that G_5 is bounded by segments of the lines $x_1 = 0$, $x_1 = d$, and $x_2 = \pm h$ and convex arcs (as shown in Figure 6).

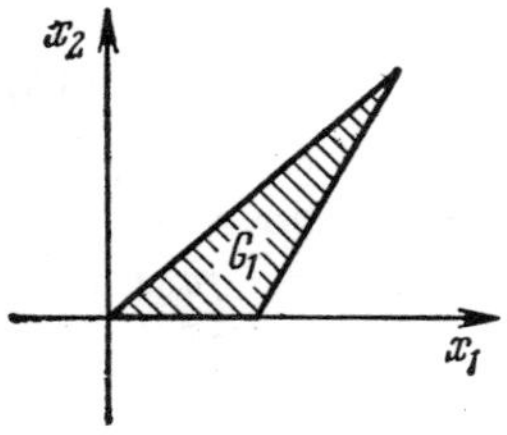

FIG. 2

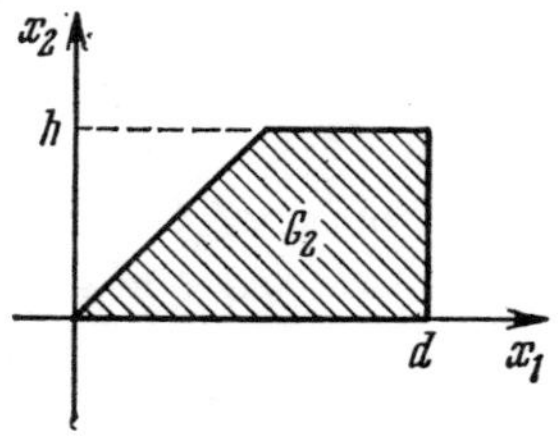

FIG. 3

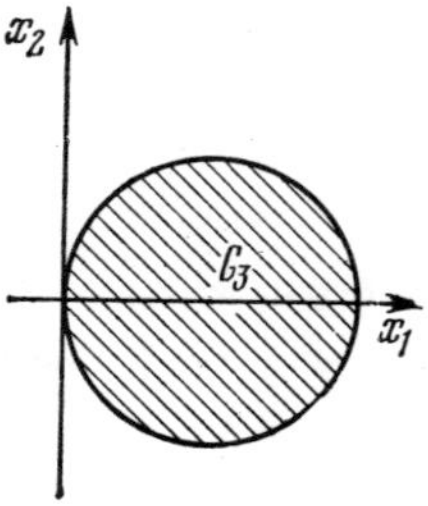

FIG. 4

First of all, let us see what vectors $s=(s_1, s_2)$ characterize those classes $\underline{A}(s, H)$ to which the regions G_i belong. Since such vectors are defined only up to a positive constant factor c, let us assume for definiteness that they are normalized in such a way that $s_1 = 1$. (The value of the second parameter H changes when we multiply s by c, but this value is not of interest to us here.) Thus, the classes of regions are characterized by vectors of the form $s=(1, s_2)$.

On the basis of the definitions of the classes $\underline{A}(s, H)$ (see §8), one can easily show that

1) $G_1 \in \underline{A}(\mathbf{1}, H)$, $\mathbf{1}=(1, 1)$, and $G \notin \underline{A}(s, H)$ if $s=(1, s_2) \neq \mathbf{1}$;

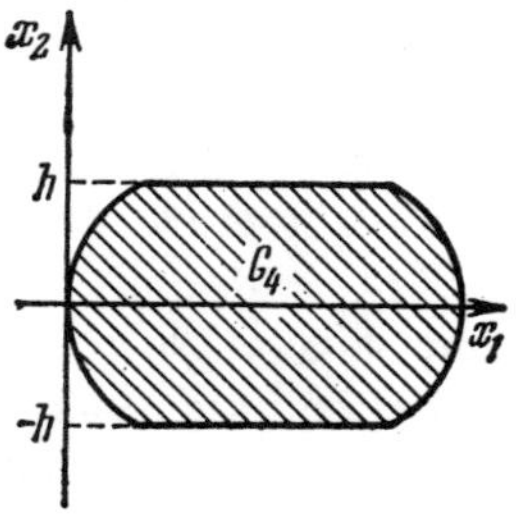

FIG. 5

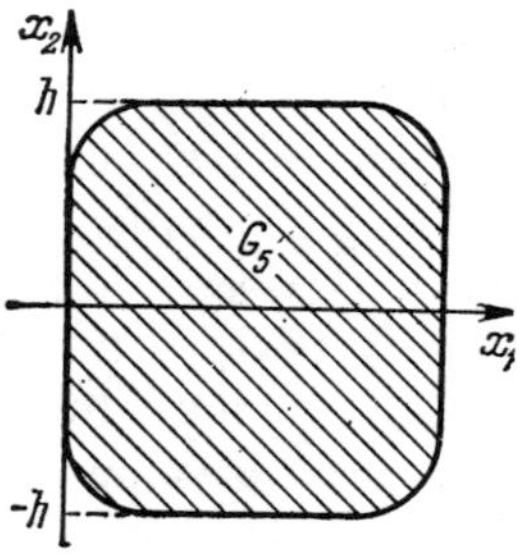

FIG. 6

2) $G_2 \in \underline{A}(s, H)$, where $s = (1, s_2)$, if and only if $0 < s_2 \leqslant 1$;

3) $G_3 \in \underline{A}(s, H)$, where $s = (1, s_2)$, if and only if $\frac{1}{2} \leqslant s_2 \leqslant 2$;

4) $G_4 \in \underline{A}(s, H)$, where $s = (1, s_2)$, if and only if $0 < s_2 \leqslant 2$;

5) $G_5 \in \underline{A}(s, H)$, where $s = (1, s_2)$, for $0 < s_2 < \infty$.

From the limits indicated for the parameter s_2 for which $G_i \in \underline{A}(s, H)$, it follows that (15) can be written (since in our case $n = 2$ and $s_1 = 1$) in the form

$$l_1 = \frac{l_2}{s_2}, \tag{15'}$$

we easily obtain relationships between l_1 and l_2 for which the imbedding Theorem 12.1 for $W_p^l(G_i)$ will be the same as for $G = E^n$. For G_1, these relationships have the form $l_1 = l_2$;

$$\begin{aligned} &\text{for } G_2, \quad l_1 \geqslant l_2; \\ &\text{for } G_3, \quad 2l_2 \geqslant l_1 \geqslant \frac{1}{2} l_2; \\ &\text{for } G_4, \quad l_1 \geqslant \frac{1}{2} l_2; \\ &\text{for } G_5, \quad l_1 > 0, \quad l_2 > 0. \end{aligned}$$

One can easily show by means of examples that these relationships are also necessary for validity of the imbedding Theorem 12.1.

We note that, for example, Theorem 12.1 takes the same form for a disk as for $G = E^n$ if and only if the differential parameters of the class of functions in question satisfy the inequalities $2l_2 \geqslant l_1 \geqslant \frac{1}{2} l_2$. This substantiates what was said in the introduction to the section to the effect that the imbedding theorems that are valid for the entire space do not always remain valid for subregions even though the boundaries of these are as smooth as one might want.

12.4. Comparison of the results for the spaces $W_p^{l1}(G)$ and $W_p^{(l)}(G)$. In this subsection, we shall modify our convention and let l denote not a vector but a positive integer.

The basic purpose of this subsection is to compare results of the type of Theorem 12.1 for the class of functions all of whose unmixed derivatives of order l, that is, derivatives of the form $D_i^l f$ (for $i = 1, \ldots, n$), exist with the corresponding results for the class of functions for which derivatives of order l, mixed and unmixed alike, exist.

Suppose, as usual, that $\mathbf{1} = (\overbrace{1, \ldots, 1}^{n})$. Then, $l\mathbf{1} = (l, \ldots, l)$ and hence $W_p^{l\mathbf{1}}(G)$ is a function space of the type considered above and characterized only by the fact that all the coordinates of the vector l_1 are equal to l. Together with $W_p^{l\mathbf{1}}(G)$ we shall consider the function space $W_p^{(l)}(G)$ introduced by Sobolev.

We denote by $W_p^{(l)}(G)$ the space of functions $f \in L_p(G)$ that have on G all generalized derivatives of order l (both mixed and unmixed) with finite norm

$$\| f \|_{W_p^{(l)}(G)} = \| f \|_{p,\, G} + \sum_{|\beta| = l} \| D^\beta f \|_{p,\, G} \qquad (\beta = (\beta_1, \ldots, \beta_n)). \tag{16}$$

Let us compare results of the type of Theorem 12.1 for functions belonging to the function spaces $W_p^{l\mathbf{1}}(G)$ and $W_p^{(l)}(G)$.

Obviously, Theorem 12.1 holds for $f \in W_p^{l\mathbf{1}}(G)$. Applying it to the case under consideration, we may write inequality (1) guaranteeing the existence of $D^\alpha f \in L_q(G)$ and the estimate (2) in the following form:

$$\delta = \frac{l}{\max\limits_{i=1, \ldots, n} s_i} - \left(\alpha + \frac{1}{p} - \frac{1}{q}, \ \frac{1}{s}\right) \geqslant 0. \tag{17}$$

We note also that if (17) is not satisfied, it follows from subsection 12.2 that there exists a region G in the class $\underline{A}(s, H)$ such that inequality (2) fails to hold for some $f \in W_p^{l\mathbf{1}}(G)$.

We now give a theorem characterizing the analogous properties of functions in the class $W_p^{(l)}(G)$.

Theorem. *Suppose that* $f \in W_p^{(l)}(G)$, *where* $G \in \underline{A}(s, H)$ *and* $1 \leqslant p \leqslant q \leqslant \infty$, *that* $\alpha = (\alpha_1, \ldots, \alpha_n)$, *where* α_i *for* $i = 1, \ldots, n$ *is a nonnegative integer, and that*

$$\bar{\delta} = \frac{l - |\alpha|}{\max\limits_{i=1,\ldots,n} s_i} - \left(\frac{1}{p} - \frac{1}{q}, \frac{1}{s}\right) \geqslant 0. \tag{18}$$

For $\bar{\delta} = 0$ *and* $|\alpha| < l$, *let us suppose that* $1 < p_n < q_n < \infty$. *Then,* $D^\alpha W_p^{(l)}(G) \hookrightarrow L_q(G)$; *more precisely, f has on G the derivative* $D^\alpha f \in L_q(G)$ *and for all h in* $(0, H]$ *there exists a constant C independent of f and h such that*

$$\| D^\alpha f \|_{q,G} \leqslant C \left(h^{-\delta_0} \| f \|_{p,G} + h^{\bar{\delta}} \sum_{|\beta| = l} \| D^\beta f \|_{p,G} \right), \tag{19}$$

where $\delta_0 = \left(\alpha + \frac{1}{p} - \frac{1}{q}, \frac{1}{s}\right)$.

This theorem is proven in the same way as Theorem 12.1.* The only significant difference, thanks to which inequality (17) is replaced with inequality (18), is that in the present case we use the integral representations 7(22) and 7(25), the right-hand members of which contain the mixed derivatives $D^{l^i} f$ for $i = 1, \ldots, n$ (instead of the unmixed derivatives $D_i^{l_i} f$ for $i = 1, \ldots, n$ that appeared in Theorem 12.1), where

$$l^i = l^i(\alpha) = (\alpha_1, \ldots, \alpha_{i-1}, l_i^i(\alpha), \alpha_{i+1}, \ldots, \alpha_n), \quad |l^i| = l \quad (i = 1, \ldots, n).$$

The validity of these representations follows from the fact that the components of the vectors l^i satisfy conditions 7(20).

Following the proof of Theorem 12.1, we see that inequality (19) holds if

*The case $\delta = 0$, $q = p$, which was examined separately in the proof of Theorem 12.1, is trivial here since we now have $|\alpha| = l$.

$$\min_{i=1,\ldots,n}\left(l^i - \alpha - \frac{1}{p} + \frac{1}{q},\ \frac{1}{s}\right) \geqslant 0. \tag{20}$$

However, since

$$\min_i\left(l^i - \alpha,\ \frac{1}{s}\right) = \min_i \frac{l_i^l - \alpha_i}{s_i} = \min_i \frac{l - |\alpha|}{s_i} = \frac{l - |\alpha|}{\max_i s_i},$$

inequality (20) coincides with inequality (18).

Thus, if $f \in W_p^{l1}(G)$, the condition guaranteeing the existence of $D^\alpha f \in L_q(G)$ and an estimate for $\| D^\alpha f \|_{q,\,G}$ in terms of the norm of f in the space $W_p^{l1}(G)$ is determined by inequality (17). If $f \in W_p^{(l)}(G)$, the analogous condition is determined by inequality (18).

One can easily see that, in the case in which all the components of the vector s are equal, that is, when $G \in \underline{A}\ (l1, H)$, the left-hand members of inequalities (17) and (18) will also be equal $(\delta = \bar{\delta})$ for arbitrary α and q. Consequently, in this case, the sets of the vectors α and q for which inequality (2) or inequality (19) is valid coincide (at least for $1 < p < \infty$).*

If the components of the vector s are not all equal, then the numbers δ and $\bar{\delta}$ are in general different. Specifically,

$$\bar{\delta} - \delta = \sum_{j=1}^{n}\left(\frac{1}{s_j} - \frac{1}{\max_i s_i}\right)\alpha_j \geqslant 0.$$

In the special case in which $q = p$, if the vector s is normalized in such a way** that $\max_i s_i = 1$, inequalities (17) and (18) take the forms

*This assertion follows from the fact that, in the present case, the norms in $W_p^{l1}(G)$ and $W_p^{(l)}(G)$ are equivalent (See Theorem 10.2).

**The remark made following Theorem 12.1 remains valid with respect to the theorem of the present subsection.

$$l-\left(\alpha, \frac{1}{s}\right) \geqslant 0 \qquad (0 < s_i \leqslant 1,\ s_{i_0}=1), \tag{17'}$$

$$l-|\alpha| \geqslant 0. \tag{18'}$$

These last inequalities imply that, for certain classes of regions G, the set of vectors α for which inequality (2) (with $l_i = l$ for $i = 1, \ldots, n$ and with $q = p$) is valid can be quite different from the set of vectors α for which inequality (19) is valid.

Suppose, for example, that $G \in \underline{A}(s, H)$, where $s = \left(1, \frac{1}{l+\varepsilon}, \ldots, \frac{1}{l+\varepsilon}\right)$ (where $\varepsilon > 0$) but $G \notin \underline{A}(s'\ H)$, where $s' = (1, s_2', \ldots, s_n')$ if at least one coordinate s_i' exceeds $1/(l+\varepsilon)$ (for $2 \leqslant i \leqslant n$).

Obviously, inequality (17′) with $s = \left(1, \frac{1}{l+\varepsilon}, \ldots, \frac{1}{l+\varepsilon}\right)$ can be satisfied only by vectors α of the form

$$\alpha = (\alpha_1, 0, \ldots, 0), \qquad 0 \leqslant \alpha_1 \leqslant l. \tag{21}$$

Therefore, on a given region G it is possible to guarantee the existence of estimates of the form (2) only for derivatives $D^\alpha f$ characterized by vectors α of the form (21). In particular, it is not possible to guarantee an estimate of the form (2) for any mixed partial derivative. Another way for us to see this is if in the example of subsection 12.2 we set $l_1 = l_2 = l$, $s_1 = 1$, and $s_2 = \frac{1}{l+\varepsilon}$. On the other hand, it follows from inequality (18′) that, no matter what class $\underline{A}(s, H)$ the region G belongs to, any function $f \in W_p^{(l)}(G)$ has on G all generalized derivatives up to order l inclusively and inequality (19) (with $q = p$) holds for them.

Everything that has been said in the present section brings out the significant influence that the properties of the domain of definition of the functions can have on the imbedding theorems and the fact that this influence can show up in different ways on functions in the different classes.

§13. Inequalities between the L_p-norms of mixed derivatives

The present section is devoted, roughly speaking, to the question as to what partial derivatives of functions of several variables can be estimated in the norm of L_p in terms of the sum of analogous norms of a given set of partial derivatives. A more precise wording of the question is as follows:

Let $\mathscr{E}$ denote a given finite set of vectors $l = (l_1, \ldots, l_n) \in E_l^n$ with nonnegative integer components. Let G denote a region contained in the space E_x^n. Suppose that $1 \leqslant p \leqslant \infty$.*

Let us denote by $\mathscr{W}_p^{\{\mathscr{E}\}}(G)$ the set of functions f belonging to $L^{\text{loc}}(G)$ that have on G generalized derivatives of the form $D^l f$ (where $l \in \mathscr{E}$) with finite seminorm

$$\sum_{l \in \mathscr{E}} \| D^l f \|_{p, G}.$$

The question arises of finding the set $\mathscr{D}$ (depending on the given set $\mathscr{E}$, the properties of the region G, and the parameter p) of those vectors $\alpha = (\alpha_1, \ldots, \alpha_n)$ (where the α_i are nonnegative integers) such that every function $f \in \mathscr{W}_p^{\{\mathscr{E}\}}(G)$ has a derivative $D^\alpha f \in L_p(G)$ and there exists a constant C independent of f such that

$$\| D^\alpha f \|_{p, G} \leqslant C \sum_{l \in \mathscr{E}} \| D^l f \|_{p, G}. \tag{1}$$

We note that the function classes $W_p^l(G)$ studied above can be regarded as classes $\mathscr{W}_p^{\{\mathscr{E}\}}(G)$ characterized by the set $\mathscr{E}$ consisting of $n+1$ vectors of the form

$$l^i = (0, \ldots, 0, l_i, 0, \ldots, 0) \quad (i = 1, \ldots, n), \quad l^0 = (0, \ldots, 0).$$

*In the present section, p will denote a number.

For functions belonging to these classes, inequalities of the form (1) were obtained in §10 (under the assumption that G satisfies a weak l-horn condition) and in §12 (under the assumption that G satisfies a weak s-horn condition). It follows from the results of §10 that, if $1 < p < \infty$ and $G \in \underline{A}(l, H)$, then the set $\mathscr{D}$ corresponding to the class $W_p^l(G)$ consists of all points (vectors) with nonnegative-integer coordinates belonging to the closed convex polyhedron spanned by the points l^i (for $i = 0, 1, \ldots, n$).

As we shall see later, in the case of an arbitrary set $\mathscr{E}$ results of the type of inequality (1) can be quite different from the results indicated for the classes $W_p^l(G)$ even, for example, when G is a rectangular parallelepiped with faces parallel to the coordinate hyperplanes (a rectangular parallelepiped satisfies any l-horn condition).

For this question, the geometric properties of the set $\mathscr{E}$ of points l are of great significance.

The value of the parameter p exerts considerable influence on the size of the set $\mathscr{D}$. (This was true also in the case of the classes $W_p^l(G)$.) In this sense, it is sufficient to distinguish the cases $p = 1$, $1 < p < \infty$, and $p = \infty$. However, in what follows we shall assume that $1 < p < \infty$.

We note that the question treated in the present section was studied in the articles by Bahvalov [1], Nikol'skiĭ [7], Golovkin [7], and Il'in [9]. In a recent paper by Boman [1], conditions for validity of inequality (1) for $f \in C_0^\infty(E^n)$ were obtained rather simply for $1 < p < \infty$ and for $p = \infty$.

13.1. Classes of regions G to be considered. Suppose that

$$H = (H_1, \ldots, H_n),\ H_i > 0, \quad \theta_i = +1 \text{ or } -1 \quad (i = 1, \ldots, n),$$

$$\square(H) = \left\{ x\colon\ 0 < \frac{x_i}{\theta_i} < H_i,\ i = 1, \ldots, n \right\} \tag{2}$$

is a rectangular parallelepiped with edges parallel to the coordinate axes. In what follows, we shall call $\square(H)$ an n-dimensional rectangle. Obviously, for given H, there exist exactly 2^n distinct rectangles of the form (2) that can be obtained by different choices of the numbers θ_i.

We shall say that a region G contained in E^n satisfies a weak rectangle condition and shall write $G \in \underline{A}(\square, H)$ if there exists a finite number Λ of open sets G_λ and rectangles $\square_\lambda(H)$ of the form (2) such that

$$G = \bigcup_{\lambda=1}^{\Lambda} G_\lambda = \bigcup_{\lambda=1}^{\Lambda} (G_\lambda + \square_\lambda(H)).$$

If $G \in \underline{A}(\square, H)$, where the first k components (for $1 \leqslant k \leqslant n$) of the vector H can be chosen arbitrarily large, we shall write $G \in \underline{A}(\square, H^{n-k})$ and $H^{n-k} = (H_{k+1}, \ldots, H_n)$ if $k < n$ or $G \in \underline{A}(\square)$ if $k = n$. For example, the strip $0 < x_i < \infty$ (for $i = 1, \ldots, k$), $0 < x_i < R$ (for $i = k+1, \ldots, n$) is a region belonging to the class $\underline{A}(\square, H^{n-k})$ and the entire space E^n is a region belonging to the class $\underline{A}(\square)$.

13.2. Notation. First of all, let us review some of the notation of subsection 7.10.

We denote by e any subset (including the empty set) of the set of natural numbers, $e^n = \{1, \ldots, n\}$. Let $e' = e^n \setminus e$ (so that $e \cup e' = e^n$), $\mathbf{1}^e = (\delta_1^e, \ldots, \delta_n^e)$, where $\delta_j^e = 1$ for $j \in e$ and $\delta_j^e = 0$ for $j \in e'$, and $|\mathbf{1}^e| = \sum_{j=1}^{n} \delta_j^e$.

If $r = (r_1, \ldots, r_n)$ is an n-dimensional vector, the notation r^e denotes either an n-dimensional vector $r^e = (r_1^e, \ldots, r_n^e)$ such that $r_j^e = r_j$ for $j \in e$ but $r_j^e = 0$ for $j \in e'$ or a vector of dimension $|\mathbf{1}^e|$ with components r_j, where j ranges over the set e. We shall also write $r = (r^e, r^{e'})$.

Suppose that $\varepsilon = (\varepsilon_1, \ldots, \varepsilon_n)$, $h = (h_1, \ldots, h_n)$, and $v = (v_1, \ldots, v_n)$, where $0 < \varepsilon_j < h_j, v_j > 0$ for $j = 1, \ldots, n$, and $\alpha = (\alpha_1, \ldots, \alpha_n)$. Then,

$$v^\alpha = v_1^{\alpha_1} \ldots v_n^{\alpha_n}, \qquad D_{v^e}^{1^e} F = \frac{\partial}{\partial v_{j_1}} \cdots \frac{\partial}{\partial v_{j_s}} F,$$

$$\int_{\varepsilon^e}^{h^e} F\, dv^e = \int_{\varepsilon_{j_1}}^{h_{j_1}} dv_{j_1} \ldots \int_{\varepsilon_{j_s}}^{h_{j_s}} F\, dv_{j_s} = \left(\prod_{k=1}^{s} \int_{\varepsilon_{j_k}}^{h_{j_k}} dv_{j_k} \right) F,$$

where F is some function and $\{j_1, \ldots, j_s\} = e$.

Let $\mathscr{E}$ denote a given finite set of vectors (points) $l = (l_1, \ldots, l_n)$ in the space E_l^n and let $\alpha = (\alpha_1, \ldots, \alpha_n)$ denote a vector.

We denote by $\mathscr{E}^e(\alpha)$ the set of all vectors $l \in \mathscr{E}$ for which $l_j \leqslant \alpha_j$ for all $j \in e'$ and we denote by $P\mathscr{E}^e(\alpha)$ the projection of the set $\mathscr{E}^e(\alpha)$ onto the coordinate plane $l^{e'} = 0$ (where $l_{j_1} = \ldots = l_{j_s} = 0$ and $\{j_1, \ldots, j_s\} = e'$); that is,

$$\mathscr{E}^e(\alpha) = \{l\colon\ l \in \mathscr{E},\ l_j \leqslant \alpha_j\ \ \forall j \in e'\},$$
$$P\mathscr{E}^e(\alpha) = \{l^e\colon\ l \in \mathscr{E}^e(\alpha)\}.$$

In particular, $\mathscr{E}^{e^n}(\alpha) = P\mathscr{E}^{e^n}(\alpha) = \mathscr{E}\ (e' = \varnothing)$, and $\mathscr{E}^{\varnothing}(\alpha)$ consists of those points $l \in \mathscr{E}$ for which $l_j \leqslant \alpha_j$ (for $j = 1, \ldots, n$). Obviously, $P\mathscr{E}^{\varnothing}(\alpha)$ either is the empty set or consists only of the coordinate origin.

We denote by $M^e(\alpha)$ the closed convex polyhedron spanned by the points $l^e \in P\mathscr{E}^e(\alpha)$, that is,

$$M^e(\alpha) = \left\{ r^e\colon r^e = \sum_{\gamma} \mu_\gamma l^{e,\gamma},\ \mu_\gamma \geqslant 0,\ \sum_{\gamma} \mu_\gamma = 1,\ l^{e,\gamma} \in P\mathscr{E}^e(\alpha) \right\}.$$

$M^e(\alpha)$ may be a polyhedron of any dimension not exceeding $|\mathbf{1}^e|$. In particular, it may consist of a single point or may even be the empty set.

13.3. The basic integral identity. The investigative method to be used in this section is based on the integral identity 7(112). If we now assume that $\varepsilon = (\varepsilon, \ldots, \varepsilon)$ (that is, all the components of the vector ε are equal), that $h = (h_1, \ldots, h_n)$ with $0 < \varepsilon < h_j$ for $j = 1, \ldots, n$, and that $f \in L^{\mathrm{loc}}(G)$, we have, by 7(112),

$$D^\alpha F(x;\ \varepsilon) = \sum_{e \subseteq e^n} (-1)^{|\alpha|} \mathscr{G}^e(x;\ \varepsilon,\ h), \tag{3}$$

where

$$F(x;\ \varepsilon) = \int_{E^n} f(x + y) \prod_{j=1}^{n} (\varepsilon^{-1}\Omega_j(y_j\varepsilon^{-1}))\, dy, \tag{4}$$

$$\mathcal{I}^e(x;\ \varepsilon,\ h) = \int_{\varepsilon^e}^{h^e} v^{-1^e}\Phi(x;\ v^e,\ h^{e'})\,dv^e, \tag{5}$$

$$\Phi(x;\ v^e,\ h^{e'}) = \int_{E^n} f(x+y)\left(\prod_{j\in e} v_j^{-1}D_{y_j}^{\alpha_j}M_j(y_jv_j^{-1})\right)\left(\prod_{j\in e'} h_j^{-1}D_{y_j}^{\alpha_j}\Omega_j(y_jh_j^{-1})\right)dy. \tag{6}$$

We note that $M_j(y_j) = D^{k_j}\mathcal{L}_j(y_j)$ for $j = 1, \ldots, n$, where the k_j are sufficiently large natural numbers and $\mathcal{L}_j(y_j)$ and $\Omega_j(y_j)$ are functions in the class $C_0^\infty(E^1)$ with supports satisfying the conditions of subsection 7.10.

If $G \in \underline{A}(\square,\ H)$, then equation (3) with $h_j \leqslant H_j$ (where $j = 1, \ldots, n$) is valid for every point x in G. However, we should note that the functions Ω_j and $\mathcal{L}_j$ are in general dependent on x although this fact is not brought out in the notation. More precisely, for points x belonging to the same subset G_λ which belongs to the covering of G in accordance with the definition of the class $\underline{A}(\square, H)$, we may assume that the kernels Ω_j and $\mathcal{L}_j$ are independent of x. For points x belonging to different G_λ, these kernels are in general different. As we switch from G_{λ_1} to G_{λ_2}, they are not continuously dependent on x. Since in what follows our treatment will be for $x \in G_\lambda$ for fixed λ, the dependence of these kernels on x does not have to be emphasized.

The basic formula (3) has the peculiarity that its right-hand member contains parameters h_j that the left-hand member does not depend on. The possible values of h_j are determined by the inequalities $\varepsilon < h_j \leqslant H_j$ (for $j = 1, \ldots, n$), where the H_j are the components of the vector H appearing in the definition of the class $\underline{A}(\square,\ H)$ to which the region G belongs.

Let us show that under certain assumptions regarding the function f and the region G, formula (3) can be simplified (the number of terms in the right-hand member can be reduced) by varying the values of h_j.

Suppose that $1 \leqslant k \leqslant n$, that $G \in \underline{A}\ (\square,\ H^{n-k})$, and that the function f has on G the generalized derivative $D^r f \in L_q(G)$ for $1 \leqslant q \leqslant \infty$. Suppose that $\frac{1}{q} + \alpha_j - r_j > 0$ for $j = 1, \ldots, n$, that is, that $r_j \leqslant \alpha_j$ if $1 \leqslant q < \infty$ or $r_j < \alpha_j$ if $q = \infty$. Under this assumption on G, the parameters $h_1, \ldots, h_k$ in formulas (3)-(6) can assume arbitrary values in the interval (ε, ∞).

Suppose that $e^k = \{1, \ldots, k\}$. Let us show now that those terms in the right-hand member of (3) that correspond to the sets e for which $e^k \cap e' \neq \varnothing$ (where $e' = e^n \setminus e$) approach zero as $h_1, \ldots, h_k \to \infty$.

It follows from formulas (5) and (6) and the definition of a generalized derivative that $\mathcal{G}^e(x;\ \varepsilon, h)$ can be represented in the form

$$\mathcal{G}^e = (-1)^{|r|}\left(\prod_{j\in e}\int_{\varepsilon}^{h_j} v^{-1-\alpha_j+r_j}\,dv_j\right)\int_{E^n} D^r f(x+y)$$

$$\times\left(\prod_{j\in e} v_j^{-1} D^{\alpha_j-r_j} M_j\left(y_j v_j^{-1}\right)\right)\left(\prod_{j\in e'} h_j^{-1-\alpha_j+r_j} D^{\alpha_j-r_j}\Omega_j\left(y_j h_j^{-1}\right)\right) dy.$$

Estimating the integral with respect to y with the aid of Hölder's inequality with exponent q and remembering that Ω_j and M_j are functions belonging to the class C_0^∞, we obtain

$$|\mathcal{G}^e| \leqslant \| D^r f\|_{q,G}\left(\prod_{j\in e}\| D^{\alpha_j-r_j}M_j\|_{q'}\right)\left(\prod_{j\in e'}\| D^{\alpha_j-r_j}\Omega_j\|_{q'}\right)$$

$$\times\left(\prod_{j\in e}\int_{\varepsilon}^{h_j} v_j^{-1-\frac{1}{q}-\alpha_j+r_j}\,dv_j\right)\left(\prod_{j\in e'} h^{-\frac{1}{q}-\alpha_j+r_j}\right)$$

$$\leqslant C(\varepsilon)\| D^r f\|_{q,G}\left(\prod_{j\in e'\setminus(e'\cap e^k)} h_j^{-\frac{1}{q}-\alpha_j+r_j}\right)\left(\prod_{j\in e'\cap e^k} h_j^{-\frac{1}{q}-\alpha_j+r_j}\right).$$

If $e' \cap e^k \neq \varnothing$ and $\frac{1}{q} + \alpha_j - r_j > 0$ (for $j = 1, \ldots, n$), then, as $h_1, \ldots, h_k \to \infty$, the last factor in the right-hand member of the last inequality approaches zero and the other factors are, for fixed $\varepsilon > 0$ and h_j (for $j = k+1, \ldots, n$), bounded. Our assertion now follows.

Thus, as we let $h_1, \ldots, h_k$ approach ∞, the only terms in the right-hand member of (3) that remain are those corresponding to sets e for which $e' \cap e^k = \varnothing$, that is, for which $e^k \subseteq e$.

Therefore, in the present case, formula (3) takes the form

$$D^{\alpha} F(x; \varepsilon) = \lim_{h_1, \ldots, h_k \to \infty} \sum_{e^k \subseteq e \subseteq e^n} (-1)^{|\alpha|} \mathcal{G}^e(x; \varepsilon, h). \quad (7)$$

In particular, if $k = n$ and $G \in \underline{A}(\square)$, we have

$$D^{\alpha} F(x; \varepsilon) = \lim_{h_1, \ldots, h_n \to \infty} (-1)^{|\alpha|} \mathcal{G}^{e^n}(x; \varepsilon, h). \quad (8)$$

We note that formula (3) can be regarded as the special case of formula (7) corresponding to $k = 0$ (i.e., $e^k = \varnothing$).

13.4. Transformation of the functions $\mathcal{G}^e(x; \varepsilon, h)$. Let us put the functions $\mathcal{G}^e(x; \varepsilon, h)$ appearing in the right-hand member of formula (7) in a form that will be more convenient for us.

13.4.1. Suppose that we are given a set $\mathcal{E}$ of points $l = (l_1, \ldots, l_n)$ and a point $\alpha = (\alpha_1, \ldots, \alpha_n)$. Let e denote any nonempty subset of the set $e^n = \{1, \ldots, n\}$ and let $M^e(\alpha)$ denote the closed convex polyhedron spanned by the points $l^e \in P\mathcal{E}^e(\alpha)$. Without loss of generality, we may assume that $e = e^m = \{1, \ldots, m\}$, where $1 \leqslant m \leqslant n$. Then, $l^e = (l_1^e, \ldots, l_m^e)$.

Let us suppose also that $\alpha^e = (\alpha_1, \ldots, \alpha_m) \in M^e(\alpha)$. Let us draw the simplex of lowest dimension s, where $0 \leqslant s \leqslant m$, with vertices at the points $l^e \in P\mathcal{E}^e(\alpha)$ that contains the point α^e (here, $s = 0$ means that α^e coincides with some point $l^e \in P\mathcal{E}^e(\alpha)$). We denote the vertices of this simplex by $l^{e, i}$ (for $i = 1, \ldots, s+1$). Then, there exist numbers $\mu_i > 0$ (for $i = 1, \ldots, s+1$) such that

$$\sum_{i=1}^{s+1} \mu_i = 1 \quad (9)$$

and

$$\sum_{i=1}^{s+1} \mu_i \left(l^{e,\,i} - \alpha^e\right) = 0. \tag{10}$$

We note that the condition on the dimension of the simplex containing the point α^e implies that any s vectors belonging to the set $\{l^{e,\,i} - \alpha^e\}_{i=1}^{s+1}$ are linearly independent.

To every point $l^{e,\,i}$, for $1 \leqslant i \leqslant s+1$, we assign a definite point $l^i = (l_1^i, \ldots, l_n^i) \in \mathscr{E}^e(\alpha)$ whose projection onto the plane $l_{m+1} = \ldots = l_n = 0$ coincides with $l^{e,\,i}$ (if there are several such points, we choose one of them). Consequently, $l_j^i = l_j^{e,\,i}$ (for $j = 1, \ldots, m$ and $i = 1, \ldots, s+1$).

Let us now write equation (10) in the form of the equivalent system of equations

$$\sum_{i=1}^{s+1} \mu_i \left(l_j^i - \alpha_j\right) = 0 \qquad (j = 1, \ldots, m), \tag{11}$$

where the l_j^i are the components of the vectors l^i (for $i = 1, \ldots, s+1$).

We define the matrix

$$A = \begin{bmatrix} l_1^1 - \alpha_1 & \ldots & l_1^{s+1} - \alpha_1 \\ l_2^1 - \alpha_2 & \ldots & l_2^{s+1} - \alpha_2 \\ \vdots & & \vdots \\ l_m^1 - \alpha_m & \ldots & l_m^{s+1} - \alpha_m \end{bmatrix}, \tag{12}$$

corresponding to the system (11). By virtue of what was said above, the rank of this matrix is equal to s. To be specific, we assume that

$$\Delta_s = \begin{vmatrix} l_1^1 - \alpha_1 & \ldots & l_1^s - \alpha_1 \\ \vdots & \vdots & \vdots \\ l_s^1 - \alpha_s & \ldots & l_s^s - \alpha_s \end{vmatrix} \neq 0. \qquad (13)$$

Let us perform s successive transformations on the matrix A that will put it into a certain special form. The sequence of transformations that we shall perform later on the function $\mathcal{J}^e$ is closely connected with this sequence of transformations.

We introduce s matrices $(m \times m)$ of the following form:

$$B_i = \begin{bmatrix} 1 & 0 & \ldots & 0 & a_{1i} & 0 & \ldots & 0 \\ 0 & 1 & \ldots & 0 & a_{2i} & 0 & \ldots & 0 \\ \cdot & \cdot & \cdot & \cdot & \cdot & \cdot & \cdot & \cdot \\ 0 & 0 & \ldots & 1 & a_{i-1,i} & 0 & \ldots & 0 \\ 0 & 0 & \ldots & 0 & 1 & 0 & \ldots & 0 \\ 0 & 0 & \ldots & 0 & a_{i+1,i} & 1 & \ldots & 0 \\ \cdot & \cdot & \cdot & \cdot & \cdot & \cdot & \cdot & \cdot \\ 0 & 0 & \ldots & 0 & a_{mi} & 0 & \ldots & 1 \end{bmatrix} \quad (i = 1, \ldots, s), \qquad (14)$$

where the a_{ij} are numbers to be determined below.

Obviously,

$$|B_i| = 1 \qquad (i = 1, \ldots, s). \qquad (15)$$

Let us define

$$B^i = B_i B_{i-1} \ldots B_1 \qquad (i = 1, \ldots, s). \qquad (16)$$

This matrix has the following form:

$$B^i = \begin{bmatrix} a_{11}^{(i)} & a_{12}^{(i)} & \ldots & a_{1i}^{(i)} & 0 & \ldots & 0 \\ a_{21}^{(i)} & a_{22}^{(i)} & \ldots & a_{2i}^{(i)} & 0 & \ldots & 0 \\ \cdot & \cdot & \cdot & \cdot & \cdot & \cdot & \cdot \\ a_{i1}^{(i)} & a_{i2}^{(i)} & \ldots & a_{ii}^{(i)} & 0 & \ldots & 0 \\ a_{i+1,i}^{(i)} & a_{i+1,2}^{(i)} & \ldots & a_{i+1,i}^{(i)} & 1 & \ldots & 0 \\ \cdot & \cdot & \cdot & \cdot & \cdot & \cdot & \cdot \\ a_{m1}^{(i)} & a_{m2}^{(i)} & \ldots & a_{mi}^{(i)} & 0 & \ldots & 1 \end{bmatrix}. \qquad (17)$$

We note that the elements of the ith column of the matrix B^i coincide with the elements of the ith column of the matrix B_i; that is,

$$a_{ji}^{(i)}=a_{ji} \quad (j=1, \ldots, m),$$

for $j = 1, \ldots, m$.

The elements of the matrices B_i (that is, the numbers a_{ji}) can be chosen in such a way that the matrices B^iA for $i = 1, \ldots, s$ will be matrices of the form

$$B^iA=\begin{bmatrix} b_{11} & b_{12} & b_{13} & \cdots & b_{1i} & c_{1,i+1} & \cdots & c_{1,s+1} \\ 0 & b_{22} & b_{23} & \cdots & b_{2i} & c_{2,i+1} & \cdots & c_{2,s+1} \\ 0 & 0 & b_{33} & \cdots & b_{3i} & c_{3,i+1} & \cdots & c_{3,s+1} \\ \cdot & \cdot & \cdot & \cdot & \cdot & \cdot & \cdot & \cdot \\ 0 & 0 & 0 & \cdots & b_{ii} & c_{i,i+1} & \cdots & c_{i,s+1} \\ 0 & 0 & 0 & \cdots & 0 & c_{i+1,i+1} & \cdots & c_{i+1,s+1} \\ \cdot & \cdot & \cdot & \cdot & \cdot & \cdot & \cdot & \cdot \\ 0 & 0 & 0 & \cdots & 0 & c_{m,i+1} & \cdots & c_{m,s+1} \end{bmatrix} \quad (i=1, \ldots, s), \tag{18}$$

where $b_{jk} \neq 0$ (for $j \leqslant k \leqslant i$). For arbitrary fixed j, the numbers b_{jk} (for $j + 1 \leqslant k \leqslant i$) have the same signs, opposite to the sign of b_{jj}, and in addition

$$|b_{jk}| \leqslant \frac{\mu_j |b_{jj}|}{\sum\limits_{t=j+1}^{s+1} \mu_t} \quad (j+1 \leqslant k \leqslant i). \tag{19}$$

We can arrange for the matrix B^iA to have these properties as follows: first of all, by virtue of (13), we may assume that the element $b_{11} = l_1^1 - \alpha_1$ of the matrix A is nonzero. Then, if we multiply A on the left by $B^1 = B_1$ (this transformation actually consists in adding the first row multiplied by the numbers a_{j1} to the rows numbered $j = 2, \ldots, m$ respectively), we arrange, by a certain choice of the numbers a_{j1} (for $j = 1, \ldots, m$) to have all the elements of the first column of the matrix B^1A other than b_{11} equal to 0.

By virtue of (13) and the nature of the transformation B^1, we can assume that the element b_{22} of the matrix $B^1 A$ (that is, the element in the second row and second column of that matrix) is nonzero. Multiplying $B^1 A$ on the left by B_2, in which the numbers a_{j2} are appropriately chosen, we arrange for all elements of the second column lying below b_{22} to be equal to zero and for the element b_{12} to be opposite in sign to b_{11} and to satisfy inequality (19).

Continuing in an analogous manner, after the ith step of the process, we arrive at the matrix (18). We mention that multiplication of the matrix $B^{i-1}A$ on the left by B_i does not change the first $i-1$ columns of the matrix $B^{i-1}A$.

Equations (11) tell us that the vector $\begin{pmatrix} \mu_1 \\ \vdots \\ \mu_{s+1} \end{pmatrix}$ is a solution of the system $A\begin{pmatrix} x_1 \\ \vdots \\ x_{s+1} \end{pmatrix} = 0$. Since transformation of the matrix A with the matrix B^i converts the original system into a system equivalent to it, the system $(B^i A)\begin{pmatrix} x_1 \\ \vdots \\ x_{s+1} \end{pmatrix} = 0$ has the same solution. A detailed writing out of this fact (with the use of the representation $B^i A$ given by formula (18)) leads us to the following equations:

$$\left.\begin{aligned} \sum_{k=j}^{i} b_{jk}\mu_k + \sum_{k=i+1}^{s+1} c_{jk}\mu_k &= 0 \quad (j = 1, \ldots, i), \\ \sum_{k=i+1}^{s+1} c_{jk}\mu_k &= 0 \quad (j = i+1, \ldots, m) \end{aligned}\right\} \quad (i = 1, \ldots, s). \tag{20}$$

The final matrix $B^s A$ has the form

$$B^s A = \begin{bmatrix} b_{11} & b_{12} & \ldots & b_{1s} & b_{1,\,s+1} \\ 0 & b_{22} & \ldots & b_{2s} & b_{2,\,s+1} \\ \cdot & \cdot & \cdot & \cdot & \cdot \\ 0 & 0 & \ldots & b_{ss} & b_{s,\,s+1} \\ 0 & 0 & \ldots & 0 & 0 \\ \cdot & \cdot & \cdot & \cdot & \cdot \\ 0 & 0 & \ldots & 0 & 0 \end{bmatrix}. \tag{21}$$

That only zeros appear at the intersections of the rows numbered $s+1, \ldots, m$ with the columns numbered $1, \ldots, s$ follows from the nature of the transformations described above. That the last elements of these rows are equal to zero is a consequence of the fact that the rank of the matrix B^sA is equal to s (since the transformations made do not change the rank).

The elements $b_{j,\,s+1}$ (for $j = 1, \ldots, s$) in the last column of the matrix, like the elements b_{jk} (for $k = j+1, \ldots, s$) in the preceding columns, are nonzero and have signs opposite to the signs of the b_{jj} (for $j = 1, \ldots, s$). This follows from equations (20) with $i = s$ and inequalities (19).

We immediately obtain from the definition of the product of the matrix B^i and the matrix A the following equations for the elements of the ith column of the matrix B^iA:

$$\begin{gathered}\sum_{j=1}^{i} a_{kj}^{(i)}(l_j^i - \alpha_j) = b_{ki} \quad (k = 1, \ldots, i),\\ \sum_{j=1}^{i} a_{kj}^{(i)}(l_j^i - \alpha_j) + (l_k^i - \alpha_k) = 0 \quad (k = i+1, \ldots, m).\end{gathered} \quad (i = 1, \ldots, s). \tag{22}$$

Analogously, for the elements of the $(s+1)$st column of the matrix B^sA,

$$\begin{gathered}\sum_{j=1}^{s} a_{kj}^{(s)}(l_j^{s+1} - \alpha_j) = b_{k,\,s+1} \quad (k = 1, \ldots, s),\\ \sum_{j=1}^{s} a_{kj}^{(s)}(l_j^{s+1} - \alpha_j) + (l_k^{s+1} - \alpha_k) = 0 \quad (k = s+1, \ldots, m).\end{gathered} \tag{23}$$

13.4.2. Let us now see about transforming the function $\mathcal{G}^e = \mathcal{G}^e(x;\, \varepsilon,\, h)$. First, let us introduce some notation.

We denote by $\chi_1(t)$ and $\chi_{-1}(t)$ the functions defined on E_+^1 (that is, the interval $0 < t < \infty$) by

$$\chi_1(t) = \begin{cases} 1 & \text{for} \quad 0 < t < 1, \\ \frac{1}{2} & \text{for} \quad t = 1, \\ 0 & \text{for} \quad t > 1, \end{cases} \tag{24}$$
$$\chi_{-1}(t) = 1 - \chi_1(t), \qquad 0 < t < \infty.$$

Let ε and h denote positive numbers such that $0 < \varepsilon < h$. Then,

$$\chi^{\varepsilon,\, h}(t) = \chi_1\left(\frac{\varepsilon}{t}\right)\chi_{-1}\left(\frac{h}{t}\right) = \begin{cases} 1 & \text{for} \quad t \in (\varepsilon,\ h), \\ \frac{1}{2} & \text{for} \quad t = \varepsilon = h, \\ 0 & \text{for} \quad t \notin [\varepsilon,\ h]. \end{cases} \tag{25}$$

Let us define

$$\begin{aligned} \theta_{ji} &= \operatorname{sign} b_{ji}, \\ \chi_{ji}(t) &= \chi_{\theta_{ji}}(t) \end{aligned} \qquad (j = 1, \ldots, s; \quad i = j, \ldots, s + 1), \tag{26}$$

where the b_{ji} are elements of the matrix $B^s A$ (see (21)).

By virtue of the properties of this matrix,

$$\theta_{jj} = -\theta_{ji} \qquad (i = j + 1, \ldots, s + 1; \quad j = 1, \ldots, s).$$

Therefore, for arbitrary t in $0 < t < \infty$, we have

$$\begin{aligned} \chi_{jj}(t) + \chi_{ji}(t) &= 1, \\ \chi_{ji}(t) &= \chi_{jk}(t) \end{aligned} \quad (i,\ k = j + 1, \ldots, s + 1; \ j = 1, \ldots, s). \tag{27}$$

Let $t_1, \ldots, t_s$ denote arbitrary positive numbers. On the basis of formulas (27), we have

$$\begin{aligned} 1 &= \chi_{11}(t_1) + \chi_{12}(t_1) = \chi_{11}(t_1) + \chi_{12}(t_1)[\chi_{22}(t_2) + \chi_{23}(t_2)] \\ &= \chi_{11}(t_1) + \chi_{12}(t_1)\chi_{22}(t_2) + \chi_{13}(t_1)\chi_{23}(t_2)[\chi_{33}(t_3) + \chi_{34}(t_3)]. \end{aligned}$$

Continuing this chain of equations, we arrive at the following identity (in $t_1, \ldots, t_s$):

$$1 = \chi_{11}(t_1) + \chi_{12}(t_1)\chi_{22}(t_2) + \chi_{13}(t_1)\chi_{23}(t_2)\chi_{33}(t) + \ldots$$
$$\ldots + \chi_{1s}(t_1)\chi_{2s}(t_2)\ldots\chi_{ss}(t_s) + \chi_{1,s+1}(t_1)\chi_{2,s+1}(t_2)\ldots\chi_{s,s+1}(t_s). \tag{28}$$

In the notation that we have adopted, we now write the function $\mathcal{G}^e$ defined by formulas (5) and (6) in the form

$$\mathcal{G}^e = \left(\prod_{j=1}^{m} \int_{E_+^1} v_j^{-1}\chi^{e,h_j}(v_j)\,dv_j\right)\Phi(x;\ v_1, \ldots, v_m, h^{e'}), \tag{29}$$

where

$$\Phi(x;\ v_1, \ldots, v_m, h^{e'})$$
$$= \int_{E^n} f(x+y)\left(\prod_{j=1}^{m} v_j^{-1} D_{y_j}^{\alpha_j} D_j^{k_j} \mathcal{L}_j(y_j v_j^{-1})\right)\left(\prod_{j=m+1}^{n} h_j^{-1} D_{y_j}^{\alpha_j}\Omega(y_j h_j^{-1})\right)dy. \tag{30}$$

Here, we took into account the facts that $e = e^m = \{1, \ldots, m\}$ and $e' = \{m+1, \ldots, n\}$.

Let us multiply the integrand in (29) by the right-hand member of (28), taking $t_1 = v_1 v_2^{-a_{21}} \ldots v_m^{-a_{m+1}}$ (where the a_{j1} are the elements of the first column of the matrix B_1 (see (14)), $t_2 = t_1^{-a_{12}} v_2 v_3^{-a_{32}} \ldots v_m^{-a_{m2}}$ (where the a_{j2} are the elements of the second column of the matrix B_2), $\ldots$, and finally

$$t_s = t_1^{-a_{1s}} t_2^{-a_{2s}} \ldots t_{s-1}^{-a_{s-1,s}} v_s v_{s+1}^{-a_{s+1,s}} \ldots v_m^{-a_{ms}}$$

(where the a_{js} are the elements of the matrix B_s). In accordance with the expansion (28), we obtain a representation of $\mathcal{G}^e$ in the form of a sum of integrals:

$$\mathcal{J}^e = \sum_{i=1}^{s+1} I_i^e. \tag{31}$$

In each of these integrals, we make a change of variable. For $i \leqslant s$, we set in I_i^e

$$\begin{aligned}
v_1 &= t_1 v_2^{a_{21}} \dots v_m^{a_{m1}}, \\
v_2 &= t_1^{a_{12}} t_2 v_3^{a_{32}} \dots v_m^{a_{m2}}, \\
&\dots\dots\dots\dots\dots\dots \\
v_i &= t_1^{a_{1i}} t_2^{a_{2i}} \dots t_{i-1}^{a_{i-1,i}} t_i v_{i+1}^{a_{i+1,i}} \dots v_m^{a_{mi}}, \\
v_{i+1} &= v_{i+1}, \\
&\dots\dots\dots \\
v_m &= v_m,
\end{aligned} \tag{32}$$

where the a_{jq} are the elements of the qth column of the matrix B_q. In I_{s+1}^e, we make the same change as in I_s^e. Obviously, the transformation (32) is the product of i elementary transformations, in the first of which we replace only v_1, in the second only v_2, ..., in the ith only v_i. The Jacobians of these transformations are equal respectively to $\frac{v_1}{t_1}, \frac{v_2}{t_2}, \dots, \frac{v_i}{t_i}$.

We note that, at the first step, the variable v_1 is replaced with $\psi_1^{(1)} = t_1 v_2^{a_{21}} \dots v_m^{a_{m1}}$; at the second step, $\psi_1^{(1)}$ and v_2 are replaced respectively with $\psi_1^{(2)}$ (since $\psi_1^{(1)}$ contains v_2) and $\psi_2^{(2)}$, where

$$\psi_j^{(2)} = \prod_{k=1}^{2} t_k^{a_{kj}^{(2)}} \prod_{k=3}^{m} v_k^{a_{kj}^{(2)}},$$

the $a_{kj}^{(2)}$ are the elements of the matrix B^2, etc. After the ith step, the variables $v_1, \dots, v_i$ are replaced respectively with $\psi_1^{(i)}, \dots, \psi_i^{(i)}$, where

$$\psi_j^{(i)} = \prod_{k=1}^{i} t_k^{a_{kj}^{(i)}} \prod_{k=i+1}^{m} v_k^{a_{kj}^{(i)}} \qquad (j = 1, \dots, i), \tag{33}$$

the $a_{kj}^{(i)}$ being elements of the matrix B^i.

After these substitutions, each of the integrals in the right-hand member of (31) acquires the form

$$I_i^e = \left(\prod_{j=1}^{q} \int_{E_+^1} t_j^{-1} \chi_{ji}(t_j)\, dt_j \right) \left(\prod_{j=q+1}^{m} \int_{E_+^1} v_j^{-1} \chi^{\varepsilon, h_j}(v_j)\, dv_j \right) \times \left(\prod_{j=1}^{q} \chi^{\varepsilon, h_j}(\psi_j^{(q)}) \right) \Phi(x;\ \psi_1^{(q)}, \ldots, \psi_q^{(q)}, v_{q+1}, \ldots, v_m, h^e), \tag{34}$$

where

$$q = \min(i, s) \qquad (1 \leqslant i \leqslant s+1) \tag{35}$$

($q = i$ for $1 \leqslant i \leqslant s$ and $q = s$ for $i = s+1$), and the $\psi_j^{(q)}$ (for $j = 1, \ldots, q$) are determined from formulas (33).

Thus, $\mathcal{I}^e$ is represented in the form of the sum of $s+1$ terms. We note that the same number of vectors $l^{e,i}$ participate in the representation of α^e by formula (10). In each term of I_i^e, we transform the function Φ in such a way that the integrand will contain not $f(x+y)$ but the derivative $D^{l^i} f(x+y)$, where $l^i = (l_1^i, \ldots, l_n^i)$ is the vector in the set $\mathscr{E}^e(\alpha)$ corresponding to the vector $l^{e,i}$ in the sense indicated in sub-subsection 13.4.1.

Assuming that the numbers k_j appearing in the definition of the kernels Ω_j are sufficiently large that

$$k_j + \alpha_j - l_j^i > 0 \qquad (j = 1, \ldots, m; \quad i = 1, \ldots, s+1) \tag{36}$$

and noting that (since $l^i \in \mathscr{E}^e(\alpha)$)

$$\alpha_j - l_j^i \geqslant 0 \qquad (j = m+1, \ldots, n; \quad i = 1, \ldots, s+1),$$

we obtain from equation (30) on the basis of the definition 6(1) of a generalized derivative

$$\Phi\left(x;\ \psi_1^{(q)},\ \ldots,\ \psi_q^{(q)},\ v_{q+1},\ \ldots,\ v_m,\ h^{e'}\right)$$

$$= \int_{E^n} f(x+y)\left(\prod_{j=1}^{q}\left[\psi_j^{(q)}\right]^{-1-\alpha_j} D_j^{\alpha_j+k_j}\mathscr{L}_j\left(\frac{y_j}{\psi_j^{(q)}}\right)\right)$$

$$\times\left(\prod_{j=q+1}^{m} v_j^{-1-\alpha_j} D_j^{\alpha_j+k_j}\mathscr{L}_j\left(y_j v_j^{-1}\right)\right)\left(\prod_{j=m+1}^{n} h_j^{-1-\alpha_j} D_j^{\alpha_j}\Omega_j\left(y_j h_j^{-1}\right)\right)dy$$

$$=(-1)^{|l^i|}\prod_{j=1}^{q}\left[\psi_j^{(q)}\right]^{l_j^i-\alpha_j}\prod_{j=q+1}^{m} v_j^{l_j^i-\alpha_j}\prod_{j=m+1}^{n} h_j^{l_j^i-\alpha_j}$$

$$\times\int_{E^n} D^{l^i} f(x+y)\left(\prod_{j=1}^{q}\left[\psi_j^{(q)}\right]^{-1} P_{ji}\left(\frac{y_j}{\psi_j^{(q)}}\right)\right)$$

$$\times\left(\prod_{j=q+1}^{m} v_j^{-1} P_{ji}\left(y_j v_j^{-1}\right)\right)\left(\prod_{j=m+1}^{n} h_j^{-1} Q_{ji}\left(y_j h_j^{-1}\right)\right) dy, \tag{37}$$

where

$$P_{ji}(y_j) = D^{k_j+\alpha_j-l_j^i}\mathscr{L}_j(y_j), \qquad Q_{ji}(y_j) = D^{\alpha_j-l_j^i}\Omega_j(y_j). \tag{38}$$

Since $\mathscr{L}_j \in C_0^\infty(E^1)$, if inequalities (36) are satisfied, the functions $P_{ji}(y_j)$ satisfy the conditions

$$\int_{E^1} P_{ji}(y_j)\, dy_j = 0 \qquad (j=1,\ \ldots,\ m;\ i=1,\ \ldots,\ s+1). \tag{39}$$

Equations (39) will be used a great deal in what follows.

The coefficient of the integral in the right-hand member of (37) is, on the basis of equations (33) and (22) (for $i \leqslant s$) or (23) (for $i = s + 1$), converted to the form

$$(-1)^{|l^i|}\prod_{j=1}^{q}\left[\psi_j^{(q)}\right]^{l_j^i-\alpha_j}\prod_{j=q+1}^{m} v_j^{l_j^i-\alpha_j}\prod_{j=m+1}^{n} h_j^{l_j^i-\alpha_j}$$

$$= C_i(h^{e'}) \prod_{j=1}^{q} \left(\prod_{k=1}^{q} t_k^{a_{kj}^{(q)}} \prod_{k=q+1}^{m} v_k^{a_{kj}^{(q)}} \right)^{l_j^i - \alpha_j} \prod_{j=q+1}^{m} v_j^{l_j^i - \alpha_j}$$

$$= C_i(h^{e'}) \prod_{k=1}^{q} t_k^{\sum_{j=1}^{q} a_{kj}^{(q)} (l_j^i - \alpha_j)} \prod_{k=q+1}^{m} v_k^{\sum_{j=1}^{q} a_{kj}^{(q)} (l_j^i - \alpha_j) + l_k^i - \alpha_k}$$

$$= C_i(h^{e'}) \prod_{k=1}^{q} t_k^{b_{ki}}, \tag{40}$$

where the b_{ki} are elements of the ith column of the matrix (21).

On the basis of (31), (34), (37), and (40), we finally get the following representation for $\mathcal{J}^e$:

$$\mathcal{J}^e(x;\ \varepsilon,\ h) = \sum_{i=1}^{s+1} I_i^e(x;\ \varepsilon,\ h), \tag{41}$$

where

$$I_i^e(x;\ \varepsilon,\ h)$$
$$= C_i(h^{e'}) \left(\prod_{j=1}^{q} \int_{E_+^1} t_j^{b_{ji}-1} \chi_{ji}(t_j)\, dt_j \right) \left(\prod_{j=q+1}^{m} \int_{E_+^1} v_j^{-1} \chi^{\varepsilon, h_j}(v_j)\, dv_j \right)$$
$$\times \left(\prod_{j=1}^{q} \chi^{\varepsilon, h_j}(\psi_j^{(q)}) \right) \Phi_i(x;\ \psi_1^{(q)}, \ldots, \psi_q^{(q)}, v_{q+1}, \ldots, v_m, h^{e'}), \tag{42}$$

$$\Phi_i = \int_{E^n} D^{l^i} f(x+y) \left(\prod_{j=1}^{q} [\psi_j^{(q)}]^{-1} P_{ji}\left(\frac{y_j}{\psi_j^{(q)}} \right) \right)$$
$$\times \left(\prod_{j=q+1}^{m} v_j^{-1} P_{ji}(y_j v_j^{-1}) \right) \left(\prod_{j=m+1}^{n} h_j^{-1} Q_{ji}(y_j h_j^{-1}) \right) dy, \tag{43}$$

q depends on i and is defined by formula (35), and s (that is, the number of terms in (41)) depends on e.

We conclude this subsection by noting that

$$\int_{E_+^1} t_j^{b_{ji}-1} \chi_{ji}(t_j)\, dt_j < \infty \quad (j=1, \ldots, q;\ i=1, \ldots, s+1). \tag{44}$$

This is true because, if $b_{ji} > 0$, then $\theta_{ji} = \operatorname{sgn} b_{ji} = 1$, $\chi_{ji}(t_j) = \chi_1(t_j)$, and

$$\int_{E_+^1} t_j^{b_{ji}-1} \chi_{ji}(t_j)\, dt_j = \int_0^1 t_j^{b_{ji}-1}\, dt_j < \infty,$$

whereas, if $b_{ji} < 0$, then $\theta_{ji} = -1$, $\chi_{ji}(t_j) = \chi_{-1}(t_j)$, and

$$\int_{E_+^1} t_j^{b_{ji}-1} \chi_{ji}(t_j)\, dt_j = \int_1^\infty t_j^{b_{ji}-1}\, dt_j < \infty.$$

13.5. Estimates of the basic integrals. In what follows, we shall look at the basic identities (7) and (41) separately for each G_λ, for $1 \leqslant \lambda \leqslant \Lambda$. As was pointed out in subsection 13.3, this has the advantage that the kernels appearing in these identities can be assumed independent of x, a fact of no small importance with various estimates.

The basic purpose of the present subsection is to obtain estimates for $\| I_i^e(\cdot, \varepsilon, h)\|_{p, G_\lambda}$, where $e \neq \varnothing$ and $1 < p < \infty$, and for analogous norms of the expressions obtained from $I_i^e(x; \varepsilon, h)$ with limiting values of the parameters h_j and ε as $h_j \to \infty$ and $\varepsilon \to 0$.

For convenience, we shall extend the function $I_i^e(x; \varepsilon, h)$ (defined on G_λ) to a function defined on the entire space E^n. We define

$$\mathcal{F}_i(x) = \begin{cases} D^{l^i} f(x), & x \in G, \\ 0, & x \in E^n \setminus G. \end{cases} \tag{45}$$

Obviously,

$$\| \mathcal{F}_i \|_p = \| \mathcal{F}_i \|_{p,\, E^n} = \left\| D^{l^i} f \right\|_{p,\, G}. \tag{46}$$

If in formulas (43) and (42) we replace the function $D^{l^i} f$ with $\mathcal{F}_i$, we obtain the functions $\tilde{\Phi}_i$ and $\tilde{I}_i^e(x;\ \varepsilon,\ h)$, which we may assume defined (by the same formulas) for all $x \in E^n$ and coinciding respectively with Φ_i and $I_i^e(x;\ \varepsilon,\ h)$ for $x \in G_\lambda$. Thus, the problem reduces to obtaining the corresponding estimates for $\tilde{I}_i^e$ in the norm of $L_p(E^n)$.

Let us fix $e \neq \varnothing$. Without loss of generality, we may assume that $e = e^m = \{1, \ldots, m\}$, where $1 \leqslant m \leqslant n$. With an eye to obtaining estimates for the functions I_i^e appearing, by virtue of (41), in the basic formula (7), we shall assume (in accordance with that formula) that the number k in $0 \leqslant k \leqslant m$ and the set $e^k = \{1, \ldots, k\}$ are given (if $k = 0$, we have $e^k = \varnothing$).

For brevity, let us assume that the vector h has the following form in formulas (3), (7), (8), and (41):

$$h = (\overbrace{\eta, \ldots, \eta}^{k}, \overbrace{\zeta, \ldots, \zeta}^{n-k}),$$

that is, that $h_j = \eta$ (for $j = 1, \ldots, k$) but $h_j = \zeta$ (for $j = k + 1, \ldots, n$).

Since the indices e, i, and q and the parameter ζ will be assumed fixed in what follows (only ε and η can be variable parameters), it will be expedient to use simpler notation.

Let us set

$$I_{\varepsilon\eta}(x) = \left(\prod_{j=1}^{q} \int_{E_+^1} t_j^{\beta_j - 1} \mu_j(t_j)\, dt_j \right) \left(\prod_{j=q+1}^{m} \int_{E_+^1} v_j^{-1}\, dv_j \right) \Psi_{\varepsilon\eta}(u)\, \Phi(x;\ u,\ \zeta), \tag{47}$$

where

$$\mu_j(t_j) = \chi_{j,i}(t_j), \quad \beta_j = b_{j,l} \qquad (j = 1, \ldots, q). \tag{48}$$

By virtue of (44),

$$\int_{E^1_+} t_j^{\beta_j-1}\mu_j(t_j)\,dt_j < \infty \qquad (j=1,\ \ldots,\ q), \tag{49}$$

$$\Psi_{\varepsilon\eta}(u) = \prod_{j=1}^{m} \chi^{\varepsilon,\,h_j}(u_j), \tag{50}$$

$$h_j = \eta \quad (j=1,\ \ldots,\ k), \qquad h_j = \zeta \quad (j=k+1,\ \ldots,\ n),$$
$$u = (\psi_1,\ \ldots,\ \psi_q,\ v_{q+1},\ \ldots,\ v_m), \tag{51}$$
$$\psi_j = \prod_{i=1}^{q} t_i^{a_{ji}} \prod_{i=q+1}^{m} v_i^{a_{ji}} \qquad (j=1,\ \ldots,\ q),$$

where the a_{ji} are real numbers,

$$\Phi(x;\ u,\ \zeta) = \int_{E^n} \mathscr{F}(x-y)\,M(y;\ u,\ \zeta)\,dy, \tag{52}$$

$$M(y;\ u,\ \zeta) = \prod_{j=1}^{m} u_j^{-1}P_j(y_ju_j^{-1}) \prod_{j=m+1}^{n} \zeta^{-1}Q_j(y_j\zeta^{-1}), \tag{53}$$

where $P_j(y_j) = -P_{ji}(-y_j)$ and $Q_j(y_j) = -Q_{ji}(-y_j)$ are functions of the class $C_0^\infty(E^1)$. Also, the functions P_j satisfy conditions (39):

$$\int_{E^1} P_j(y_j)\,dy_j = 0 \qquad (j=1,\ \ldots,\ m). \tag{54}$$

One can easily see that the function $I_{\varepsilon\eta}(x)$ takes the same form as $\tilde{I}_i^e(x;\ \varepsilon,\ h)$ (we replace y with $-y$ in the integral defining the function $\tilde{\Phi}_i$). Therefore, it will be sufficient to obtain estimates for $I_{\varepsilon\eta}$.

Let us also set

$$\Psi_\varepsilon(u) = \Psi_{\varepsilon\infty}(u) = \prod_{j=1}^{m} \chi_1(\varepsilon u_j^{-1}) \prod_{j=k+1}^{m} \chi_{-1}(\zeta u_j^{-1}) \tag{55}$$

and let us define the function $I_\varepsilon(x)$ by formula (47) with $\Psi_{\varepsilon\eta}(u)$ replaced with $\Psi_\varepsilon(u)$. We note that, if $k = 0$, we have $I_{\varepsilon\eta}(x) = I_\varepsilon(x)$.

13.5.1. Lemma. *If* $\mathscr{F} \in L_p(E^n)$, *where* $1 < p < \infty$, *then, for arbitrary fixed* ε *and* η *such that* $0 < \varepsilon < \eta$, *the functions* $I_{\varepsilon\eta}(x)$ *and* $I_\varepsilon(x)$ *exist and*

$$\lim_{\eta\to\infty} \sup_{x\in E^n} | I_\varepsilon(x) - I_{\varepsilon\eta}(x) | = 0. \tag{56}$$

PROOF. Let us first show that $I_{\varepsilon\eta}(x)$ and $I_\varepsilon(x)$ are bounded by a constant independent of x.

Let us use Hölder's inequality to estimate the function Φ appearing in the right-hand member of (47). We recall that P_j and Q_j are bounded and of compact support. Then, on the basis of (52) and (53), we obtain

$$| \Phi | \leqslant C_1 \| \mathscr{F} \|_p \zeta^{\frac{m-n}{p}} \prod_{j=1}^{m} u_j^{-\frac{1}{p}}. \tag{57}$$

Furthermore, by virtue of the inequalities

$$|\chi_{-1}(h_j u_j^{-1})| \leqslant 1, \qquad \left| u^{-\frac{1}{p}} \chi_1(\varepsilon u_j^{-1}) \right| \leqslant \varepsilon^{-\frac{1}{p}}, \tag{58}$$

we have

$$\begin{aligned} \left| \Psi_{\varepsilon\eta} \prod_{j=1}^{m} u_j^{-\frac{1}{p}} \right|, \ \left| \Psi_\varepsilon \prod_{j=1}^{m} u_j^{-\frac{1}{p}} \right| &\leqslant \left| \prod_{j=1}^{m} \chi_1(\varepsilon u_j^{-1})\, u_j^{-\frac{1}{p}} \right| \\ &= \left| \prod_{j=1}^{q} \psi_j^{-\frac{1}{p}} \chi_1(\varepsilon \psi_j^{-1}) \right| \left| \prod_{j=q+1}^{m} v_j^{-\frac{1}{p}} \chi_1(\varepsilon v_j^{-1}) \right| \\ &\leqslant \varepsilon^{-\frac{q}{p}} \left| \prod_{j=q+1}^{m} v_j^{-\frac{1}{p}} \chi_1(\varepsilon v_j^{-1}) \right|. \end{aligned} \tag{59}$$

It follows from (47), (57), (59), and (49) that

$$
\begin{aligned}
| I_{\varepsilon\eta}(x) |, \ | I_{\varepsilon}(x) | & \\
\leqslant C_2 \| \mathcal{F} \|_p \zeta^{\frac{m-n}{p}} \varepsilon^{-\frac{q}{p}} & \left(\prod_{j=1}^{q} \int_{E_+^1} t_j^{\beta_j - 1} \mu_j(t_j)\, dt_j \right) \\
& \times \left(\prod_{j=q+1}^{m} \int_{E_+^1} v_j^{-1-\frac{1}{p}} \chi_1(\varepsilon v_j^{-1})\, dv_j \right) \\
\leqslant C_3 \| \mathcal{F} \|_p \zeta^{\frac{m-n}{p}} \varepsilon^{-\frac{q}{p}} & \left(\prod_{j=q+1}^{m} \int_{\varepsilon}^{\infty} v_j^{-1-\frac{1}{p}}\, dv_j \right) \leqslant C_4 \| \mathcal{F} \|_p \zeta^{\frac{m-n}{p}} \varepsilon^{-\frac{m}{p}},
\end{aligned}
$$

where C_4 is a constant independent of x, $\mathcal{F}$, ζ, η, and ε.

Let us now prove (56). For $k = 0$, it is obvious since we then have $I_{\varepsilon\eta}(x) = I_{\varepsilon}(x)$. For $k \geqslant 1$, let us look at the difference

$$
\Psi_{\varepsilon} - \Psi_{\varepsilon\eta} = \prod_{j=1}^{m} \chi_1(\varepsilon u_j^{-1}) \prod_{j=k+1}^{m} \chi_{-1}(\zeta u_j^{-1}) \left[1 - \prod_{j=1}^{k} \chi_{-1}(\eta u_j^{-1}) \right]. \tag{60}
$$

Obviously,

$$
\begin{aligned}
1 - \prod_{j=1}^{k} \chi_{-1}(\eta u_j^{-1}) &= [1 - \chi_{-1}(\eta u_1^{-1})] + \chi_{-1}(\eta u_1^{-1})[1 - \chi_{-1}(\eta u_2^{-1})] + \\
&\quad \dots + \chi_{-1}(\eta u_1^{-1}) \dots \chi_{-1}(\eta u_{k-1}^{-1})[1 - \chi_{-1}(\eta u_k^{-1})] \\
&= \sum_{i=1}^{k} \chi_{-1}(\eta u_1^{-1}) \dots \chi_{-1}(\eta u_{i-1}^{-1}) \chi_1(\eta u_i^{-1})
\end{aligned} \tag{61}
$$

since, by virtue of (24),

$$
1 - \chi_{-1}(\eta u_i^{-1}) = \chi_1(\eta u_i^{-1}), \qquad 0 < u_i < \infty.
$$

From (60) and (61), on the basis of the first of inequalities (58) and the fact that

$$\chi_1(\varepsilon u_i^{-1})\chi_1(\eta u_i^{-1})=\chi_1(\eta u_i^{-1}), \qquad 0<\varepsilon<\eta,\ 0<u_i<\infty,$$

we obtain*

$$|\Psi_\varepsilon-\Psi_{\varepsilon\eta}|\leqslant\sum_{i=1}^{k}\chi_1(\eta u_i^{-1})\prod_{j=1}^{m}{}^{(i)}\chi_1(\varepsilon u_j^{-1}). \tag{62}$$

If we now represent the difference $I_\varepsilon(x)-I_{\varepsilon\eta}(x)$ with the aid of formula (47) and use estimates (57) and (62), we get

$$|I_\varepsilon(x)-I_{\varepsilon\eta}(x)|\leqslant C_5\|\mathcal{F}\|_p\zeta^{\frac{m-n}{p}}\sum_{i=1}^{k}N_i, \tag{63}$$

where

$$N_i=\left(\prod_{j=1}^{q}\int_{E_+^1}t_j^{\beta_j-1}\mu_j(t_j)\,dt_j\right)\left(\prod_{j=q+1}^{m}\int_{E_+^1}v_j^{-1}\,dv_j\right)$$
$$\times u_i^{-\frac{1}{p}}\chi_1(\eta u_i^{-1})\prod_{j=1}^{m}{}^{(i)}u_j^{-\frac{1}{p}}\chi_1(\varepsilon u_j^{-1}).$$

If $i\leqslant q$, we have, by virtue of the second of inequalities (58),

$$u_i^{-\frac{1}{p}}\chi_1(\eta u_i^{-1})\prod_{j=1}^{m}{}^{(i)}\chi_1(\varepsilon u_j^{-1})u_j^{-\frac{1}{p}}\leqslant\eta^{-\frac{1}{p}}\varepsilon^{-\frac{q-1}{p}}\prod_{j=q+1}^{m}v_j^{-\frac{1}{p}}\chi_1(\varepsilon v_j^{-1}).$$

Therefore, if we now apply inequalities (49), we get

$$*\ \prod_{j=1}^{m}{}^{(i)}a_j=\prod_{\substack{j=1\\ j\neq i}}^{m}a_j.$$

$$N_i \leqslant C_6 \eta^{-\frac{1}{p}} \varepsilon^{-\frac{q-1}{p}} \left(\prod_{j=q+1}^{m} \int_{E_+^1} v_j^{-1-\frac{1}{p}} \chi_1 (\varepsilon v_j^{-1})\, dv_j \right) \leqslant C_7 \eta^{-\frac{1}{p}} \varepsilon^{-\frac{m-1}{p}}.$$

If $i > q$, we have $u_i = v_i$ and

$$u_i^{-\frac{1}{p}} \chi_1 (\eta u_i) \prod_{j=1}^{m}{}^{(i)} u_j^{-\frac{1}{p}} \chi_1 (\varepsilon u_j^{-1}) \leqslant \varepsilon^{-\frac{q}{p}} v_i^{-\frac{1}{p}} \chi_1 (\eta v_i^{-1}) \prod_{j=q+1}^{m}{}^{(i)} v_j^{-\frac{1}{p}} \chi_1 (\varepsilon v_j^{-1}).$$

Consequently,

$$N_i \leqslant C_8 \varepsilon^{-\frac{q}{p}} \left(\int_{E_+^1} v_i^{-1-\frac{1}{p}} \chi_1 (\eta v_i^{-1})\, dv_i \right) \times \left(\prod_{j=q+1}^{m}{}^{(i)} \int_{E_+^1} v_j^{-1-\frac{1}{p}} \chi_1 (\varepsilon v_j^{-1})\, dv_j \right) \leqslant C_9 \eta^{-\frac{1}{p}} \varepsilon^{-\frac{m-1}{p}}.$$

It follows from the estimates for N_i and inequality (63) that

$$| I_\varepsilon (x) - I_{\varepsilon\eta} (x) | \leqslant C_{10} \| \mathcal{F} \|_p \zeta^{\frac{m-1}{p}} \varepsilon^{-\frac{m-1}{p}} \eta^{-\frac{1}{p}},$$

where C_{10} is independent of $\mathcal{F}$, x, ε, ζ, and η. Equation (56) now follows. This completes the proof of the lemma.

To prove the basic theorem, we need yet another lemma.

13.5.2. Lemma. *Suppose that* $\mathcal{F} \in C_0^\infty (E^n)$, *where* $1 < p < \infty$, *and that* $0 < \varepsilon_1 < \varepsilon_2 < 1$. *Then, there exists a constant* $A_{\mathcal{F}}$ *dependent on* $\mathcal{F}$ *but independent of* ε_1 *and* ε_2 *such that*

$$\| I_{\varepsilon_1} - I_{\varepsilon_2} \|_p \leqslant A_{\mathcal{F}} \varepsilon_2. \tag{64}$$

PROOF. Let us represent the difference $I_{\varepsilon_1}(x) - I_{\varepsilon_2}(x)$ in accordance with formula (47), replacing $\Psi_{\varepsilon\eta}(u)$ with $\Psi_{\varepsilon}(u)$ in it on the basis of the definition of $I_{\varepsilon}(x)$. Then, by virtue of Minkowski's generalized inequality 2(10), we obtain

$$\| I_{\varepsilon_1} - I_{\varepsilon_2} \|_p \leqslant \left(\prod_{j=1}^{q} \int_{E_+^1} t_j^{\beta_j - 1} \mu_j(t_j)\, dt_j \right)$$
$$\times \left(\prod_{j=q+1}^{m} \int_{E_+^1} v_j^{-1}\, dv_j \right) | \Psi_{\varepsilon_1}(u) - \Psi_{\varepsilon_2}(u) | \, \| \Phi(\cdot\,,\, u,\, \zeta) \|_p. \quad (65)$$

Let us first estimate $| \Psi_{\varepsilon_1}(u) - \Psi_{\varepsilon_2}(u) |$. By virtue of (55) and the first of inequalities (58), we obtain

$$| \Psi_{\varepsilon_1} - \Psi_{\varepsilon_2} | = \left| \prod_{j=k+1}^{m} \chi_{-1}(\zeta u_j^{-1}) \left[\prod_{j=1}^{m} \chi_1(\varepsilon_1 u_j^{-1}) - \prod_{j=1}^{m} \chi_1(\varepsilon_2 u_j^{-1}) \right] \right|$$
$$\leqslant \left| \prod_{j=1}^{m} \chi_1(\varepsilon_1 u_j^{-1}) - \prod_{j=1}^{m} \chi_1(\varepsilon_2 u_j^{-1}) \right|$$
$$= \left| \sum_{i=1}^{m} [\chi_1(\varepsilon_1 u_i^{-1}) - \chi_1(\varepsilon_2 u_i^{-1})] \prod_{j=1}^{i-1} \chi_1(\varepsilon_1 u_j^{-1}) \prod_{j=i+1}^{m} \chi_1(\varepsilon_2 u_j^{-1}) \right|$$
$$\leqslant \sum_{i=1}^{m} | \chi_1(\varepsilon_1 u_i^{-1}) - \chi_1(\varepsilon_2 u_i^{-1}) | \prod_{j=1}^{m}{}^{(i)} \chi_1(\varepsilon_{ji} u_j^{-1}), \quad (66)$$

where $\varepsilon_{ji} = \varepsilon_1$ for $j = 1, \ldots, i-1$ but $\varepsilon_{ji} = \varepsilon_2$ for $j = i+1, \ldots, m$.

Let $e^{m(i)}$ denote the set $\{1, \ldots, i-1, i+1, \ldots, m\}$ and let σ denote an arbitrary subset (possibly the empty one) of the set $e^{m(i)}$. Let $\sigma' = e^{m(i)} \setminus \sigma$. Then,

$$\prod_{j=1}^{m}{}^{(i)} \chi_1(\varepsilon_{ji} u_j^{-1}) = \sum_{\sigma \subseteq e^{m(i)}} \left(\prod_{j \in \sigma} \chi_1(\varepsilon_{ji} u_j^{-1}) \chi_{-1}(u_j^{-1}) \right) \left(\prod_{j \in \sigma'} \chi_1(u_j^{-1}) \right), \quad (67)$$

which we obtain by representing every factor on the left in the form

$$\chi_1(\varepsilon_{ji}u_j^{-1}) = \chi_1(\varepsilon_{ji}u_j^{-1})\chi_{-1}(u_j^{-1}) + \chi_1(u_j^{-1}),\ 0 < \varepsilon_{ji} < 1.$$

On the basis of (65)–(67), we obtain

$$\| I_{\varepsilon_1} - I_{\varepsilon_2} \|_p \leqslant \sum_{i=1}^{m} \sum_{\sigma \subseteq e^{m(i)}} \mathcal{J}_{i,\sigma}, \tag{68}$$

where

$$\mathcal{J}_{i,\sigma} = \left(\prod_{j=1}^{q} \int_{E_+^1} t_j^{\beta_j - 1} \mu_j(t_j)\, dt_j \right) \left(\prod_{j=q+1}^{m} \int_{E_+^1} v_j^{-1} dv_j \right) |\chi_{i,\sigma}| \, \| \Phi(\cdot, u, \zeta \|_p, \tag{69}$$

$$\chi_{i,\sigma} = [\chi_1(\varepsilon_1 u_i^{-1}) - \chi_1(\varepsilon_2 u_i^{-1})] \left(\prod_{j \in \sigma} \chi_1(\varepsilon_{ji} u_j^{-1}) \chi_{-1}(u_j^{-1}) \right) \left(\prod_{j \in \sigma'} \chi_1(u_j^{-1}) \right). \tag{70}$$

Obviously, to prove the lemma, we need only obtain an estimate for $\mathcal{J}_{i,\sigma}$ that holds for arbitrary but fixed $i \in [1, m]$ and $\sigma \subseteq e^{m(i)}$.

Fixing i and σ, let us first estimate $\| \Phi \|_p$. In finding this estimate, we shall use the notation $x = (x_i, x^\sigma, x^{\sigma'}, x^{e'})$, where $e' = \{m + 1, \ldots, n\}$ and σ and σ' are the sets referred to above.

Let us represent the function $M(y;\ u,\ \zeta)$ appearing in the definition of the function $\Phi(x;\ u,\ \zeta)$ (see (52) and (53)) in the form

$$M(y;\ u,\ \zeta) = M_1(y_i,\ y^\sigma,\ y^{e'};\ u,\ \zeta)\, M_2(y^{\sigma'};\ u),$$

where

$$M_1(y_i,\ y^\sigma,\ y^{e'};\ u,\ \zeta) = u_i^{-1} P_i(y_i u_i^{-1}) \left(\prod_{j \in \sigma} u_j^{-1} P_j(y_j u_j^{-1}) \right) \left(\prod_{j \in e'} \zeta^{-1} Q_j(y_j \zeta^{-1}) \right),$$

$$M_2(y^{\sigma'};\ u) = \prod_{j \in \sigma'} u_j^{-1} P_j(y_j u_j^{-1}).$$

Let us set

$$\Delta^{1\sigma}_{-y^\sigma}\varphi(x)=\Delta_{-y_{r_1}}\ldots\Delta_{-y_{r_s}}\varphi(x),\ \Delta_{-y_j}\varphi(x)=\varphi(x-y_je_j)-\varphi(x)$$

if $\sigma=\{r_1, \ldots, r_s\}$.

Using property (54) of the kernels P_j, we can write the function Φ (see (52)) in the following form:

$$\Phi(x;\ u,\ \zeta)$$
$$=\int_{E^n}\mathcal{F}(x_i-y_i, x^\sigma-y^\sigma, y^{\sigma'}, x^{e'}-y^{e'})\,M_1(y_i, y^\sigma, y^{e'}; u, \zeta)$$
$$\times M_2(x^{\sigma'}-y^{\sigma'};\ u)\,dy=\int_{E^n}\Delta_{-y_i}\Delta^{1\sigma}_{-y^\sigma}\mathcal{F}(x_i,\ x^\sigma,\ y^{\sigma'},\ x^{e'}-y^{e'})$$
$$\times M_1(y_i,\ y^\sigma,\ y^{e'};\ u,\ \zeta)\,M_2(x^{\sigma'}-y^{\sigma'};\ u)\,dy.$$

Then, with the aid of Minkowski's generalized inequality, we obtain

$$\|\Phi(\,\cdot\,;\ u,\ \zeta)\|_p\leqslant\int|M_1(y_i,\ y^\sigma,\ y^{e'};\ u,\ \zeta)|\,dy_i\,dy^\sigma\,dy^{e'}$$
$$\times\int dy^{\sigma'}\Big(\int\Big|\Delta_{-y_i}\Delta^{1\sigma}_{-y^\sigma}\mathcal{F}(x_i,\ x^\sigma,\ y^{\sigma'},\ x^{e'}-y^{e'})\Big|^p$$
$$\times\,dx_i\,dx^\sigma\,dx^{e'}\int|M_2(x^{\sigma'}-y^{\sigma'};\ u)|^p\,dx^{\sigma'}\Big)^{1/p}.$$

Obviously,

$$\int|M_2(y^{\sigma'};u)|^p\,dy^{\sigma'}=\prod_{j\in\sigma'}u_j^{-p}\int_{E'}|P_j(y_ju_j^{-1})|^p\,dy_j\leqslant C_1\prod_{j\in\sigma'}u_j^{1-p}$$

where C_1 is independent of u_j.

Furthermore, since $\mathcal{F}$ is an infinitely differentiable function with compact support, there exists a constant $N_{\mathcal{F}}$ such that

$$\int dy^{\sigma'}\left(\int \left|\Delta_{-y_i}\Delta^{1\sigma}_{-y^\sigma}\mathcal{F}(x_i,\ x^\sigma,\ y^{\sigma'},\ x^{e'}-y^{e'})\right|^p dx_i\,dx^\sigma\,dx^{e'}\right)^{1/p} \leqslant N_{\mathcal{F}}|\,y_i\,|\prod_{j\in\sigma}|\,y_j\,|.$$

Consequently,

$$\|\Phi\|_p \leqslant C_1N_{\mathcal{F}}\prod_{j\in\sigma'}u_j^{\frac{1-p}{p}}\int|y_i|\prod_{j\in\sigma}|\,y_j||\,M_1(y_i,y^\sigma,y^{e'};u,\zeta)|\,dy_i dy^\sigma\,dy^{e'}$$
$$\leqslant C_2N_{\mathcal{F}}u_i\prod_{j\in\sigma}u_j\prod_{j\in\sigma'}u_j^{\frac{1-p}{p}}\int|\,y_i\,||\,P_i\,(y_i)\,|\prod_{j\in\sigma}|\,y_j\,||\,P_j\,|(y_j)\,|$$
$$\times\prod_{j\in e'}|\,Q\,(y_j)\,|\,dy_i\,dy^\sigma\,dy^{e'}\leqslant C_3N_{\mathcal{F}}u_i\prod_{j\in\sigma}u_j\prod_{j\in\sigma'}u_j^{\frac{1-p}{p}}, \qquad (71)$$

where C_3 is independent of $\mathcal{F}$ and u_j.

It now follows from (70) and (71) that

$$|\chi_{i,\,\sigma}|\|\Phi\|_p\leqslant C_3N_{\mathcal{F}}\,|\,u_i\,(\chi_1\,(\varepsilon_1u_i^{-1})-\chi_1\,(\varepsilon_2u_i^{-1}))\,|$$
$$\times\left(\prod_{\substack{j\in\sigma\\1\leqslant j\leqslant q}}\psi_j\chi_1\,(\varepsilon_{ji}\psi_j^{-1})\,\chi_{-1}\,(\psi_j^{-1})\right)\left(\prod_{\substack{j\in\sigma'\\1\leqslant j\leqslant q}}\psi_j^{\frac{1-p}{p}}\chi_1\,(\psi_j^{-1})\right)$$
$$\times\left(\prod_{\substack{j\in\sigma\\q+1\leqslant j\leqslant m}}v_j\chi_1\,(\varepsilon_{ji}v_j^{-1})\,\chi_{-1}\,(v_j^{-1})\right)\left(\prod_{\substack{j\in\sigma'\\q+1\leqslant j\leqslant m}}v_j^{\frac{1-p}{p}}\chi_1\,(v_j^{-1})\right) \qquad (72)$$

since u_j is equal to ψ_j for $j=1,\ldots,q$ but equal to v_j for $j=q+1,\ldots,m$ (see (51)).

From the definition of the functions χ_1 and χ_{-1} we get immediately the inequalities

$$|\,u_i\,(\chi_1\,(\varepsilon_1u_i^{-1})-\chi_1\,(\varepsilon_2u_i^{-1}))\,|\leqslant\varepsilon_2,\quad 0<\varepsilon_1<\varepsilon_2,$$
$$|\,\psi_j\chi_1\,(\varepsilon_{ji}\psi_j^{-1})\,\chi_{-1}\,(\psi_j^{-1})\,|\leqslant 1,$$
$$\left|\psi_j^{\frac{1-p}{p}}\chi_1\,(\psi_j^{-1})\right|\leqslant 1 \qquad (p>1). \qquad (73)$$

On the basis of (69), (72), and (73), we now obtain

$$\mathcal{I}_{i,\sigma} \leqslant C_3 N_{\mathcal{F}} \left(\prod_{\substack{j=1 \\ j \neq i}}^{q} \int_{E_+^1} t_j^{\beta_j - 1} \mu_j(t_j)\, dt_j \right) \times \left(\prod_{\substack{j \in \sigma \\ q+1 \leqslant j \leqslant m}} \int_{E_+^1} \chi_1(\varepsilon_{ji} v_j^{-1}) \chi_{-1}(v_j^{-1})\, dv_j \right) \times \left(\prod_{\substack{j \in \sigma' \\ q+1 \leqslant j \leqslant m}} \int_{E_+^1} v_j^{-1 - \frac{p-1}{p}} \chi_1(v_j^{-1})\, dv_j \right) \mathcal{L}_i,$$

where, for $i \leqslant q$,

$$\mathcal{L}_i = \int_{E_+^1} t_i^{\beta_i - 1} \mu_i(t_i) \left| \psi_i \left[\chi_1(\varepsilon_1 \psi_i^{-1}) - \chi_1(\varepsilon_2 \psi_i^{-1}) \right] \right| dt_i \leqslant C_4 \varepsilon_2$$

by virtue of the first of inequalities (73) and inequality (49), but, for $i > q$,

$$\mathcal{L}_i = \int_{E_+^1} \left[\chi_1(\varepsilon_1 v_i^{-1}) - \chi_1(\varepsilon_2 v_i^{-1}) \right] dv_i = \int_{\varepsilon_1}^{\varepsilon_2} dv_i < \varepsilon_2.$$

Noting now that

$$\int_{E_+^1} \chi_1(\varepsilon_{ji} v_j^{-1}) \chi_{-1}(v_j^{-1})\, dv_j \leqslant \int_0^1 dv_j = 1$$

and

$$\int_{E_+^1} v_j^{-1 - \frac{p-1}{p}} \chi_1(v_j^{-1})\, dv_j \leqslant \int_1^\infty v_j^{-1 - \frac{p-1}{p}}\, dv_j = \frac{p}{p-1},$$

we finally obtain

$$\mathcal{J}_{l,\sigma} \leqslant CN_{\mathcal{F}}\varepsilon_2, \tag{74}$$

where C is independent of ε_1 and ε_2. The assertion of the lemma follows from inequalities (68) and (74).

The chief purpose of the present subsection is to prove the following theorem:

13.5.3. Theorem. *Suppose that* $\mathcal{F} \in L_p(E^n)$, *where* $1 < p < \infty$. *Then,*

1) *there exists a constant* C_p *independent of* $\mathcal{F}$, ε, *and* η *such that*

$$\| I_{\varepsilon\eta} \| \leqslant C_p \| \mathcal{F} \|_p, \quad 0 < \varepsilon < \eta < \infty, \tag{75}$$

$$\| I_\varepsilon \|_p \leqslant C_p \| \mathcal{F} \|_p, \quad \varepsilon > 0, \tag{76}$$

2) *there exists a function* $I(x)$ *such that*

$$\lim_{\varepsilon \to 0} \| I_\varepsilon - I \|_p = 0, \quad \| I \|_p \leqslant C_p \| \mathcal{F} \|_p. \tag{77}$$

PROOF. First of all, let us prove inequality (75). We set

$$K_{\varepsilon\eta}(x)$$

$$= \left(\prod_{j=1}^{q} \int_{E_+^1} t_j^{\beta_j - 1} \mu_j(t_j)\, dt_j \right) \left(\prod_{j=q+1}^{m} \int_{E_+^1} v_j^{-1}\, dv_j \right) \Psi_{\varepsilon\eta}(u) M(x; u, \zeta). \tag{78}$$

It follows from the proof of Lemma 13.5.1 that the integrand in (47) is absolutely integrable. Therefore, using Fubini's theorem to interchange the order of integration, we can write $I_{\varepsilon\eta}(x)$ in the form of the convolution

$$I_{\varepsilon\eta}(x) = \int_{E^n} \mathcal{F}(x - y) K_{\varepsilon\eta}(y)\, dy. \tag{79}$$

The next step involves the following theorem by Lizorkin (see [1]) on $(L_p \to L_p)$-multipliers of Fourier integrals, which we state without proof.

Theorem (Lizorkin [1]). *Suppose that the function $T(x)$, its mixed nth derivative $\frac{\partial^n T(x)}{\partial x_1 \ldots \partial x_n}$, and all its lower-order derivatives are continuous off the coordinate planes (that is, for $|x_j| > 0$ for $j = 1, \ldots, n$). Suppose that all these derivatives satisfy the inequality*

$$\left| x_1^{l_1} \ldots x_n^{l_n} \frac{\partial^{|l|} T(x)}{\partial x_1^{l_1} \ldots \partial x_n^{l_n}} \right| \leqslant B, \tag{80}$$

where B is a constant and l_j is either zero or 1 for $j = 1, \ldots, n$ and satisfies the relation

$$0 \leqslant |l| = \sum_{j=1}^{n} l_j \leqslant n.$$

Then, if $\mathcal{F} \in L_p(E^n)$, the function $T(x)\tilde{\mathcal{F}}(x)$ (where $\tilde{\mathcal{F}}$ is the Fourier transform of $\mathcal{F}$) is the Fourier transform of some function $\mathcal{I}(x)$ belonging to $L_p(E^n)$, and there exists a constant $A_{p,n}$ depending only on p and n such that

$$\| \mathcal{I} \|_p \leqslant A_{p,n} B \| \mathcal{F} \|_p, \qquad 1 < p < \infty.$$

Let us denote by $\tilde{K}_{\varepsilon\eta}(x)$ the Fourier transform of $K_{\varepsilon\eta}(x)$.

By virtue of the theorem on the Fourier transformation of a convolution and Lizorkin's theorem, to prove inequality (75), it will be sufficient to show that the function $T(x) = \tilde{K}_{\varepsilon\eta}(x)$ satisfies the conditions of Lizorkin's theorem with constant B independent of ε and η.

By (78), we have

$$\tilde{K}_{\varepsilon\eta}(x) = \int_{E^n} K_{\varepsilon\eta}(y)\, e^{-2\pi i (x, y)}\, dy$$

$$= \left(\prod_{j=1}^{q} \int_{E_+^1} t_j^{\beta_j - 1} \mu_j(t_j)\, dt_j \right) \left(\prod_{j=q+1}^{m} \int_{E_+^1} v_j^{-1}\, dv_j \right) \Psi_{\varepsilon\eta}(u)$$

$$\times \int_{E^n} M(y;\ u,\ \zeta)\, e^{-2\pi i (x,\, y)}\, dy. \qquad (81)$$

The change of order of integration that we made here is valid by virtue of the absolute summability of the integrand (this follows easily from the definitions of $\Psi_{\varepsilon\eta}$ and M).

Let us use formula (53) to rewrite the inner integral in the form

$$\int_{E^n} M(y;\ u,\ \zeta)\, e^{-2\pi i (x,\, y)}\, dy$$

$$= \prod_{j=1}^{m} \left(u_j^{-1} \int_{E^1} P_j(y_j u_j^{-1})\, e^{-2\pi i x_j y_j}\, dy_j \right)$$

$$\times \prod_{j=m+1}^{n} \left(\zeta^{-1} \int_{E^1} Q_j(y_j \zeta^{-1})\, e^{-2\pi i x_j y_j}\, dy_j \right) = \prod_{j=1}^{n} \mathscr{L}_j. \qquad (82)$$

Let us show that

$$|\mathscr{L}_j|,\ \left| x_j \frac{\partial \mathscr{L}_j}{\partial x_j} \right| \leqslant \begin{cases} C_j \min\left(|u_j x_j|,\ \dfrac{1}{|u_j x_j|} \right) & \text{for } j = 1, \ldots, m, \\ C_j & \text{for } j = m+1, \ldots, n, \end{cases} \qquad (83)$$

where the C_j (for $j = 1, \ldots, n$) are constants independent of x_j, u_j, and ζ.

Suppose first that $1 \leqslant j \leqslant m$. We have

$$\mathscr{L}_j = u_j^{-1} \int_{E^1} P_j(y_j u_j^{-1})\, e^{-2\pi i x_j y_j}\, dy_j = \int_{E^1} P_j(z_j)\, e^{-2\pi i x_j z_j u_j}\, dz_j. \qquad (84)$$

If $x_j = 0$, then, by virtue of property (54) of the kernel P_j, we have $\mathcal{L}_j = 0$. Suppose that $x_j \neq 0$. Then, on the one hand, by again using (54) and remembering that $P_j \in C_0^\infty(E^1)$, we have

$$|\mathcal{L}_j| = \left| \int_{E^1} P_j(z_j)\left[e^{-2\pi i x_j z_j u_j} - 1\right] dz_j \right| \leqslant 2\pi |x_j u_j| \int_{E^1} |P_j(z_j) z_j| dz_j \leqslant C_j |x_j u_j|.$$

On the other hand,

$$|\mathcal{L}_j| = \left| \int_{E^1} P_j(z_j) e^{-2\pi i x_j z_j u_j} dz_j \right| = \left| \frac{1}{2\pi i x_j u_j} \int_{E^1} \frac{d}{dz_j}(P_j(z_j)) e^{-2\pi i x_j z_j u_j} dz_j \right| \leqslant \frac{1}{2\pi |x_j u_j|} \int_{E^1} \left| \frac{d}{dz_j} P_j(z_j) \right| dz_j \leqslant \frac{C_j}{|x_j u_j|}.$$

The estimate (83) for $|\mathcal{L}_j|$ now follows from the inequalities that we have obtained.

Also, it follows from (84) that

$$\frac{\partial \mathcal{L}_j}{\partial x_j} = -2\pi i u_j \int_{E^1} P_j(z_j) z_j e^{-2\pi i x_j z_j u_j} dz_j. \qquad (85)$$

If $x_j = 0$, we obviously have $x_j \dfrac{\partial \mathcal{L}_j}{\partial x_j} = 0$. Suppose that $x_j \neq 0$. On the one hand, by virtue of (85), we have

$$\left| x_j \frac{\partial \mathcal{L}_j}{\partial x_j} \right| \leqslant 2\pi |u_j x_j| \int_{E^1} |P_j(z_j) z_j| dz_j \leqslant C_j |u_j x_j|. \qquad (86)$$

On the other hand, by twice integrating by parts, we obtain from (85)

$$\left| x_j \frac{\partial \mathscr{L}_j}{\partial x_j} \right| = \left| 2\pi u_j x_j \int_{E^1} P_j(z_j)\, z_j e^{-2\pi i x_j z_j u_j}\, dz_j \right|$$

$$= \frac{1}{2\pi |u_j x_j|} \left| \int_{E^1} \frac{d^2}{dz^2} (P_j(z_j)\, z_j)\, e^{-2\pi i x_j z_j u_j}\, dz_j \right|$$

$$\leqslant \frac{1}{2\pi |u_j x_j|} \int_{E^1} \left| \frac{d^2}{dz^2} (P_j(z_j)\, z_j) \right| dz_j \leqslant \frac{C_j}{|u_j x_j|}. \quad (87)$$

The estimate (83) for $\left| x_j \dfrac{\partial \mathscr{L}_j}{\partial x_j} \right|$ follows from (86) and (87).

Suppose now that $m+1 \leqslant j \leqslant n$. Noting again that $Q_j \in C_0^\infty(E^1)$, we see that

$$|\mathscr{L}_j| = \left| \int_{E^1} Q_j(z_j)\, e^{-2\pi i x_j z_j \zeta}\, dz_j \right| \leqslant \int_{E^1} |Q_j(z_j)|\, dz_j \leqslant C_j$$

and

$$\left| x_j \frac{\partial \mathscr{L}_j}{\partial x_j} \right| = \left| 2\pi \zeta x_j \int_{E^1} Q_j(z_j)\, z_j e^{-2\pi i x_j z_j \zeta}\, dz_j \right|$$

$$= \left| \int_{E^1} \frac{d}{dz_j} (Q_j(z_j)\, z_j)\, e^{-2\pi i x_j z_j \zeta}\, dz_j \right| \leqslant \int_{E^1} \left| \frac{d}{dz_j} (Q_j(z_j)\, z_j) \right| dz_j \leqslant C_j.$$

Thus, inequalities (83) are proven. The estimates (80) follow immediately from them.

Let us estimate, for example, $\left| x_1 \ldots x_s \dfrac{\partial^s \widetilde{K}_{\varepsilon\eta}(x)}{\partial x_1 \ldots \partial x_s} \right|$. On the basis of inequalities (49) and (83) and the obvious estimate

$$|\Psi_{\varepsilon\eta}(u)| \leqslant 1,$$

we get from (81) and (82)

$$\left| x_1 \ldots x_s \frac{\partial \tilde{K}_{\varepsilon\eta}(x)}{\partial x_1 \ldots \partial x_s} \right|$$

$$\leqslant \left(\prod_{j=1}^{q} \int_{E_+^1} t_j^{\beta_j - 1} \mu_j(t_j)\, dt_j \right) \left(\prod_{j=q+1}^{m} \int_{E_+^1} v_j^{-1}\, dv_j \right) |\Psi_{\varepsilon\eta}(u)|$$

$$\times \left| \prod_{j=1}^{s} x_j \frac{\partial \mathscr{L}_j}{\partial x_j} \right| \left| \prod_{j=s+1}^{n} \mathscr{L}_j \right|$$

$$\leqslant C_0 \prod_{j=q+1}^{m} \left(\int_{E_+^1} v_j^{-1} \min\left(|v_j x_j|, \frac{1}{|v_j x_j|} \right) dv_j \right)$$

$$\leqslant C_0 \left(\int_0^\infty u^{-1} \min(u, u^{-1})\, du \right)^{m-q} = B,$$

where B is independent of ε and η.

One can also easily see that the derivatives of $\tilde{K}_{\varepsilon\eta}(x)$ required in Lizorkin's theorem are continuous. Consequently, all the conditions in that theorem regarding $T(x) = \tilde{K}_{\varepsilon\eta}(x)$ are satisfied. Inequality (75) then follows, as was mentioned above.

To prove inequality (76), we note that, by virtue of Lemma 13.5.1, $I_{\varepsilon\eta}(x) \to I_\varepsilon(x)$ uniformly on E^n as $\eta \to \infty$. Therefore, if in the inequality

$$\| I_{\varepsilon\eta} \|_{p, |x| \leqslant R} \leqslant C_p \| \mathscr{F} \|_p$$

we take the limit first as $\eta \to \infty$ and then as $R \to \infty$, we obtain (76).

Let us now prove assertion 2) of the theorem. By virtue of (76), we can, for different ε in the interval (0, 1), regard the I_ε as operators mapping the space $L_p(E^n)$ into itself. The norms of these operators are bounded uniformly. Also, it follows from Lemma 13.5.2 that, if $\mathscr{F} \in C_0^\infty(E^n)$, then

$$\lim_{\varepsilon_1, \varepsilon_2 \to 0} \| I_{\varepsilon_1}(\mathscr{F}) - I_{\varepsilon_2}(\mathscr{F}) \|_p = 0,$$

that is, the operations I_ε are Cauchy-convergent on a set of functions that is dense in $L_p(E^n)$. On the basis of the Banach-Steinhaus Theorem 1.8, we now conclude that, for any function $\mathscr{F} \in L_p(E^n)$, there exists a function $I(\mathscr{F}) \in L_p(E^n)$ such that

$$\lim_{\varepsilon \to 0} \| I_\varepsilon(\mathscr{F}) - I(\mathscr{F}) \|_p = 0$$

and

$$\| I(\mathscr{F}) \|_p \leqslant C_p \| \mathscr{F} \|_p,$$

where C_p is the constant of inequality (76).

This completes the proof of Theorem 13.5.3.

13.5.4. Consequences of the results of subsection 13.5. Let us look at formula (7) for $x \in G_\lambda$ (for $1 \leqslant \lambda \leqslant \Lambda$), where $k \geqslant 1$ and

$$h = (\eta^k, \zeta^{n-k}) = (\overbrace{\eta, \ldots, \eta}^{k}, \overbrace{\zeta, \ldots, \zeta}^{n-k}).$$

Let us suppose that $\alpha^e \in M^e(\alpha)$ for all e such that $e^k \subseteq e \subseteq e^n$. Then, every function $\mathscr{I}^e$ appearing in the identity (7) can be represented by formulas (41)-(43). All the results obtained for the function $I_{\varepsilon\eta}(x)$ in the preceding sections remain valid for the function $\tilde{I}_i^e$ defined on E^n and coinciding with I_i^e on G_λ. In particular, on the basis of Lemma 13.5.1, we have, for $x \in G_\lambda$ (where $h = (\eta^k, \zeta^{n-k})$), the relation*

$$\lim_{\eta \to \infty} I_i^e(x; \varepsilon, h) = \check{I}_i^e(x; \varepsilon), \tag{88}$$

where $\check{I}_i^e(x; \varepsilon)$ is obtained from $I_i^e(x; \varepsilon, h)$ by taking $h_1 = \ldots = h_k = \eta = \infty$ in it.

*In the notation for the function $\check{I}_i^e$, we omit the parameter ζ since we are assuming it to be fixed.

By virtue of (41) and (88), we can now write the basic formula (7) in the form

$$D^{\alpha}F(x;\varepsilon)=\sum_{e^k\subseteq e\subseteq e^n}(-1)^{|\alpha|}\breve{\mathcal{J}}^{e}(x;\varepsilon),\tag{89}$$

where

$$\breve{\mathcal{J}}^{e}(x;\varepsilon)=\sum_{i=1}^{s+1}\breve{I}_i^{e}(x;\varepsilon).\tag{90}$$

Theorem 13.5.3 enables us to obtain from formulas (89) and (90) some important consequences for later use.

By virtue of assertion 2) of that theorem, for every function $\breve{I}_i^{e}(x;\varepsilon)$ there exists a function $T_i^{e}(x)$ defined at least on G_λ such that

$$\lim_{\varepsilon\to 0}\|\breve{I}_i^{e}(\,\cdot\,,\varepsilon)-T_i^{e}\|_{p,\,G_\lambda}=0\tag{91}$$

and, by virtue of (77), (45), and (46),

$$\|T_i^{e}\|_{p,\,G_\lambda}\leqslant C_p'\|\mathcal{F}_i\|_p=C_p'\|D^{l^i}f\|_{p,\,G},\quad l^i\in\mathscr{E}^{e}(\alpha),\tag{92}$$

where C_p' is a constant independent of f.

Let us set

$$T(x)=\sum_{e^k\subseteq e\subseteq e^n}\sum_{i=1}^{s^e+1}(-1)^{|\alpha|}T_i^{e}(x)\,\cdot$$

(The number s in formula (90) depends in general on e, which we take into account by writing s^e.) It then follows from (89)–(92) that

$$\lim_{\varepsilon\to 0}\|D^{\alpha}F(\,\cdot\,,\varepsilon)-T\|_{p,\,G_\lambda}=0\tag{93}$$

and

$$\| T \|_{p, G_\lambda} \leqslant \sum_{e^k \subseteq e \subseteq e^n} C'_p \sum_{l \in \mathscr{C}^e(\alpha)} \| D^l f \|_{p, G} \leqslant C''_p \sum_{l \in \mathscr{C}} \| D^l f \|_{p, G}, \tag{94}$$

where C''_p is independent of f.

Let us suppose now that $D^\alpha F(x;\ \varepsilon)$ is represented on G_λ by formula (3), that is, that $k = 0$ ($e^k = \varnothing$), $h = (\zeta, \ldots, \zeta)$, where ζ is a fixed number. Let us suppose also that $\alpha^e \in M^e(\alpha)$ for all $e \subseteq e^n$ (including $e = \varnothing$). Let us write formula (3) in the form

$$D^\alpha F(x;\ \varepsilon) = (-1)^{|\alpha|} \mathscr{I}^\varnothing(x) + R(x;\ \varepsilon), \tag{95}$$

where

$$\mathscr{I}^\varnothing(x) = \int_{E^n} f(x+y) \prod_{j=1}^{n} \left(\zeta^{-1} D_{y_j}^{\alpha_j} \Omega_j\left(y_j \zeta^{-1}\right)\right) dy,$$

$$R(x;\ \varepsilon) = (-1)^{|\alpha|} \sum_{\substack{e \subseteq e^n \\ e \neq \varnothing}} \breve{\mathscr{I}}^e(x;\ \varepsilon),$$

$$\breve{\mathscr{I}}^e(x;\ \varepsilon) = \mathscr{I}^e(x;\ \varepsilon,\ h) \qquad (\forall e \subseteq e^n).$$

Just as above, we can show that there exists a function $\tilde{T}(x)$ defined on G_λ such that

$$\lim_{\varepsilon \to 0} \| R(\cdot,\ \varepsilon) - \tilde{T} \|_{p, G_\lambda} = 0 \tag{96}$$

and

$$\| \tilde{T} \|_{p, G_\lambda} \leqslant \tilde{C}_p \sum_{l \in \mathscr{C}} \| D^l f \|_{p, G}. \tag{97}$$

Furthermore, since $\alpha^\varnothing \in M^\varnothing(\alpha)$, there exists a vector $l \in \mathscr{E}$ such that $l_j \leqslant \alpha_j$ for $j = 1, \ldots, n$. Therefore, by virtue of the definition of a generalized derivative, we can put $\mathscr{I}^\varnothing(x)$ in the form

$$\mathcal{J}^{\varnothing}(x) = (-1)^{|l|} \int_{E^n} D^l f(x+y) \prod_{j=1}^{n} \left(\zeta^{-1-\alpha_j+l_j} D^{\alpha_j - l_j} \Omega_j (y_j \zeta^{-1})\right) dy.$$

From this it easily follows that

$$\|\mathcal{J}^{\varnothing}\|_{p,\, G_\lambda} \leqslant C \|D^l f\|_p, \qquad l \in \mathscr{E}. \tag{98}$$

Let us now set

$$T(x) = (-1)^{|\alpha|} \mathcal{J}^{\varnothing}(x) + \tilde{T}(x).$$

One can easily show on the basis of (95)-(98) that the function $T(x)$ satisfies equation (93) and inequality (94).

Combining what has been said, we arrive at the following result:

If $D^\alpha F(x;\, \varepsilon)$ *is represented by formula* (7) (*resp. by formula* (3)) *and* $\alpha^e \in M^e(\alpha)$ *for all* e *such that* $e^k \subseteq e \subseteq e^n$ (*resp. for all* $e \subseteq e^n$), *then, for every* λ *in* $[1, \Lambda]$, *there exists a function* $T_\lambda(x)$ *defined on* G_λ *for which equation* (93) *and inequality* (94) *are valid.*

13.6. The basic results. In this subsection, we shall formulate the basic results of the present section. First, let us prove theorems providing sufficient conditions for existence of the generalized derivative $D^\alpha f$ and validity of inequality (1) for an arbitrary function $f \in \mathscr{W}_p^{\{\mathscr{E}\}}(G)$ under various assumptions regarding the region G. We shall then show that the conditions imposed in these theorems on the vector α are in a certain sense also necessary.

13.6.1. Theorem. *Suppose that* $1 < p < \infty$ *and* $G \in \underline{A}(\square, H)$. *Let* $\alpha = (\alpha_1, \ldots, \alpha_n)$ *denote a vector with nonnegative integer-valued coefficients that satisfies the condition*

$$\alpha^e \in M^e(\alpha) \quad \forall e \subseteq e^n \qquad (\text{including } e = \varnothing). \tag{99}$$

Then, $D^\alpha \mathscr{W}_p^{\{\mathscr{E}\}}(G) \hookrightarrow L_p(G)$; *that is, for* $f \in \mathscr{W}_p^{\{\mathscr{E}\}}(G)$, *there exists on* G *a generalized derivative* $D^\alpha f \in L_p(G)$ *and there exists a constant* C *independent of* f *such that inequality* (1), *that is, the inequality*

$$\| D^\alpha f \|_{p,\,G} \leqslant C \sum_{l \in \mathscr{E}} \| D^l f \|_{p,\,G},$$

holds.

PROOF. Let $\{G_\lambda\}_1^\Lambda$ denote a collection of open sets forming a covering G in accordance with the definition of the class $\underline{A}(\square, H)$. To prove the theorem, we obviously need only prove the existence of $D^\alpha f$ on each G_λ (see subsection 6.1) and estimate $\| D^\alpha f \|_{p,\,G_\lambda}$ (for $\lambda = 1, \ldots, \Lambda$).

Let us fix λ. Let us now use the identity (3) for the function f, this identity being valid for all $x \in G_\lambda$. Since $\alpha^e \in M^e(\alpha)$ for all $e \subseteq e^n$, there exists, by virtue of the result of 13.5.4, a function $T_\lambda(x)$ defined on G_λ and satisfying equation (93) and inequality (94), that is, a function $T_\lambda(x)$ such that

$$\lim \| D^\alpha F(\cdot\,;\,\varepsilon) - T_\lambda \|_{p,\,G_\lambda} = 0$$

and

$$\| T_\lambda \|_{p,\,G_\lambda} \leqslant C \sum_{l \in \mathscr{E}} \| D^l f \|_{p,\,G}, \tag{100}$$

where C is a constant independent of f.

Thus, $D^\alpha F(x;\,\varepsilon) \to T_\lambda(x)$ in $L_p(G_\lambda)$ as $\varepsilon \to 0$. On the other hand, since $f \in L^{\mathrm{loc}}(G)$ (by virtue of the remark to Lemma 5.2), it follows that $F(x;\,\varepsilon) \to f(x)$ as $\varepsilon \to 0$ in the sense of convergence in $L^{\mathrm{loc}}(G_\lambda)$.

It then follows on the basis of Lemma 6.2 that the generalized derivative $D^\alpha f = T_\lambda$ exists on G_λ and, by virtue of (100),

$$\| D^\alpha f \|_{p,\,G_\lambda} \leqslant C \sum_{l \in \mathscr{E}} \| D^l f \|_{p,\,G}.$$

Since this last inequality holds for every λ in $[1, \Lambda]$, this completes the proof of the theorem.

13.6.2. Theorem. *Suppose that* $f \in \mathscr{W}^{\{\mathscr{E}\}}(G)$, $1 < p < \infty$, $G \in \underline{A}(\square, H^{n-k})$, $1 \leqslant k \leqslant n$, *and* $\alpha = (\alpha_1, \ldots, \alpha_n)$ (*where the* α_j, *for* $j = 1, \ldots, n$, *are nonnegative integers) satisfies the condition*

$$\alpha^e \in M^e(\alpha) \quad \forall e\colon\ e^k \subseteq e \subseteq e^n \qquad (e^k = \{1, \ldots, k\}). \qquad (101)$$

Suppose also that there exists $D^r f \in L_q(G)$, *where* $0 \leqslant r_j \leqslant \alpha_j$ *(for* $j = 1, \ldots, n$*) and* $1 \leqslant q < \infty$.

Then, the function f has, on G, the generalized derivative $D^\alpha f \in L_p(G)$, *and inequality* (1) *with constant C independent of f holds for it.*

Proof of this theorem differs from the proof of Theorem 13.6.1 only in that here we apply not the identity (3) for the function f but the identity (7). By virtue of the results of subsection 13.3, to generalize that identity we need only note that $G \in A(\square, H^{n-k})$, that $D^r f \in L_q(G)$ exists on G, and that, since $r \leqslant \alpha$ and $q < \infty$, we have $\frac{1}{q} + \alpha_j - r_j > 0$ for $j = 1, \ldots, n$.

We note in particular the special case of Theorem 13.6.2 corresponding to $k = n$.

13.6.2′. Theorem. *Suppose that* $f \in \mathscr{W}_p^{\{\mathscr{E}\}}(G)$, *where* $1 < p < \infty$ *and* $G \in \underline{A}(\square)$, *that* $\alpha \in M^{e^n}$, *and that there exists* $D^r f \in L_q(G)$, *where* $r \leqslant \alpha$ *and* $1 \leqslant q < \infty$. *Then, the function f has on G a derivative $D^\alpha f$ and inequality* (1) *holds for it.*

13.6.3. Remark. In contrast with Theorem 13.6.1, we make in Theorem 13.6.2 the supplementary assumption that the function f has a derivative $D^r f \in L_q(G)$ although $\|D^r f\|_{q,G}$ does not appear in the right-hand member of inequality (1). Simple examples show that assumptions of this nature are necessary.

We note that if, for a given vector α, there exists a vector $l^0 \leqslant \mathscr{E}$ such that $l^0 \leqslant \alpha$, then the role of the derivative $D^r f$ in Theorem 13.6.2 can be played by $D^{l^0} f$ (with $q = p$) and then the only requirement we need to make on f is that $f \in \mathscr{W}_p^{\{\mathscr{E}\}}(G)$.

13.6.4. The necessity of conditions (99) and (101) in Theorems 13.6.1 and 13.6.2. Let us show that conditions (99) and (101) are significant for validity of the theorems.

Suppose that $0 \leqslant k \leqslant n$. We denote by $\square^{(k)}(R)$ the strip in the space E_x^n characterized by the inequalities $0 < x_i < \infty$ (for $i =$

$1, \ldots, k$) and $0 < x_i < R$ (for $i = k+1, \ldots, n$). In particular, $\square^{(0)}(R)$ is a rectangular parallelepiped with vertex at the coordinate origin.

Lemma. *Let $\mathscr{E}$ denote a given finite set of vectors $l \in E_l^n$ with nonnegative integer-valued components. Let α denote a vector with nonnegative integer-valued components. If for some e we have $e^k \subseteq e \subseteq e^n$ but $\alpha^e \notin M^e(\alpha)$, then, for any positive constant C, there exists a function*

$$f \in \mathscr{W}_p^{\{\mathscr{E}\}}(\square^{(k)}(R)) \cap C^\infty(\square^{(k)}(R))$$

such that $D^r f \in L_q(\square^{(k)}(R))$, where $r \leqslant \alpha$ and $1 \leqslant q < \infty$, for which inequality (1) (*with $G = \square^{(k)}(R)$*) *does not hold for that value of C.*

PROOF. Without loss of generality, we may assume that $e = e^m = \{1, \ldots, m\}$, where $k \leqslant m \leqslant n$. Since $\alpha^e \notin M^e(\alpha)$, there exists an $(m-1)$-dimensional plane separating the point α^e from $M^e(\alpha)$. Suppose that its equation is

$$(l^e, \varkappa^e) = l_1^e \varkappa_1 + \ldots + l_m^e \varkappa_m = d, \qquad d \geqslant 0.$$

Let us set

$$(\alpha^e, \varkappa^e) = \delta.$$

Then, either

$$\min_{l^e \in M^e(\alpha)} (l^e, \varkappa^e) > d > \delta \tag{102}$$

or

$$\max_{l^e \in M^e(\alpha)} (l^e, \varkappa^e) < d < \delta. \tag{103}$$

Let us suppose that inequalities (102) hold. Let us consider on $\square^{(k)}(R)$ the following sequence of functions belonging to the class $C^\infty(\square^{(k)}(R))$:

$$f_s(x) = s^{-\frac{\varkappa_1 + \dots + \varkappa_n}{p}} e^{-\left(s^{-\varkappa_1} x_1 + \dots + s^{-\varkappa_n} x_m\right)} x_{m+1}^{\alpha_{m+1}} \dots x_n^{\alpha_n}$$
$$(s = 1, 2, \dots).$$

Let us estimate the semi-norms of these functions in $\mathscr{W}_p^{\{\mathscr{E}\}}(\square^{(k)}(R))$. To do this, we represent $\mathscr{E}$ as the union of two sets:

$$\mathscr{E} = \mathscr{E}^e(\alpha) \cup C\mathscr{E}^e(\alpha),$$

where $C\mathscr{E}^e(\alpha)$ is the complement of the set $\mathscr{E}^e(\alpha)$ with respect to $\mathscr{E}$.

If $l \in C\mathscr{E}^e(\alpha)$, then at least one of its coordinates l_j, where $m + 1 \leqslant j \leqslant n$, is greater than the corresponding coordinate α_j of the vector α. Therefore,

$$D^l f_s(x) = 0 \quad \text{for} \quad l \in C\mathscr{E}^e(\alpha) \qquad (s = 1, 2, \dots).$$

Suppose that $l = (l^e, l^{e'}) \in \mathscr{E}^e(\alpha)$. Simple calculation shows that

$$\| D^l f_s \|_{p, \square^{(k)}(\mathscr{R})} = s^{-(l^e, \varkappa^e)} C_1(R, p) \prod_{i=k+1}^{m} \left(1 - e^{-Rps^{-\varkappa_i}}\right)^{1/p}$$
$$\leqslant s^{-d} C_1(R, p) \prod_{i=k+1}^{m} \left(1 - e^{-Rps^{-\varkappa_i}}\right)^{1/p} \qquad (s = 1, 2, \dots),$$

where $C_1(R, p)$ is a constant depending on R and p but not on s.

It follows from what has been said that

$$\sum_{l \in \mathscr{E}} \| D^l f_s \|_{p, \square^{(k)}(R)} \leqslant s^{-d} C_2(R, p) \prod_{i=k+1}^{m} \left(1 - l^{-Rps^{-\varkappa_i}}\right)^{1/p} \tag{104}$$
$$(s = 1, 2, \dots).$$

Obviously, for arbitrary $r \geqslant 0$ and $q \geqslant 1$, we have $D^r f_s \in L_q(\square^{(k)}(R))$.

Thus, the functions f_s, for $s = 1, 2, \dots$, satisfy all the conditions of the lemma.

On the other hand, one can easily show that

$$\| D^{\alpha} f_s \|_{p,\, \square^{(k)}(R)} = s^{-\delta} C_3(R, p) \prod_{i=k+1}^{m} \left(1 - e^{-Rps^{-\varkappa_i}}\right)^{1/p} \tag{105}$$
$$(s = 1,\ 2,\ \ldots),$$

where $C_3\ (R,\ p)$ is positive and independent of s.

Since $\delta < d$, it follows from (104) and (105) that, for any constant $C > 0$, there exists a number s_0 such that inequality (1) does not hold with that constant and f_s if $s \geqslant s_0$. The assertion of the lemma is proven under conditions (102).

If inequalities (103) are satisfied, we arrive at an analogous result by considering the following sequence of functions:

$$f_s(x) = s^{\frac{\varkappa_1 + \ldots + \varkappa_m}{p}} e^{-\left(s^{\varkappa_1} x_1 + \ldots + s^{\varkappa_m} x_m\right)} x_{m+1}^{\alpha_{m+1}} \ldots x_n^{\alpha_n}$$
$$(s = 1,\ 2,\ \ldots).$$

13.7. Examples. Let us first look at a simple example of an application of Theorem 13.6.1 and let us show with it that the requirement that the region G belong to the class $\underline{A}(\square,\ H)$ is needed.

If $n = 2$, the set $\mathscr{E}$ consists of three vectors

$$l^1 = (1,\ 0), \quad l^2 = (0,\ 1), \quad l^3 = (2,\ 2),$$

and $\alpha = (1,\ 1)$.

Suppose that $e_0 = \varnothing$, $e_1 = \{1\}$, $e_2 = \{2\}$, and $e_3 = \{1,\ 2\}$. These are all subsets of $e^2 = \{1,\ 2\}$. Then (see 13.2) $\mathscr{E}^{e_0}(\alpha) = \{l^1,\ l^2\}$, $\mathscr{E}^{e_1}(\alpha) = \{l^1,\ l^2\}$, $\mathscr{E}^{e_2}(\alpha) = \{l^1,\ l^2\}$, $\mathscr{E}^{e_3} = \mathscr{E} = \{l^1,\ l^2,\ l^3\}$, $M^{e_0}(\alpha)$ consists of the point 0 (the coordinate origin), $M^{e_1}(\alpha) = [0,\ l^1]$ is the segment on the l_1-axis connecting the points 0 and l^1, $M^{e_2}(\alpha) = [0,\ l^2]$ is a segment on the l_2-axis, and $M^{e_3}(\alpha) = M^{e^2}$ is the closed triangle with vertices at the points l^i (for $i = 1,\ 2,\ 3$).

One can easily see that $\alpha^e \in M^e(\alpha)$ for all $e \subseteq e^2$, that is, that the vector α satisfies conditions (99) of Theorem 13.6.1. It follows from this theorem that, if $G \in \underline{A}(\square, H)$, where $H = (H_1, H_2)$, and $f \in \mathcal{W}_p^{\{\mathcal{E}\}}(G)$, where $1 < p < \infty$, then $D^\alpha f = \dfrac{\partial^2 f}{\partial x_1\, \partial x_2}$ exists on G and there exists a constant C independent of f such that

$$\left\| \frac{\partial^2 f}{\partial x_1\, \partial x_2} \right\|_{p,\, G} \leqslant C \left(\left\| \frac{\partial f}{\partial x_1} \right\|_{p,\, G} + \left\| \frac{\partial f}{\partial x_2} \right\|_{p,\, G} + \left\| \frac{\partial^4 f}{\partial x_1^2\, \partial x_2^2} \right\|_{p,\, G} \right). \tag{106}$$

(Inequality (106) was mentioned in subsection 7.10.)

Consider now the region G bounded by the lines $y = 0, y = 1/2$, and $x = 1$ and the curve $x = y^a$, where $1 \leqslant a < \infty$. Obviously, for no $H_i > 0$ (for $i = 1, 2$) is G a member of $\underline{A}(\square, H)$.

Suppose that

$$f_s(x_1, x_2) = x_2 e^{-s x_1} \qquad (s = 1, 2, \ldots)$$

is a sequence of functions defined on G. Obviously, $f_s \in \mathcal{W}_p^{\{\mathcal{E}\}}(G)$ for $s = 1, 2, \ldots$, and, as one can easily show,

$$\left\| \frac{\partial f_s}{\partial x_1} \right\|_{p,\, G} + \left\| \frac{\partial f_s}{\partial x_2} \right\|_{p,\, G} + \left\| \frac{\partial^4 f_s}{\partial x_1^2\, \partial x_2^2} \right\|_{p,\, G} = O\left(s^{1 - \frac{1}{p} - \frac{1}{ap} - \frac{1}{a}} \right).$$

At the same time,

$$\left\| \frac{\partial^2 f_s}{\partial x_1\, \partial x_2} \right\|_{p,\, G} = O\left(s^{1 - \frac{1}{p} - \frac{1}{ap}} \right).$$

Comparing these relationships, we see that for no single positive constant C is inequality (106) satisfied for all f_s (where $s = 1, 2, \ldots$). This example shows that the requirement that the region G belong to the class $\underline{A}(\square, H)$ is in a certain sense necessary for validity of (106).

Let us now look at some examples of application of the theorems of subsection 13.6 for the case $n = 2$.

In Figures 7–13, the points l^i represent the points of the corresponding sets $\mathscr{E}$.

We denote by $\Omega_1(\mathscr{E})$ the set of points α that have nonnegative-integer-valued components α_j for $j = 1, 2, \ldots, n$, that belong to the closed polyhedron M spanned by the points of the set $\mathscr{E}$, and that satisfy conditions (99) of Theorem 13.6.1, that is, points such that $\alpha^e \in M^e(\alpha)$ for all $e \subseteq e^2 = \{1, 2\}$.

We denote by $\Omega_2(\mathscr{E})$ the set of points $\alpha \in M$ satisfying the conditions of Theorem 13.6.2 with $e^k = \{1\}$ and $r \in \mathscr{E}$, that is, points satisfying the following two conditions: 1) $\alpha^e \in M^e(\alpha)$ for all e such that $\{1\} \subseteq e \subseteq \{1, 2\}$ (such e are $e = \{1\}$ and $e = \{1, 2\}$); 2) for every point $\alpha \in \Omega_2(\mathscr{E})$ there exists a point $r_\alpha = l^i \in \mathscr{E}$ for which $l_j^i \leqslant \alpha_j$ (for $j = 1, 2$).

We denote by $\Omega_{2'}(\mathscr{E})$ the set of points α satisfying the conditions of Theorem 13.6.2′ with $r \in \mathscr{E}$, that is, the set of points $\alpha \in M$ for each of which there exists a point $r'_\alpha = l^i \in \mathscr{E}$ such that $l_j^i \leqslant \alpha_j$ for $j = 1, 2$.

We note that $\Omega_1(\mathscr{E}) \subseteq \Omega_2(\mathscr{E}) \subseteq \Omega_{2'}(\mathscr{E})$. The second of these inclusion relations is obvious, and the first follows from the fact that, for every point $\alpha \in \Omega_1(\mathscr{E})$, the set $\mathscr{E}^{\varnothing}(\alpha)$ is nonempty (in the opposite case, we would have $\alpha^{\varnothing} \notin M^{\varnothing}(\alpha)$) and any point $l \in \mathscr{E}^{\varnothing}(\alpha)$ can be taken for r_α in the definition of points of the set $\Omega_2(\mathscr{E})$ since its coordinates satisfy the inequalities $l_j \leqslant \alpha_j$ for $j = 1, 2$.

In each of Figures 7–10, the set Ω_1 is shaded with vertical lines, its complement with respect to the set Ω_2 is shaded with horizontal lines, and the complement of Ω_2 with respect to $\Omega_{2'}$ is shaded by slanting lines. We note that, in Figure 9, Ω_1 consists of points α lying on the segments l^1N and Nl^3 and of the point l^2. In Figure 10, Ω_1 consists only of points in l^1, l^2, and l^3. All the other sets are clear from the drawings.

In Figures 11–13, the sets $\mathscr{E}$ consist of two points, and the corresponding polyhedrons M are degenerate, consisting of segments. In Figure 11, the segment l^1l^2 is parallel to the l_1-axis. In this case,

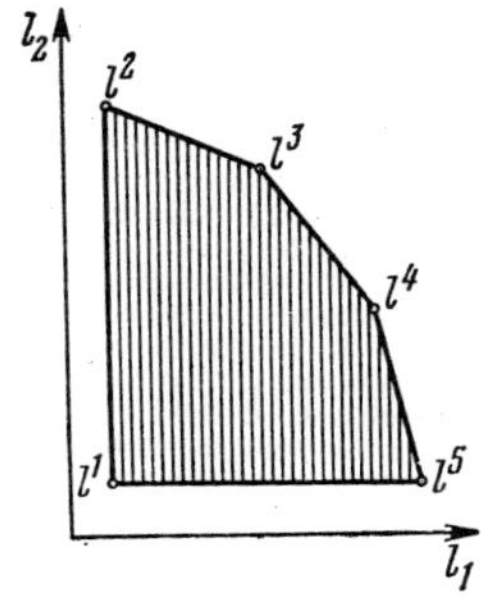

FIG. 7

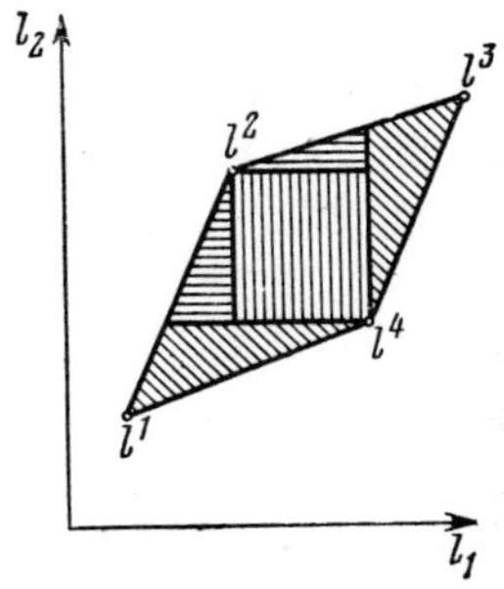

FIG. 8

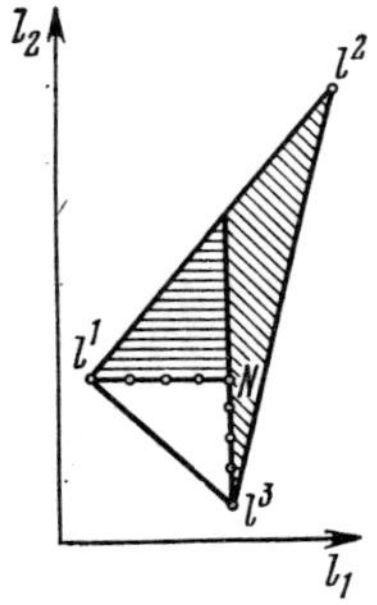

FIG. 9

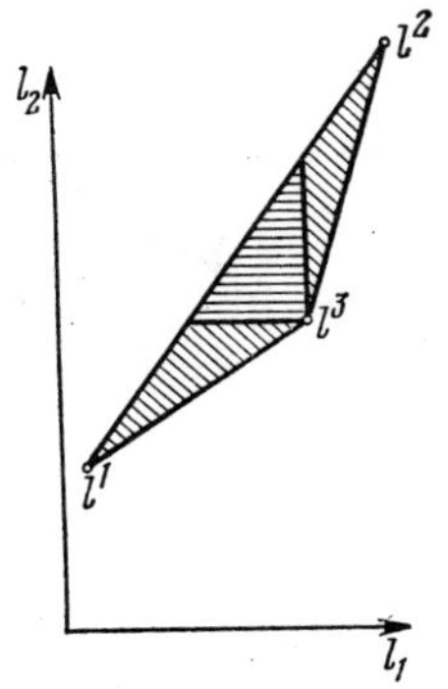

FIG. 10

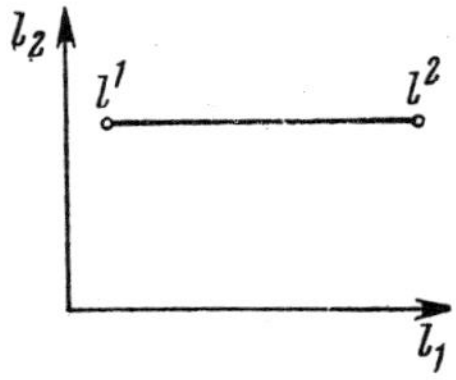

FIG. 11

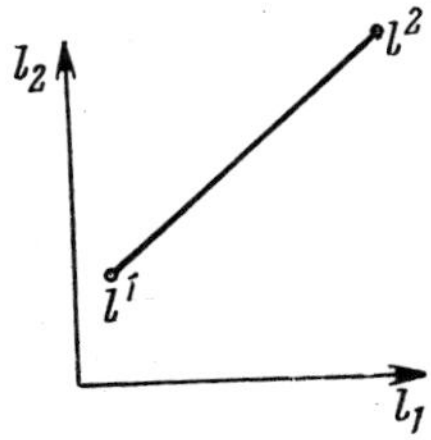

FIG. 12

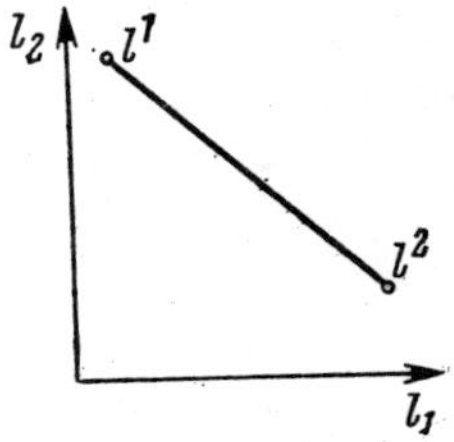

FIG. 13

$\Omega_1 = \Omega_2 = \Omega_{2'}$, and all points of the segment $l^1 l^2$ with integer-valued coordinates belong to these sets. In Figure 12, Ω_1 and Ω_2 consist only of the points l^1 and l^2, and Ω_2' consists of all points α of the segment $l^1 l^2$. For each of them we can take the point l^1 as r_α. In Figure 13, the three sets Ω_1, Ω_2, and $\Omega_{2'}$ all consist only of the points l^1 and l^2.

Suppose that $f \in \mathscr{W}_p^{\{\mathscr{E}\}}(G)$. Then, on the basis of theorems 13.6.1 and 13.6.2, we conclude that, if $G \in \underline{A}(\square, H)$, where $H = (H_1, H_2)$, then inequality (1) will hold for $\alpha \in \Omega_1(\mathscr{E})$. If $G \in \underline{A}(\square, H_2)$, then (1) holds for $\alpha \in \Omega_2(\mathscr{E})$. If $G \in \underline{A}(\square)$, then (1) holds for $\alpha \in \Omega_2$ $(\mathscr{E})$.

We note also that, in the case of the set $\mathscr{E}$ shown in Figure 13, inequality (1) for the functions in the class $\mathscr{W}_p^{\{\mathscr{E}\}}(G)$ (with constant C independent of f) does not hold for any α different from l^1 and l^2 even if $G \in \underline{A}(\square)$. However, if we consider the class of functions $f \in \mathscr{W}_p^{\{\mathscr{E}\}}(G)$, with derivative $D^r f \in L_q(G)$, where $r \leqslant l^1$, l^2 and $1 \leqslant q < \infty$, then, for $G \in \underline{A}(\square)$, inequality (1) holds for arbitrary α belonging to the segment $l^1 l^2$ although $\|D^r f\|_{q,G}$ does not belong to that segment. In this case, the conditions of Theorem 13.6.2$'$ are satisfied.

§14. The behavior of functions in W_p^l at ∞ and the density of C_0^∞ in W_p^l

In this section, we shall show that, when certain relationships between the parameters l, p, and n are satisfied, then functions in $W_p^l(E^n)$ tend to polynomials at infinity (more accurately, are in a certain sense approximated by polynomials).

For a function $f(x)$ with finite seminorm

$$\sum_{i=1}^n \left\| \frac{\partial}{\partial x_i} f \right\|_{p,\,E^n},$$

where $p < n$, Uspenskiĭ [3] has established the existence of a constant $c = c(f)$ that $f(x)$ approaches along all rays and he has given an estimate for $f(x) - c$ in terms of the derivatives $\frac{\partial}{\partial x_i} f$. Sobolev [3] has studied the general case connected with the seminorm

$$\sum_{|\alpha|=m} \|D^\alpha f\|_{p,\,E^n}.$$

In the same case, Sedov [1] then established a property of functions that tend to polynomials in terms of estimates of their norms in L_q on E^n and planes of different dimensions.

Connected with these properties of the functions is the possibility of approximating them with infinitely differentiable functions of compact support (see Sobolev [3]).

Here, we shall study the behavior, at infinity, of functions with various differentiability properties with respect to different directions and we shall investigate approximation of them by means of functions in C_0^∞.

In this section, we shall examine an unbounded region G contained in E^n and including 0 for which there exists an l-horn $V = V(l, \infty)$ such that $G + V(l, \infty) = G$. In our subsequent estimates, we shall assume that $1 \leqslant p \leqslant q \leqslant \infty$.

14.1. Again, let $W_p^l(G, \square)$ denote a function space with norm 9(3) (a cube $\square \subset G$)

$$\|f\|_{W_p^l(G,\,\square)} = \|f\|_{p,\,\square} + \|f\|_{L_p^l(G)}, \tag{1}$$

where the seminorm

$$\|f\|_{L_p^l(G)} = \sum_{i=1}^{n} \|D_i^{l_i} f\|_{p,\,G}. \tag{2}$$

Let us partition the space $W_p^l(G, \square)$ into classes of functions, considering two functions to belong to the same class if and only if their difference coincides almost everywhere on G with a polynomial $P_{l-1}(x)$ of degree $l_i - 1$ in x_i for each i. We denote by $L_p^l(G)$ the space of these classes with norm (2) (calculated with respect to any representative). Thus, the space $L_p^l(G)$ is a factor space of the space $W_p^l(G)$ (with respect to the space of polynomials P_{l-1}) and hence a Banach space (see Kantorovič and Akilov [1]).

Suppose that $\lambda_i = \frac{1}{l_i}$ and $|\lambda| = |1:l| = \sum_1^n \lambda_i$. For $|\alpha : l| \leqslant 1$ and $0 < \varepsilon < h < \infty$, we have on the basis of 7(13), 7(22), and 7(11),

$$D^\alpha f_{h^\lambda}(x) - D^\alpha f_{\varepsilon^\lambda}(x)$$

$$= -(-1)^{|\alpha|} \int_\varepsilon^h \sum_{i=1}^n v^{-|\lambda|-(\alpha,\lambda)}\, dv \int_{E^n} D_i^{l_i} f(x+y)\, \mathscr{L}_i^{(\alpha)}(y : v^\lambda)\, dy$$

$$= D^\alpha f_{\varepsilon, h}(x), \tag{3}$$

where

$$f_{v^\lambda}(x) = v^{-|\lambda|} \int_{E^n} f(x+y)\, \Omega(y : v^\lambda)\, dy. \tag{4}$$

In (3) and (4), we shall assume that $x \in G$, that the kernels $\Omega(y)$ belong to C_0^∞, and that the $\mathscr{L}_i(y)$ belong to C_0^∞ and are concentrated in the horn $V(l)$, so that the support of the representation (3), (4) is contained in the translated horn $x + V(l)$. By virtue of 7(26),

$$\int \mathscr{L}_i^{(\alpha)}(y)\, dy = 0 \qquad (i = 1, \ldots, n). \tag{5}$$

In connection with the representation (3), (4), let us examine the properties of the integral

$$\mathscr{I}_{\varepsilon, h}(x) = \int_\varepsilon^h \int v^{-|\lambda|-(\alpha,\lambda)} g(x+y)\, \mathscr{L}_i^{(\alpha)}(y : v^\lambda)\, dy\, dv, \tag{6}$$

$$0 < \varepsilon < h < \infty,$$

where $g(x) \in L_p(E^n)$ and $g(x) = 0$ on $E^n \setminus G$.

In what follows, we shall use the following notation (where $q \geqslant p$):

$$\varkappa = \left(\frac{1}{p} - \frac{1}{q} + \alpha, \lambda\right) = \left|\left(\frac{1}{p} - \frac{1}{q} + \alpha\right) : l\right|, \quad \varkappa_\infty = \left(\frac{1}{p} + \alpha, \lambda\right). \tag{7}$$

Lemma. For $\varkappa_\infty > 1$, the integral (6) converges uniformly on E^n as $h \to \infty$. If $\varkappa = 1$ and if one of the following conditions

$$1 < p = q < \infty,$$
$$1 < p_n < q_n < \infty,$$
$$1 = p_n < q_n = \infty$$

holds, then the integral (6) converges in $L_q(E^n)$ as $\varepsilon \to 0$.

PROOF. By virtue of Minkowski's generalized inequality 2(12) and Young's inequality 2(18) for $q = \infty$ and $\frac{1}{p} + \frac{1}{r} = 1$, we see that

$$\| \mathcal{I}_{\varepsilon, h} \|_\infty \leqslant \int_\varepsilon^h \| g \|_p v^{-|\lambda| - (\alpha, \lambda)} \| \mathcal{L}_i^{(\alpha)}(\cdot, v^\lambda) \|_r \, dv$$
$$\leqslant C \| g \|_p \int_\varepsilon^h v^{-\left(\frac{1}{p} + \alpha, \lambda\right)} dv \to 0$$

as ε and h approach ∞.

In the case $1 < p = q < \infty$, the second assertion of the lemma follows from Theorem 4.5. In the other two cases, this assertion is contained in the proof of Lemma 10.1.

14.2. Lemma. *Suppose that* $f \in W_p^l(G, \square)$ *and* $\varkappa_\infty > 1$. *Then,* $D^\alpha f_{h^\lambda}(x)$ *converges uniformly on* G *as* $h \to \infty$. *If in addition* $\alpha_j \geqslant l_j$ *for some* j, *then* $D^\alpha f_{h^\lambda}(x) \to 0$.

Proof of the first assertion of the lemma follows from the representation (3) and the convergence of $D^\alpha f_{\varepsilon, h}(x)$ as $h \to \infty$, which follows from Lemma 14.1 with $g(x) = D_i^{l_i} f(x)$ for $x \in G$ and $g(x) = 0$ for $x \in E^n \setminus G$.

Let us prove the second part of the lemma. Let us set $\alpha = (0, \ldots, 0, l_j, 0, \ldots, 0) + \beta$. Then,

$$D^\alpha f_{h^\lambda}(x) = (-1)^{|\beta|} h^{-|\lambda| - (\beta, \lambda)} \int D_j^{l_j} f(x+y) \Omega^{(\beta)}(y : h^\lambda) \, dy,$$

and Hölder's inequality leads to the estimate

$$\| D^\alpha f_{h^\lambda} \|_{\infty, G} \leqslant C h^{-(\beta, \lambda) - \left(\frac{1}{p}, \lambda\right)} \| D_j^{l_j} f \|_{p, G} = C h^{1 - \varkappa_\infty} \| D_j^{l_j} f \|_{p, G},$$

from which the assertion of the lemma follows.

14.3. Lemma. *For every function* $f \in W_p^l(G, \square)$, *there exists a polynomial*

$$Q_{l-1}(x; f) = Q_{l-1}(x; f; p) = \sum_{\alpha \leqslant l-1} C_\alpha x^\alpha, \tag{8}$$

$$C_\alpha = 0 \quad \text{for} \quad \left(\frac{1}{p} + \alpha, \lambda\right) \leqslant 1, \tag{9}$$

such that, for $\varkappa_\infty > 1$,

$$D^\alpha f_{h^\lambda}(x) \to D^\alpha Q_{l-1}(x, f) \tag{10}$$

uniformly on G *as* $h \to \infty$.

PROOF. Let us set

$$C_\alpha = \lim_{h \to \infty} \frac{1}{\alpha!} D^\alpha f_{h^\lambda}(0) \qquad \text{for} \qquad \varkappa_\infty > 1. \tag{11}$$

By Lemma 14.2, these limits exist.

For $\varkappa_\infty = \left(\frac{1}{p} + \alpha, \lambda\right) > 1$, let us expand $D^\alpha f_{h^\lambda}(x)$ according to Taylor's formula in powers of x keeping terms up to those of sufficiently high power. By virtue of (11), all the coefficients in the expansion approach limiting values as $h \to \infty$. If α is no less than l, then, by Lemma 14.2, $D^\alpha f_{h\lambda}(x)$ approaches 0 uniformly on G. It obviously follows that the remainder term of Taylor's formula approaches 0 uniformly on every bounded subset of G.

Thus, (10) holds uniformly on every bounded subset of G. By virtue of Lemma 14.2, (10) is also satisfied uniformly on G.

14.4. Theorem. *Suppose that* $f(x) \in W_p^l(G, \square)$, *that* $\varkappa = 1$, *and that we have one of the following relations*

$$1 < p = q < \infty,$$
$$1 < p_n < q_n < \infty,$$
$$1 = p_n < q_n = \infty.$$

Then, equations (8), (9), *and* (11) *imply the existence of a constant* C *independent both of* f *and of* ε *such that*

$$\left\| D^\alpha \left[f - Q_{l-1}(\cdot\,;\, f) \right] \right\|_{q,\,G} \leqslant C \| f \|_{L_p^l(G)} \tag{12}$$

and, for arbitrary $\varepsilon > 0$,

$$\left\| D^\alpha \left[f_{\varepsilon\lambda} - Q_{l-1}(\cdot\,;\, f) \right] \right\|_{q,\,G} \leqslant C \| f \|_{L_p^l(G)}. \tag{13}$$

On the other hand, if $\varkappa > 1$ *and* $1 \leqslant p \leqslant q \leqslant \infty$, *then there exists a constant* $C(\varepsilon, \alpha)$ *independent of* f *such that, for arbitrary* $\varepsilon > 0$,

$$\left\| D^\alpha \left[f_{\varepsilon\lambda} - Q_{l-1}(\cdot\,;\, f) \right] \right\|_{q,\,G} \leqslant C(\varepsilon, \alpha) \| f \|_{L_p^l(G)}. \tag{14}$$

PROOF. For $\varkappa = 1$, we obtain from (3) on the basis of Theorem 4.5 and inequality 10(6) the estimate

$$\left\| D^\alpha f_{h\lambda} - D^\alpha f_{\varepsilon\lambda} \right\|_{q,\,G} \leqslant C \sum_{i=1}^{n} \| D_i^{l_i} f \|_{p,\,G} \qquad (0 < \varepsilon < h < \infty),$$

in which C is independent of both ε and h. If we take the limit as

$h \to \infty$ in the left-hand member of this estimate, we obtain on the basis of Lemma 14.3 inequality (13). To obtain the esitmate (12), we need now only take the limit as $\varepsilon \to 0$ since

$$D^{\alpha} f_{\varepsilon\lambda} \to D^{\alpha} f$$

in $L_p^{\text{loc}}(G)$ as $\varepsilon \to 0$.

This last relation is valid by virtue of Lemma 6.2, the conditions of which are satisfied. Specifically, $f_{\varepsilon\lambda} \to f$ in the sense of $L_p^{\text{loc}}(G)$ by a property of averages. At the same time, by virtue of the representation (3) and Lemma 14.1, $D^{\alpha} f_{\varepsilon\lambda} - D^{\alpha} f_{h\lambda} \to 0$ in $L_q^{\text{loc}}(G)$ as ε and h approach 0. The estimate (14) is proven in a manner analogous to the proof of (13). Instead of Theorem 4.5 and the estimate 10(6), we need only apply Minkowski's generalized inequality 2(12) and Young's inequality 2(18).

This theorem shows that the defined derivatives of $f(x)$ or $f_{\varepsilon\lambda}(x)$ are close "at infinity" to the corresponding derivatives of the polynomial $Q_{l-1}(x; f)$ in the sense of finiteness of the left-hand members of (12) and (14). *For brevity, we shall say that the function $f(x)$ tends to the polynomial $Q_{l-1}(x; f)$ at infinity.*

We note that, for a given function $f(x)$, there exists no more than one polynomial of the form

$$Q_{l-1}(x) = \sum_{\alpha \leqslant l-1} c_{\alpha} x^{\alpha}, \quad c_{\alpha} = 0 \quad \text{for} \quad \left(\frac{1}{p} + \alpha, \lambda\right) \leqslant 1,$$

for which the left-hand members of (12) and (14) are finite even if we assume that in them $q = q^{(\alpha)}$ is a single-valued function of α defined for $|\alpha : l| > 1 - \left|\frac{1}{p} : l\right|$, where $\alpha \leqslant l - 1$.

To see this, suppose that there exist two such polynomials $Q_{l-1}(x)$ and $Q^{*}_{l-1}(x)$. Then, we have

$$D^{\alpha}\left[Q_{l-1}(x) - Q^{*}_{l-1}(x)\right] \in L_{q^{(\alpha)}}(G),$$

which is possible only if their coefficients c_{α} and c^{*}_{α} are equal for all $\alpha \leqslant l - 1$.

14.5. Let us look at the set $\overset{*}{W}{}_p^l(G, \square)$ of functions in $W_p^l(G, \square)$ that vanish at infinity. Let us show that this set is a subspace of the space $W_p^l(G, \square)$ with finite norm

$$\|f\|_{\overset{*}{W}{}_p^l(G,\square)} = \|f\|_{W_p^l(G,\square)} + \sum_{\alpha \in A_p} \left\| D^\alpha f_{\varepsilon\lambda} \right\|_{\infty, G}, \qquad \varepsilon > 0, \quad (15)$$

where

$$A_p = \left\{\alpha:\ \alpha \leqslant l-1,\ |\alpha : l| > 1 - \left|\frac{1}{p} : l\right|\right\}.$$

(For certain values of p and l, the set A_p is empty; ε is any positive number and we may assume, for example, that $\varepsilon = 1$.)

For every function $f \in W_p^l(G, \square)$ that tends to a polynomial at infinity, the norm (15) is, by virtue of (14), finite. Conversely, every function f with finite norm (15) must, since it is a function in $W_p^l(G, \square)$, tend to the polynomial $Q_{l-1}(x; f)$ at infinity. The uniqueness of such a polynomial implies that it coincides with the zero polynomial since the left-hand member of (14) is, by virtue of the finiteness of the norm (15), finite for $Q_{l-1} \equiv 0$.

Thus, *the set* $\overset{*}{W}{}_p^l(G, \square)$ *of functions in* $W_p^l(G, \square)$ *that approach zero at infinity is a Banach space with norm* (15) (*its completeness is obvious*). *By virtue of* (14) *the norms* (1) *and* (15) *are equivalent on the set* $\overset{*}{W}{}_p^l(G, \square)$, *so that* $\overset{*}{W}{}_p^l(G, \square)$ *is a subspace of the space* $W_p^l(G, \square)$.

We mention in passing that, instead of the norm (15), we could just as well introduce on $\overset{*}{W}{}_p^l(G, \square)$ the norm

$$\|f\|_{W_p^l(G,\square)} + \sum_{\alpha \in A_p} \left\| D^\alpha f_{\varepsilon\lambda} \right\|_{q^{(\alpha)}, G}, \qquad \varepsilon > 0, \quad (16)$$

where $q^{(\alpha)}$ is a single-valued function of $\alpha \in A_p$ for which $q^{(\alpha)} \geqslant p$, $\varkappa \geqslant 1$, and the estimates (13) and (14) are satisfied. If, for some $\alpha \in A_p$, the estimate (12) is satisfied for $q^{(\alpha)}$, then, in the corresponding terms of the norm (16), we can assume $\varepsilon = 0$.

We note also that the number of terms in the summation $\sum_{\alpha \in A_p}$ in the norm (15) or the norm (16) can in some cases be decreased if in so doing we can preserve the uniqueness, which is important to us: if a polynomial P_{l-1} is such that $\| P_{l-1} \|_{\overset{*}{W}{}_p^l(G, \square)} < \infty$, then it has the form

$$P_{l-1}(x) = \sum_{|\alpha : l| \leqslant 1 - \left|\frac{1}{p} : l\right|} c_\alpha x^\alpha.$$

Since a function of compact support $\varphi(x) \in C_0^\infty(E^n)$ vanishes at infinity, we conclude on taking the limit in (14) that every function $f(x)$ that can be approximated with an arbitrary degree of accuracy in the norm $W_p^l(G, \square)$ by functions in $C_0^\infty(E^n)$ also vanishes, that is, belongs to $\overset{*}{W}{}_p^l(G, \square)$. We shall show also that, conversely, every function $f(x) \in \overset{*}{W}{}_p^l(G, \square)$, for $1 < p < \infty$, can be approximated with an arbitrary degree of accuracy in the norm $W_p^l(G, \square)$ by functions in $C_0^\infty(E^n)$. In so doing, we shall show that $\overset{*}{W}{}_p^l(G, \square)$ *coincides with the closure, in* $W_p^l(G, \square)$, *of the set of functions* $C_0^\infty(E^n)$ *for* $1 < p < \infty$.

14.6. Lemma. *Every monomial* x^α *such that* $|\alpha : l| \leqslant 1 - \left|\frac{1}{p} : l\right|$, *where* $1 < p_n < \infty$, *can be approximated by functions in* $C_0^\infty(E^n)$ *with an arbitrary degree of accuracy in* $\overset{*}{W}{}_p^l(G, \square)$.

PROOF. Consider an infinitely differentiable function defined on E^1 such that

$$\psi(t) = \begin{cases} 1, & t \leqslant 1, \\ 0, & t \geqslant 2. \end{cases}$$

Let us show that the norm in $W_p^l(E^n, \square)$ of the difference

$$g_\eta(x) = x^\alpha - x^\alpha \psi\left(\frac{\ln \sum_1^n x_j^{2l_j}}{\ln \eta}\right)$$

approaches zero as $\eta \to +\infty$. When we do, this will prove the lemma.

For sufficiently large η, we have $\| g_\eta \|_{p, \square} = 0$, so that we need only to estimate the terms $\sum_{i=1}^n \left\| D_i^{l_i} g_\eta \right\|_{p, E^n}$ of the norm (1), (2). We note that $D_i^{l_i} x^\alpha = 0$. By virtue of Leibnitz's formula for differentiating a product and the properties of the function $\psi(t)$, we have, for $\eta > 2$,

$$\left| D_i^{l_i}\left[x^\alpha \psi\left(\frac{\ln \sum x_j^{2l_j}}{\ln \eta}\right)\right]\right|$$

$$\leqslant C \left| \frac{x^\alpha}{x_i^{\alpha_i}} \right| \left| \sum_{k=0}^{\alpha_i} \right| x_i^{\alpha_i - k} \frac{\partial^{l_i - k}}{\partial x_i^{l_i - k}} \psi\left(\frac{\ln \sum x_j^{2l_j}}{\ln \eta}\right)\Bigg|$$

$$\leqslant C_1 \frac{|x^\alpha| \, |x_i|^{l_i}}{\ln \eta \sum_1^n x_j^{2l_j}} \sum_{k=0}^{\alpha_i} \left| \psi^{(l_i - k)}\left(\frac{\ln \sum x_j^{2l_j}}{\ln \eta}\right)\right|.$$

Remembering that the right-hand member of this inequality can be nonzero only for $\eta \leqslant \sum_1^n x_j^{2l_j} \leqslant \eta^2$, let us estimate it by

$$C_2 \frac{\left(\sum_1^n x_j^{2l_j}\right)^{\frac{1}{2}|\alpha : l| - \frac{1}{2}}}{\ln \sum_1^n x_j^{2l_j}} \leqslant C_3 \frac{\left(\sum_1^n x_j^{2l_j} + \eta\right)^{\frac{1}{2}|\alpha : l| - \frac{1}{2}}}{\ln\left(x_n^{2l_n} + \eta\right)} = \bar{g}_\eta(x).$$

The norm $\| \bar{g}_\eta \|_{p, E^n}$ is estimated by making the substitution $x_j =$

$\eta^{\lambda_j l/2} y_j$ for $(j = 1, \ldots, n)$ and then calculating the integrals with respect to $y_1, \ldots, y_n$, which under the conditions of the lemma converge. The last of them converges because $p_n > 1$. We finally obtain the result that, as $\eta \to +\infty$,

$$\sum_{i=1}^{n} \| D_i^{l_i} g_\eta \|_{p, E^n} \leqslant n \| \bar{g}_\eta \|_{p, E^n}$$

$$\leqslant C_4 \eta^{\frac{1}{2}\left(\alpha + \frac{1}{p}, \lambda\right) - \frac{1}{2}} \left\{ \int_0^\infty \frac{1}{y_n + 1} \left[\frac{1}{\ln \left(y_n^{2l_n} + 1\right) + \ln \eta} \right]^{p_n} dy_n \right\}^{1/p_n} \to 0.$$

We have thus proven the possibility of approximating the monomial x^α to an arbitrary degree of accuracy in $W_p^l(G, \square)$, where $1 < p_n < \infty$. Since the norms $W_p^l(G, \square)$ and $\overset{*}{W}{}_p^l(G, \square)$ are equivalent on functions of compact support, the same holds true in $\overset{*}{W}{}_p^l(G, \square)$.

We note that, for $|\alpha : l| < 1 - \left|\frac{1}{p} : l\right|$, we can take as cut-off function the simpler function

$$\psi\left(\frac{\sum_1^n x_j^{2l_j}}{\eta}\right).$$

14.7. Lemma. *Suppose that* $1 \leqslant p < \infty$, $p \neq 1$, *and* $g(x) \in L_p(E^n)$. *Then,* $\| g_{h\lambda} \|_{p, E^n} \to 0$ *as* $h \to +\infty$.

PROOF. By virtue of the fact that the set C_0^∞ is dense in $L_p(E^n)$ (see Theorem 1.6), we can, on the basis of the estimate 7(8), assume that

$$|g(x)| \leqslant N \quad \text{on} \quad E^n, \quad g(x) = 0 \quad \text{for} \quad |x| \geqslant R.$$

But in this case,

$$|g_{h\lambda}(x)| \leqslant C h^{-|\lambda|} \int |g(x+y)|\, dy \leqslant C_1 h^{-|\lambda|} R^n,$$

and $g_{h^\lambda}(x)=0$ outside the parallelepiped $\{x: |x_i|<R+ah^{\lambda_i}, i=1, \ldots, n\}$, where the positive constant a is determined by the position of the support of the averaging kernel Ω.

Then, by direct calculation, we obtain

$$\|g_{h^\lambda}\|_p \leqslant C_1 h^{-|\lambda|} R^n \prod_{i=1}^{n} (2R+2ah^{\lambda_i})^{\frac{1}{p_i}} \leqslant$$
$$\leqslant C(R)\, h^{-\left(1-\frac{1}{p},\, \lambda\right)} \to 0 \qquad (h\to+\infty),$$

as we needed to show.

14.8. Lemma. *For a sufficiently smooth function $f(x)$ defined on the cube*

$$\square=\{x:\ a_i\leqslant x_i\leqslant a_i+\delta,\quad i=1, \ldots, n\},$$

there exists a polynomial

$$P(x;\ f)=\sum_{|\alpha:l|\leqslant 1-\left|\frac{1}{p}:l\right|} c_\alpha x^\alpha \tag{17}$$

linearly dependent on f such that

$$\|f-P(\cdot\,;\ f)\|_{p,\,\square}\leqslant C \sum_{1<\left|\left(\alpha+\frac{1}{p}\right):l\right|\leqslant 2} \max_{x\in\square} |D^\alpha f(x)|. \tag{18}$$

PROOF. For $P(x;\ f)$ that satisfies (17) and (18), the operator

$$\left(\prod f\right)(x)=P(x;\ f)$$

is a projection operator defined on the space of functions $f(x)$ defined on $\square$ with continuous derivatives $D^\alpha f$, for $|\alpha:l|\leqslant 2-\left|\frac{1}{p}:l\right|$.

As will be seen from the construction, we can arrange for the operator $\prod f$ to map, for example, the space $L_1(\square)$ into itself. However, this last requirement will play no role in the present subsection.

Suppose that the function $g(t)$ has a generalized derivative of order $s + 1$ that is summable on (a, b). Then, the lower-order derivatives are continuous on (a, b) and Taylor's formula with a remainder in integral form is valid:

$$g(t) = \sum_{k=0}^{s} \frac{1}{k!} g^{(k)}(t^{(0)})(t - t^{(0)})^k + \frac{(t - t^{(0)})^{s+1}}{s!} \int_0^1 (1 - \xi)^s g^{(s+1)}(t^{(0)} + (t - t^{(0)})\xi)\, d\xi. \quad (19)$$

$$(t \in (a, b),\ t^{(0)} \in (a, b))$$

Let us multiply this equation by the function $\eta(t^{(0)}) \in C_0^\infty(E^1)$ (for which supp $\eta \subset (a, b)$ and $\int \eta(t^{(0)})\, dt^{(0)} = 1$) and let us integrate with respect to $t^{(0)}$. Obviously,

$$g(t) = P_s(t;\ g) + R_s(t;\ g^{(s+1)}), \quad (20)$$

where $P_s(t;\ g)$ is an sth-degree polynomial in t of the form

$$P_s(t; g) = \sum_{k=0}^{s} \frac{(-1)^k}{k!} \int g(t^{(0)}) \frac{d^k}{dt^{(0)\,k}} [(t - t^{(0)})^k \eta(t^{(0)})]\, dt^{(0)} \quad (21)$$

and

$$R_s(t;\ g^{(s+1)}) = \frac{1}{s!} \int \int_0^1 \eta(t^{(0)})(t - t^{(0)})^{s+1} (1 - \xi)^s g^{(s+1)}(t^{(0)} + (t - t^{(0)})\xi)\, d\xi\, dt^{(0)}. \quad (22)$$

Formula (19) and formula (20) both give in the one-dimensional case an expansion of the function $g(t)$ with the aid of the projection operator. In the expansion (20), the projection operator (21) is defined on $L_1(a, b)$.

For a function $f(x) = f(x_1, \ldots, x_n)$ of n variables, the expansion (20) can be made with respect to any one of them, with the others treated as parameters. The so-obtained "elementary" expansion with the aid of an "elementary" projection operator makes it possible to construct more complicated projection operators in the multi-dimensional case. In this connection, see Sobolev [2], [3], Besov [5], [8], Il'in [14], Portnov [1], [2], and subsection 7.5 of the present book.

However, we shall proceed along a different path, using the smoothness of the function referred to in the lemma and relying on the simpler expansion (19).

Thus, the function $f(x) = f(x_1, \ldots, x_n)$ defined on the cube $\square$ is assumed to have continuous derivatives $D^\alpha f(x)$ on $\square$ of all orders that will be necessary. Suppose that $x^{(0)} = (x_1^{(0)}, \ldots, x_n^{(0)}) \in \square$. Singling out the variable x_1, let us apply the expansion (19) in powers $(x_1 - x_1^{(0)})^{\alpha_1}$ with the exponents α_1 ranging over all values for which $\alpha_1 : l_1 \leqslant 1 - \left|\frac{1}{p} : l\right|$. In the coefficients of the polynomial obtained, let us apply the expansion (19) in powers $(x_2 - x_2^{(0)})^{\alpha_2}$ in such a way that in the final expansion in products of the form $(x_1 - x_1^{(0)})^{\alpha_1}(x_2 - x_2^{(0)})^{\alpha_2}$ the exponents α_1 and α_2 range over all possible values satisfying the inequality $\frac{\alpha_1}{l_1} + \frac{\alpha_2}{l_2} \leqslant 1 - \left|\frac{1}{p} : l\right|$. In the coefficients in the expansion obtained, let us apply the expansion (19) in powers $(x_3 - x_3^{(0)})^{\alpha_3}$, etc. The end result is an expansion of the form

$$f(x) = P(x;\ f) + \sum_{1 < \left|\left(\alpha + \frac{1}{p}\right) : l\right| \leqslant 2} R_\alpha(x;\ D^\alpha f), \tag{23}$$

where the polynomial $P(x;\ f)$ depends linearly on f, has the form

(17), and obviously satisfies the estimate (18).

14.9. Theorem. *The functions in* $C_0^\infty(E^n)$ *constitute a dense set in the space* $\overset{*}{W}{}_p^l(G, \square)$, *where* $1 \leqslant p < \infty$ *and* $p_n > 1$.

PROOF. Suppose that $f \in \overset{*}{W}{}_p^l(G, \square)$, that is, that f is a member of $W_p^l(G, \square)$ and vanishes at infinity. Since the norms in the sense of $\overset{*}{W}{}_p^l(G, \square)$ and $W_p^l(G, \square)$ are equivalent on functions belonging to $\overset{*}{W}{}_p^l(G, \square)$, it will be sufficient to approximate the function f in the norm of the space $W_p^l(G, \square)$. By virtue of 5(9), $f_{\varepsilon\lambda}(x)$ approximates the function $f(x)$ in the norm $W_p^l(G, \square)$ with a prestated accuracy for sufficiently small $\varepsilon > 0$. Thus, it will be sufficient to resolve the question of approximation of the function $f_{\varepsilon\lambda}$ in the norm of $W_p^l(G, \square)$.

By virtue of (3), we have $f_{\varepsilon\lambda} = f_{h\lambda} - f_{\varepsilon, h}$. Let us show that such an approximation is possible for $f_{\varepsilon, h}$. Suppose that $f_{\varepsilon, h}^{[R]}(x) \in C_0^\infty(E^n)$ differs from $f_{\varepsilon, h}(x)$ only by replacement of $D_i^{l_i} f(x)$ in the integrands with the function $\chi_R(x) D_i^{l_i} f(x)$, defined equal to zero on the rest of E^n, where $\chi_R(x)$ is the characteristic function of the intersection of the region G and the ball $|x| < R$.

With the aid of Minkowski's generalized inequality 2(12) and Young's inequality 2(18), we obtain

$$\| f_{\varepsilon, h} - f_{\varepsilon, h}^{[R]} \|_{W_p^l(G, \square)} \leqslant C(\varepsilon, h) \sum_{i=1}^{n} \| (1 - \chi_R) D_i^{l_i} f \|_{p, G} \to 0 \quad \text{as } R \to \infty.$$

Thus, it remains to show that $f_{h\lambda}(x)$ can be approximated with an arbitrary degree of accuracy in $W_p^l(G, \square)$ by functions belonging to $C_0^\infty(E^n)$ for sufficiently large h.

Suppose that $P(x; f_{h\lambda})$ is the projection polynomial (17) corresponding to the function $f_{h\lambda}$. By virtue of (18) and (10) and Lemma 14.7,

$$\| f_{h\lambda} - P(\cdot\,; f_{h\lambda}) \|_{W_p^l(G, \square)} = \sum_{i=1}^{n} \| D_i^{l_i} f_{h\lambda} \|_{p, G} + \| f_{h\lambda} - P(\cdot\,; f_{h\lambda}) \|_{p, \square}$$

$$\leqslant \sum_{i=1}^{n} \| D_i^{l_i} f_{h\lambda} \|_{p,\, G} + C \sum_{1 < \left| \left(\alpha + \frac{1}{p}\right) : l \right| \leqslant 2} \max_{x \in \square} \left| D^{\alpha} f_{h\lambda}(x) \right| \to 0$$

as $h \to \infty$.

Thus, the difference $f_{h\lambda}(x) - P(x;\ f_{h\lambda})$ is nicely approximated by zero and the problem is reduced to approximating the polynomial $P(x;\ f_{h\lambda})$, which, by virtue of Lemma 14.6, is possible. This completes the proof of the lemma.

14.10. Similar questions on the behavior of the functions at infinity and approximation of them by means of infinitely differentiable functions of compact support can be solved in an analogous manner in the more general case in which we are considering not the space $W_p^l(G,\ \square)$ but the space with norm

$$\| f \|_{p,\, \square} + \sum_{j=1}^{N} \| P_j(D) f \|_{p,\, G},$$

where the l-polynomials $P_j(\xi)$ do not have a common complex root other than $\xi = 0$ (see section 11). However, we shall confine ourselves to the special case of the Sobolev space $W_p^{(m)}(G,\ \square)$ with norm

$$\| f \|_{W_p^{(m)}(G,\, \square)} = \| f \|_{p,\, \square} + \sum_{|\alpha| = m} \| D^{\alpha} f \|_{p,\, G} = \| f \|_{p,\, \square} + \| f \|_{L_p^{(m)}(G)}, \tag{24}$$

where m is a natural number and $1 < p < \infty$. The case of scalar $p \geqslant 1$ was examined by Sobolev in [3].

Reasoning analogously to the above and setting $l = (m, \ldots, m)$, we see that, for $|\alpha| > m$, $D^{\alpha} f_{h\lambda}(x) \to 0$ uniformly on G as $h \to \infty$. It follows that there exists a polynomial

$$Q_{m-1}(x;\ f) = \sum_{m - \left| \frac{1}{p} \right| < |\alpha| \leqslant m-1} c^{\alpha} x^{\alpha}$$

such that

$$D^{\alpha} f_{h^{\lambda}}(x) \to D^{\alpha} Q_{m-1}(x;\ f), \qquad |\alpha| > m - \left|\frac{1}{p}\right|,$$

uniformly on G as $h \to +\infty$. For $|\alpha| > m - \left|\frac{1}{p}\right|$, we have the estimate

$$\| D^{\alpha} [f_{\varepsilon} - Q_{m-1}(\cdot\,;\ f)]_{\infty,\, G} \leqslant C \| f \|_{L_p^{(m)}(G)}, \quad \varepsilon > 0,$$

which shows that the function $f(x)$ tends to the polynomial $Q_{m-1}(x;\ f)$. The set of functions in $W_p^{(m)}(G,\ \square)$ that vanish at infinity coincides with the space $\overset{*}{W}{}_p^{(m)}(G\ \square)$ with norm

$$\| f \|_{\overset{*}{W}{}_p^{(m)}(G,\ \square)} = \| f \|_{W_p^{(m)}(G,\ \square)} + \sum_{m - \left|\frac{1}{p}\right| < |\alpha| \leqslant m-1} \left\| D^{\alpha} f_{\varepsilon^{\lambda}} \right\|_{\infty,\, G},$$

$$\varepsilon > 0, \quad \frac{1}{\lambda} = l = (m,\ \ldots,\ m).$$

14.11. Lemma. *On the space $W_p^l(G,\ \square)$, where $1 \leqslant p \leqslant \infty$, there exists a linear projection operator*

$$\left(\prod f\right)(x) = P_{l-1}(x;\ f) = \sum_{\alpha \leqslant l-1} c_{\alpha}(f)\, x^{\alpha}$$

such that, for an arbitrary compact subset K of G,

$$\| f - P_{l-1}(\cdot\,;\ f) \|_{p,\, K} \leqslant C(K) \| f \|_{L_p^l(G)}.$$

PROOF. Let us write the expansion (17) applied to the variable x_i on $(a_i,\ b_i)$ with $s = l_i - 1$ in the form

$$f(x) = \left(\prod{}_i f\right)(x) + \left(\prod{}_i^* f\right)(x) \tag{25}$$

in such a way that $\prod_i f$ and $\prod_i^* f$ are calculated respectively from formulas (21) and (22). By applying the expansion (25) with respect to x_1, then with respect to x_2 in the polynomial $\prod_1 f$, then with respect to x_3 in the polynomial $\prod_2\prod_1 f$, and proceeding in this way, we obtain

$$f=\prod_1 \dots \prod_n f+\prod_1 \dots \prod_{n-1}\prod_n^* f+ \\ +\prod_1 \dots \prod_{n-2}\prod_{n-1}^* f+ \dots +\prod_1^* f. \quad (26)$$

The first term in the right-hand member is a polynomial

$$\prod_1 \dots \prod_n f=P_{l-1}(x;\ f),$$

and the remaining terms contain $D_i^{l_i} f$ under the integral sign.

From the expansion (26), we now obtain the estimate

$$\| f-P_{l-1}(\cdot\,;\ f)\|_{p,\,\square}\leqslant C\| f\|_{L_p^l(\square)}\leqslant C\| f\|_{L_p^l(G)}, \quad (27)$$

where

$$\square=\{x:\ a_i<x_i<b_i;\ i=1,\ \dots,\ n\},\quad \bar{\square}\subset G.$$

The assertion of the lemma now follows from the estimate 9(4) applied to the function $f^*=f-P_{l-1}$.

14.12. Theorem. *The space* $W_p^l(G,\square)$, *for* $1<p<\infty$, *admits a linear bounded estension to the space* $W_p^l(E^n,\ \square)$.

PROOF. We are required to construct a linear bounded operator from $W_p^l(G,\square)$ into $W_p^l(E^n,\ \square)$:

$$W_p^l(G,\ \square)\ni f\to\tilde{f}\in W_p^l(E^n,\ \square),\quad \tilde{f}\,|_G=f.$$

Consider the integral operator

$$\tilde{f}_{\varepsilon,\,h}(x)=\int_{\varepsilon}^{h}\sum_{i=1}^{n}v^{-|\lambda|}dv\int_{E^n}f_i(x+y)\,\mathscr{L}_i(y:v^{\lambda})\,dy,$$

where $f_i(x) = D_i^{l_i} f(x)$ for $x \in G$ but $f_i(x) = 0$ outside G, so that $f_i \in L_p(E^n)$.

By Lemma 14.1, the derivatives $D_i^{l_i} \tilde{f}_{\varepsilon, h}(x)$ (for $i = 1, \ldots, n$) converge in the sense of $L_p(\square_N)$, where $\square_N = \{x: |x_i| < N;\ i = 1, \ldots, n\}$, for every N as $\varepsilon \to 0$ and $h \to \infty$. By virtue of (3) and 5(9) and Lemma 14.7,

$$D_i^{l_i} \tilde{f}_{\varepsilon, h}(x) \to D_i^{l_i} f(x) \quad \text{in} \quad L_p(G) \qquad (i = 1, \ldots, n).$$

By virtue of Lemma 14.11,

$$\tilde{f}_{\varepsilon, h}(x) - P_{l-1}(x;\ \tilde{f}_{\varepsilon, h}) \to f^*(x) \quad \text{in} \quad W_p^l(\square_N)$$

for every N.

Since $D_i^{l_i} f(x) = D_i^{l_i} f^*(x)$ on G, we have

$$f(x) = f^*(x) + P_{l-1}^*(x), \qquad x \in G.$$

The function $\tilde{f}(x) = f^*(x) + P_{l-1}^*(x)$, which is defined on E^n, satisfies all the assertions of the theorem when we remember that, by virtue of Theorem 4.5,

$$\sum_1^n \| D_i^{l_i} \tilde{f} \|_{p,\, E^n} \leqslant \sup_{0 < \varepsilon < h < \infty} \sum_1^n \| D^{l_i} \tilde{f}_{\varepsilon, h} \|_{p,\, E^n} \leqslant C \sum_{i=1}^n \| D_i^{l_i} f \|_{p,\, G}.$$

14.13. Let us establish a number of propositions that we shall use in §15. Suppose that $l = (l_1, \ldots, l_n)$ is a vector whose components are natural numbers. Consider the spaces of the functions $L^l_{p^{(1)}, \ldots, p^{(n)}}$ and $W^l_{p^{(0)};\, p^{(1)}, \ldots, p^{(n)}}$ defined on E^n with seminorm and norm respectively

$$\| f \|_{L^l_{p^{(1)}, \ldots, p^{(n)}}} = \sum_{j=1}^n \| D_j^{l_j} f \|_{p^{(j)}},$$
$$\| f \|_{W^l_{p^{(0)};\, p^{(1)}, \ldots, p^{(n)}}} = \| f \|_{p^{(0)}} + \sum_{j=1}^n \| D_j^{l_j} f \|_{p^{(j)}}. \tag{28}$$

We denote by $B^{(j)}_{\delta_j, H_j}$ (where $0<\delta_j<H_j<\infty$ for $j=1, \ldots, n$) the operator that assigns to an infinitely differentiable function $f(x)$, where $x=(x_1, \ldots, x_n)$, the difference between its average with respect to the variable x_j with parameter δ_j and with parameter H_j:

$$B^{(j)}_{\delta_j, H_j} f(x) = \int \Omega_j(t) f(x+\delta_j t e_j)\, dt - \int \Omega_j(t) f(x+H_j t e_j)\, dt$$

$$= \int_{\delta_j}^{H_j} \int v^{l_j-2} \mathscr{L}_j\left(\frac{t}{v}\right) D_j^{l_j} f(x+te_j)\, dt\, dv, \qquad (29)$$

where $\Omega_j(t) \in C_0^\infty(E^1)$ and $\mathscr{L}_j(t) \in C_0^\infty(E^1)$ are taken from the one-dimensional representation 7(13).

Let us define the operator $B_{\delta, H}$, where $\delta=(\delta_1, \ldots, \delta_n)$, $H=(H_1, \ldots, H_n)$, and $0<\delta_i<H_i<\infty$, on the set of functions $F(x) \in C^\infty(E^n)$ by

$$B_{\delta, H} F(x) = \int_{\delta_1}^{H_1} \cdots \int_{\delta_n}^{H_2} \int_{E^n} \prod_1^n h_j^{l_j-2} \mathscr{L}_j\left(\frac{y_j-x_j}{h_j}\right) F(y)\, dy\, dh_1 \ldots dh_n, \qquad (30)$$

so that if $F(x)=D^l f(x)$, we have on the basis of (29)

$$B_{\delta, H} D^l f(x) = \left(\prod_{j=1}^n B^{(j)}_{\delta_j, H_j}\right) f(x). \qquad (31)$$

Lemma. *Suppose that* $1<p^{(j)}<\infty$ *and that*

$$D^l f \in L_{p^{(j)}} \quad (j=0, 1, \ldots, n),$$

where

$$f \in L^l_{p^{(1)}, \ldots, p^{(n)}} \cap C^\infty$$

or

$$f \in W^{l}_{p^{(0)};\, p^{(1)}, \ldots,\, p^{(n)}} \cap C^{\infty}.$$

Then, f can be approximated with an arbitrary degree of accuracy in $L^{l}_{p^{(1)}, \ldots,\, p^{(n)}}$ or in $W^{l}_{p^{(0)};\, p^{(1)}, \ldots,\, p^{(n)}}$ respectively by the functions $B_{\delta, H} D^{l} f$ and the functions $B_{\delta, H} \chi_R D^{l} f \in C_0^{\infty}$ for a suitable choice of δ, H, and R, where $\chi_R(x)$ is the characteristic function of the ball $|x| < R$.

PROOF. We stop only for the case $f \in W^{l}_{p^{(0)};\, p^{(1)}, \ldots,\, p^{(n)}}$ since the case $f \in L^{l}_{p^{(1)}, \ldots,\, p^{(n)}}$ is treated analogously and more simply. Let us show that, for $j = 1, \ldots, n$,

$$\left\| f - B^{(j)}_{\delta_j,\, H_j} f \right\|_{W^{l}_{p^{(0)};\, p^{(1)}, \ldots,\, p^{(n)}}} \to 0 \text{ as } \delta_j \to 0 \text{ and } H_j \to \infty. \tag{32}$$

The first term in the middle member of (29) approaches f in $W^{l}_{p^{(0)};\, p^{(1)}, \ldots,\, p^{(n)}}$ because of the reversibility of the orders of differentiation and averaging and the continuity in the wide sense in $L_{p^{(j)}}$ (see Theorem 1.5 and compare with Lemma 5.2). Thus, it remains to show that for $g \in L_p(E^n)$, where $1 < p < \infty$, we have

$$\left\| \int \Omega_j(t)\, g(\cdot + Hte_j)\, dt \right\|_p \to 0 \quad \text{as} \quad H \to \infty \tag{33}$$

and to use this result for $g = f$, $p = p^{(0)}$ and also for $g = D_i^{l_i} f$, $p = p^{(i)}$. We shall not stop to prove (33) since its proof does not differ greatly from that of the lemma of 14.5. Thus, the validity of (32) is established.

Applying (32) successively for $j = 1, 2, \ldots, n$, we see that every function $f \in W^{l}_{p^{(0)};\, p^{(1)}, \ldots,\, p^{(n)}} \cap C^{\infty}$ can be approximated with an arbitrary degree of accuracy in $W^{l}_{p^{(0)};\, p^{(1)}, \ldots,\, p^{(n)}}$ by functions of the form $B_{\delta, H} D^{l} f$.

By virtue of 1(6),

$$\chi_R D^{l} f \to D^{l} f \quad \text{in} \quad L_{p^{(j)}} \quad \text{as} \quad R \to \infty \quad (j = 0, 1, \ldots, n),$$

so that

$$B_{\delta, H}\chi_R D^l f \to B_{\delta, H} D^l f \quad \text{in} \quad W^l_{p^{(0)};\, p^{(1)}, \ldots, p^{(n)}}$$

as $R \to \infty$, which completes the proof of the lemma.

14.14. Theorem. *The set* $C_0^\infty(E^n)$ *is dense in the space* $W^l_{p^{(0)};\, p^{(1)}, \ldots, p^{(n)}}$, *where* $l = (l_1, \ldots, l_n)$ *is a vector whose components are natural numbers and* $1 < p^{(j)} < \infty$ *(for* $j = 0, 1, \ldots, n$*).*

PROOF. An arbitrary function f in $W^l_{p^{(0)};\, p^{(1)}, \ldots, p^{(n)}}$ can be approximated in the norm of that space with an arbitrary degree of accuracy by a function f_ε, which is its average with kernel in the space C_0^∞ and with sufficiently small averaging parameter ε (see Lemma 5.2). Obviously, all the conditions of Lemma 14.13 are satisfied for the function f_ε and it remains only to use its conclusion.

§15. Multiplicative inequalities for L_p-norms of derivatives

Hadamard, Šilov, Kolmogorov, Rodov, and other mathematicians have written papers devoted to multiplicative estimates for norms of the derivatives of a function of a single variable. All of them dealt with the case of the same metric C for all the factors. For the case of several variables, Gagliardo [3] and Nirenberg [2] obtained the estimate

$$\sum_{|\alpha|=r} \| D^\alpha f \|_p \leqslant C \| f \|_{p_1}^{1-\theta} \Big(\sum_{|\alpha|=l} \| D^\alpha f \|_{p_2} \Big)^\theta \tag{1}$$

under suitable relationships between the numerical parameters n, r, l, p, p_1, p_2, and θ, of which we shall speak below.

We mention that, under certain special assumptions regarding p_1, p_2, and p, for example, $p_1 = p_2$ or $p_1 \leqslant p$ and $p_2 \leqslant p$, inequality (1) and its analogue for the special case of a mixed L_p-norm has been established by V. P. Il'in (candidate's dissertation, 1951, the result published in 1957 and 1959 [2]), Erling (1954 for $p = 2$), and

Nirenberg (1955). Recently, Solonňikov [4] obtained an analogue of inequality (1) for the anisotropic case (assigning differential properties to a function) in L_{p_2}.

The multiplicative inequality (1) is equivalent to the corresponding additive inequality with arbitrary parameter $\varepsilon > 0$:

$$\sum_{|\alpha|=r} \| D^{\alpha} f \|_{p} \leqslant C \left(\varepsilon^{-\frac{1}{1-\theta}} \| f \|_{p_1} + \varepsilon^{\frac{1}{\theta}} \sum_{|\alpha|=l} \| D^{\alpha} f \|_{p_2} \right) \quad \forall \varepsilon > 0. \qquad (1')$$

Inequality (1′) is obtained by introducing factors $1/\varepsilon$ and ε into the right-hand side of (1) and considering the two cases when one of the terms in the right-hand member of (1′) is greater than the other. Inequality (1) follows from (1′) for a choice of ε that equates the two terms in the right-hand member of (1′). We note that inequality (1′) is equivalent to its special case ($\varepsilon = 1$) and is obtained by applying this case to the function $f_{\varepsilon}(x) = f(\varepsilon x)$.

Analogous relations are obtained in the anisotropic case. Here, the additive inequality with arbitrary parameter of the type (1′) follows from the corresponding additive inequality by applying the latter to $f(\varepsilon^{\beta} x)$, where $\varepsilon^{\beta} x = \left(\varepsilon^{\beta_1} x_1, \ldots, \varepsilon^{\beta_n} x_n\right)$ for suitably chosen positive $\beta_1, \ldots, \beta_n$.

Below, we shall establish some estimates generalizing and refining estimates of the type (1). Instead of *sums* of norms of the derivatives, we shall have norms of the individual derivatives. This will enable us, for example, to generalize the estimate (1) to the anisotropic case. The right-hand member of the multiplicative inequality to be obtained will contain an arbitrary number, $N \geqslant 2$, of factors. The inequality itself will be derived for mixed $L_{\mathbf{p}}$-norms. However, what will be important for the method that we shall use will be restrictions of the type $1 < p < \infty$ or $\mathbf{1} < \mathbf{p} < \boldsymbol{\infty}$ although in a number of cases the inequality remains valid without them (see, for example, the works cited by Gagliardo, Nirenberg, and Solonnikov).

15.1. We state (without proof) the Gagliardo–Nirenberg result:

Theorem. *Inequality* (1) *is valid for* $1 \leqslant p_1 \leqslant \infty, 1 \leqslant p_2 \leqslant \infty$, $0 \leqslant r < l$,

$$\frac{n}{p} - r = (1-\theta)\frac{n}{p_1} + \theta\left(\frac{n}{p_2} - l\right), \qquad \frac{r}{l} \leqslant \theta \leqslant 1$$

with the following exceptions:

a) *If* $r=0$, $l<\frac{n}{p_2}$, *and* $p_1=\infty$, *we assume in addition that either* $f\to 0$ *as* $x\to\infty$ *or* $f\in L_q$ *for some finite* $q>0$;

b) *if* $1<p_2<\infty$ *and* $l-r-\frac{n}{p_2}$ *is a nonnegative integer, then* (1) *does not hold for* $\theta=1$.

Here, we point out that the lower bound for $\theta\left(\frac{r}{l}\leqslant\theta\right)$ is established by considering the functions

$$f(x)=\varphi(x)\sin\lambda x_1,\quad\text{where}\quad \varphi\in C_0^\infty,\quad \lambda\to\infty,$$

and noting that, for a bounded region G with a sufficiently smooth boundary, an inequality similar to (1) with the addition of a term $\|f\|_{q,\,G}$ for some positive q in the right-hand member is valid.

15.2. In what follows, we shall need some notation and some auxiliary propositions.

For a function $f(x)$, where $x=(x_1,\ldots,x_n)$, that is 2π-periodic with respect to all the variables, we denote by $\|f\|_p^*$, where $1\leqslant p\leqslant\infty$, its L_p-norm with respect to the cube $(0,2\pi)^n$. For the vector $k=(k_1,\ldots,k_n)$ with integer-valued coordinates, we denote by $\square_k$ the rectangular parallelepiped with edges parallel to the coordinate axes that has as its projection onto the jth coordinate axis the segment with ends at

$$2^{|k_j|-1}\operatorname{sign}k_j,\ (2^{|k_j|}-1)\operatorname{sign}k_j\qquad(j=1,\ldots,n).$$

The parallelepiped $\square_k$ is n-dimensional in the general case and it degenerates to a parallelepiped of lower dimension if any of the k_j are equal to zero.

Let $m=(m_1,\ldots,m_n)$ and $k=(k_1,\ldots,k_n)$ denote vectors with integer-valued coordinates. Let c_m denote a multi-indexed sequence. Then, let us define

$$c_{m,k}=\begin{cases} c_m & \text{for} \quad m\in\square_k,\\ 0 & \text{for} \quad m\notin\square_k.\end{cases} \tag{2}$$

For

$$f(x)=\sum c_m e^{imx}\in L_p^* \tag{3}$$

and a vector $r=(r_1, \ldots, r_n)$ with nonnegative components, let us set

$$f^{(r)}(x)=\sum (im_1)^{r_1}\ldots(im_n)^{r_n} c_m e^{imx},$$

where

$$(im_j)^{r_j}=|m_j|^{r_j}\exp\left\{\frac{1}{2}\pi i r_j \operatorname{sign} m_j\right\},$$

$$\delta_k(f)=\sum_m c_{m,k}e^{imx},$$

$$\delta_k^{(r)}(f)=\begin{cases} 2^{\bar{k}r}\delta_k(f) & \text{for} \quad \operatorname{supp} k\supset \operatorname{supp} r,\\ 0 & \text{for} \quad \operatorname{supp} k\not\supset \operatorname{supp} r,\end{cases}$$

where $\bar{k}=(|k_1|, \ldots, |k_n|)$. By the support of a vector we mean the set of the numbers of those components that are nonzero. (Translator's note. For example, the support of the vector (0, 1, 0, 7) is the set $\{2,4\}$.)

For $f(x)=\sum c_m e^{imx}$, let us define

$$\|f^{(r)}\|_{p}^{**}=\left\|\left(\sum_k |\delta_k^{(r)}(f)|^2\right)^{1/2}\right\|_p^*, \qquad 1\leqslant p\leqslant\infty.$$

Lemma.* For $1<p<\infty$,

$$\|f^{(r)}\|_p^{**}\sim\|f^{(r)}\|_p^*. \tag{4}$$

*This was established by N. S. Nikol'skaya for $r\neq 0$ and $p=(p,\ldots,p)$.

PROOF. Let us make clear first of all that equivalence in (4) means that finiteness of either side of (4) implies finiteness of the other side and that there exist two positive constants (dependent on p and r but not on f) such that (4) holds. Let us suppose first that $r = 0$. For the usual $L_{\boldsymbol{p}}^*$-norm, this case constitutes the content of the Littlewood-Paley theorem, the multidimensional variant of which can be found in the book by Nikol'skiĭ [9], §1.5.2. The Littlewood-Paley theorem can be generalized to the case of a mixed $L_{\boldsymbol{p}}^*$-norm following the outline of the proof given there but with the following additions.

Inequalities* 1.5.2(8) for vector $\boldsymbol{p}$ are obtained by successive n-fold application of the one-dimensional case. Instead of the chain of inequalities 1.5.2(10) with the use of Minkowski's generalized inequality, we have, for

$$p_{\min} = \min_{1 \leqslant j \leqslant n} p_j, \quad p_{\max} = \max_{1 \leqslant j \leqslant n} p_j, \quad \Omega = \{\theta = (\theta_1, \ldots, \theta_n) : 0 \leqslant \theta_j \leqslant 1 \ (j = 1, \ldots, n)\},$$

$$\begin{aligned}
\|f\|_{\boldsymbol{p}}^* &= \left\{ \int_\Omega \left(\|f\|_{\boldsymbol{p}}^* \right)^{p_{\max}} d\theta \right\}^{\frac{1}{p_{\max}}} \\
&\leqslant C_1 \left\{ \int_\Omega \left(\left\| \sum \omega_k(\theta)\, \delta_k(f) \right\|_{\boldsymbol{p}}^* \right)^{p_{\max}} d\theta \right\}^{\frac{1}{p_{\max}}} \\
&\leqslant C_2 \left\| \left(\int_\Omega \left| \sum \omega_k(\theta)\, \delta_k(f) \right|^{p_{\max}} d\theta \right)^{\frac{1}{p_{\max}}} \right\|_{\boldsymbol{p}}^* \\
&\leqslant C_3 \left\| \left(\sum \delta_k^2(f) \right)^{\frac{1}{2}} \right\|_{\boldsymbol{p}}^* \leqslant C_4 \left\| \left(\int_\Omega \left| \sum \omega_k(\theta)\, \delta_k(f) \right|^{p_{\min}} d\theta \right)^{\frac{1}{p_{\min}}} \right\|_{\boldsymbol{p}}^* \\
&\leqslant C_5 \left\{ \int_\Omega \left(\|f\|_{\boldsymbol{p}}^* \right)^{p_{\min}} d\theta \right\}^{\frac{1}{p_{\min}}} \leqslant C_6 \|f\|_{\boldsymbol{p}}^*,
\end{aligned}$$

where the constants $C_1, \ldots, C_6$ depend only on $\boldsymbol{p}$.

*Translation editor's note. The references in this paragraph to 1.5.2(8), (10), (13) are to Nikol'skiĭ's book [9].

The proof of inequality 1.5.2(13) for vector $\boldsymbol{p}$ differs from the proof given by Nikol'skiǐ in [9] for scalar p only in that the part dealing with the induction should be replaced with n-fold application of the corresponding one-dimensional inequality.

Suppose now that $r \neq 0$. The equivalence relation (4) is established with the aid of the same Littlewood–Paley inequality for vector $\boldsymbol{p}$ and the following theorem of Marcinkewicz regarding the coefficients in Fourier series.

15.3. Marcinkewicz's theorem. *Let $\{\lambda_m\}$ denote a multi-indexed sequence. Suppose that there exists a number M such that*

$$\sum_m |\Delta_1 \ldots \Delta_n \lambda_{m,k}| \leqslant M \qquad \forall k, \tag{5}$$

where the $\lambda_{m,k}$ are constructed from the λ_m with the aid of formula (2). Suppose that the difference $\Delta_j \lambda_{m,k} = \lambda_{m+e_j,k} - \lambda_{m,k}$, where e_j is the jth unit vector and $\Delta_1 \ldots \Delta_n \lambda_{m,k}$ is the multi-indexed difference corresponding to the subscripts. Then, for $f(x) = \sum c_m e^{imx} \in L^*_{\boldsymbol{p}}$ (where $1 < \boldsymbol{p} < \infty$) we have

$$F(x) = \sum \lambda_m c_m e^{imx} \in L^*_{\boldsymbol{p}},$$

and there exists a constant C_p independent of f such that

$$\| F \|^*_{\boldsymbol{p}} \leqslant C_{\boldsymbol{p}} M \| f \|^*_{\boldsymbol{p}},$$

The proof for scalar p can be found, for example, in the book by Nikol'skiǐ [9], §1.5.3. Let us show how to generalize it to the case of vector $\boldsymbol{p}$.

Following the proof given in Zygmund's book [1] (Vol. II, Chapter XV, Theorem 4.14), let us first generalize Lemma 2.10 (Zygmund [1], Vol. II, p. 224) to the case of several variables and vector $\boldsymbol{p}$ such that $1 \leqslant \boldsymbol{p} < \infty$. To do this, let us derive a weakened form of inequality (2.11) (see Zygmund [1], Vol. II, p. 225). This derivation can be carried over in an obvious way to the case of a mixed norm for functions of several variables. Here, in place of M in (2.11) we need to put $A_{\boldsymbol{p}}M$, where $A_{\boldsymbol{p}}$, for $1 \leqslant \boldsymbol{p} < \infty$, is independent of the number n.

Suppose that

$$\varphi(x) = \sum_1^n \varphi_k(x)\, z_k$$

and

$$\psi(x) = \sum_1^n \psi_k(x)\, z_k,$$

where $z \in E^n$, and that

$$\Phi(x) = \sqrt{\sum_1^n \varphi_k^2(x)}$$

and

$$\Psi(x) = \sqrt{\sum_1^n \psi_k(x)}.$$

In addition to the notation and reasoning of Zygmund ([1], Vol. II, p. 225), we use Hölder's and Minkowski's inequalities, getting

$$\int_{|z|<1} |z_1|\, dz \left\{ \int_a^b |\Psi(x)|^p\, dx \right\}^{\frac{1}{p}} = \left\{ \int_a^b \left[\int_{|z|<1} |\psi(x)|\, dz \right]^p dx \right\}^{\frac{1}{p}}$$

$$\leqslant \int_{|z|<1} \left\{ \int_a^b |\psi(x)|^p\, dx \right\}^{\frac{1}{p}} dz \leqslant M \int_{|z|<1} \left\{ \int_a^b |\varphi(x)|^p\, dx \right\}^{\frac{1}{p}} dz$$

$$\leqslant \left\{ \int_{|z|<1} dz \right\}^{\frac{1}{q'}} \left\{ \int_{|z|<1} \left[\int_a^b |\varphi(x)|^p\, dx \right]^{\frac{q}{p}} dz \right\}^{\frac{1}{q}}$$

$$\leqslant \left\{ \int_{|z|<1} dz \right\}^{\frac{1}{q'}} \left\{ \int_a^b \left[\int_{|z|<1} |\varphi(x)|^q\, dz \right]^{\frac{p}{q}} dx \right\}^{\frac{1}{p}}$$

$$= \mathcal{I}_1(n)\, \mathcal{I}_2(n) \left\{ \int_a^b |\Phi(x)|^p\, dx \right\}^{\frac{1}{p}},$$

where $1 \leqslant p < q = 2k + 1 < \infty$ (where k is a natural number), $\frac{1}{q} + \frac{1}{q'} = 1$, and

$$\mathcal{I}_1^{q'}(n) = \int\limits_{|z|<1} dz, \qquad \mathcal{I}_2^{q}(n) = \int\limits_{|z|<1} |z_1|^q \, dz.$$

To complete the proof, it remains to show that

$$\mathcal{I}_1(n)\,\mathcal{I}_2(n) \leqslant A_p \int\limits_{|z|<1} |z_1|\, dz, \quad n = 2, 4, 6, \ldots .$$

Let us look at each of these three integrals:

$$\mathcal{I}(n) = \int\limits_{|z|<1} |z_1|\, dz = 2\int\limits_0^1 z_1 c_{n-1}\left(1 - z_1^2\right)^{\frac{n-1}{2}} dz_1 = c_{n-1}\frac{2}{n+1},$$

where c_{n-1} is the volume of a unit ball in E^{n-1}.

$$\mathcal{I}_1^{q'}(n) = 2c_{n-1}\int\limits_0^1 \left(1 - z_1^2\right)^{\frac{n-1}{2}} dz = 2c_{n-1}\int\limits_0^{\frac{\pi}{2}} \sin^n t\, dt =$$

$$= \pi c_{n-1}\frac{(n-1)!!}{n!!} = \frac{\pi c_n n!}{2^n\left[\left(\frac{n}{2}\right)!\right]^2}.$$

(Translator's note: the notation $n!!$ denotes the product of n and all smaller positive integers of like parity. Thus, $5!!$ means $5 \cdot 3 \cdot 1$ and $6!!$ means $6 \cdot 4 \cdot 2$.)

Now, with the aid of Stirling's formula ($n! = \sqrt{2\pi n}\left(\frac{n}{e}\right)^n \varepsilon_n$, where $\varepsilon_n \to 1$ as $n \to +\infty$), we get

$$\mathcal{I}_1^{q'}(n) = \frac{c_{n-1}\sqrt{2\pi}}{\sqrt{n}}\,\tilde{\varepsilon}_n^{q'}, \quad \tilde{\varepsilon}_n \to 1 \quad \text{as} \quad n \to \infty.$$

For $I_{n,q} = \int\limits_0^{\pi/2} \sin^n t \cos^q t\, dt$, where $q = 2k + 1$, we get with the aid of the formulas for lowering the order

$$I_{n,q}=\frac{(q-1)(q-3)\ldots 2}{(n+q)(n+q-2)\ldots(n+3)}I_{n,1},\qquad I_{n,1}=\frac{1}{n+1}.$$

Therefore,

$$\mathcal{J}_2^q(n)=2c_{n-1}\int_0^1|z_1|^q\left(1-z_1^2\right)^{\frac{n-1}{2}}dz_1=2c_{n-1}I_{n,q}$$
$$\leqslant c_{n-1}2(q-1)!!\frac{1}{n^{k+1}}=c_{n-1}2(q-1)!!.$$

Consequently,

$$\mathcal{J}_1(n)\,\mathcal{J}_2(n)\leqslant c_{n-1}(2\pi)^{\frac{1}{2q'}}[2(q-1)!!]^{\frac{1}{q}}\left(\frac{1}{n}\right)^{\frac{1}{2q'}}\left(\frac{1}{n}\right)^{\frac{1}{2}+\frac{1}{2q}}\tilde{\varepsilon}_n=$$
$$=c_{n-1}(2\pi)^{\frac{1}{2q'}}[2(q-1)!!]^{\frac{1}{q}}\frac{\tilde{\varepsilon}_n}{n}\leqslant(2\pi)^{\frac{1}{2q'}}[2(q-1)!!]^{\frac{1}{q}}\tilde{\varepsilon}_n\mathcal{J}(n),$$

as we needed to show.

It remains now to generalize Lemma 2.15 in Zygmund's book [1], Vol. II, Chapter XV to the case of a mixed L_p-norm, where $1<p<\infty$. We mention that actually we need this generalization only for trigonometric polynomials, that is, for functions $f(x)=\sum c_m e^{imx}$ with a finite number of nonzero terms. Following the proof of Lemma 2.15 and remembering that the convolution with respect to x_j with $\cos k_j x_j$ is a uniformly (with respect to k_j) bounded operator in $L_{\boldsymbol{p}}^*$, we reduce the proof of Lemma 2.15 with vector $\boldsymbol{p}$ to the estimate (the one-dimensional case of Riesz' theorem)

$$\|\tilde{f}_j\|_{\boldsymbol{p}}^* \leqslant A_{\boldsymbol{p}}\|f\|_{\boldsymbol{p}}^*,\qquad 1<\boldsymbol{p}<\infty,$$

where $\tilde{f}_j(x)$ is constructed from $f(x)$ by the (trigonometric) conjugate operator taken with respect to the variable x_j:

$$\tilde{f}_j(x)=-\frac{1}{\pi}\int_0^{\pi}\frac{f(x+te_j)-f(x-te_j)}{2\tan\frac{t}{2}}\,dt.$$

Obviously, it will be sufficient if we assume that $n=j$. In this case, the function (trigonometric polynomial) $f(x) = f(x', x_n)$, where $x' = (x_1, \ldots, x_{n-1})$ can be regarded as an (abstract) function of the variable x_n with values in the space $L^*_{p'}$, where $p' = (p_1, \ldots, p_{n-1})$. Since the difference

$$\varepsilon(t) = \frac{1}{2\tan\frac{t}{2}} - \frac{1}{t}$$

is continuous on $(0, \pi)$ and can be extended as a continuous function to $[0, \pi]$, our problem reduces, by virtue of the estimate in L^*_p, to the estimate

$$\| f_{(n)} \|_p \leqslant B_p \| f \|_p, \qquad 1 < p < \infty,$$

where

$$f_{(n)}(x) = -\frac{1}{\pi} \int_0^\pi \frac{f(x + te_n) - f(x - te_n)}{t}\, dt =$$

$$= -\frac{1}{\pi} \lim_{\varepsilon \to 0} \int_{\varepsilon < |t| < \pi} \frac{f(x + te_n)}{t}\, dt.$$

For $n = 1$, this estimate is contained in Theorem 4.9. The case for $n > 1$ and scalar p in $(1, \infty)$ is an obvious consequence of the one-dimensional case. The general case $n \geqslant 1$, $1 < p < \infty$ follows from Theorem 3.13, from which we get the uniform boundedness (with respect to ε) of the operator with kernel

$$K_\varepsilon(u) = \begin{cases} \dfrac{1}{u} & \text{for} \quad 0 < \varepsilon < |u| < \pi, \\ 0 & \text{for} \quad |u| \notin (\varepsilon, \pi). \end{cases}$$

The passage to the limit as $\varepsilon \to 0$ is elementary and is performed just as in the proof of Theorem 4.9.

In our case, Marcinkewicz's theorem is applied to the multipliers $\{\lambda_m\}$, which for all $m \in \square_k$, where supp $k \supset$ supp r, are equal to

$$\lambda_{m,k} = \frac{|m_1|^{r_1} \ldots |m_n|^{r_n}}{2^{kr}}$$

or

$$\lambda_{m,k} = \frac{2^{kr}}{|m_1|^{r_1} \ldots |m_n|^{r_n}}.$$

Condition (5) is easily verified on the basis of the monotonicity.

15.4. Lemma. *Let r and $r^{(j)}$ for $j = 1, \ldots, N$ denote n-dimensional vectors with nonnegative components. Suppose that* $1 \leqslant p^{(j)} \leqslant \infty$,

$$0 < \mu_j < 1, \quad \sum_1^N \mu_j = 1, \quad r = \sum_1^N \mu_j r^{(j)}, \quad \frac{1}{p} = \sum_1^N \frac{\mu_j}{p^{(j)}}.$$

Then,

$$\| f^{(r)} \|_p^* \leqslant \prod_1^N \| f^{(r^{(j)})} \|_{p^{(j)}}^{*\mu_j}. \tag{6}$$

Proof. Since supp $r \supset$ supp $r^{(j)}$ for all $j = 1, \ldots, N$, we have

$$|\delta_k^{(r)}(f)|^2 \leqslant \prod_{j=1}^N |\delta_k^{(r^{(j)})}(f)|^{2\mu_j},$$

so that, by virtue of Hölder's inequality,* we have

$$\sum_k |\delta_k^{(r)}(f)|^2 \leqslant \prod_{j=1}^N \Big(\sum_k |\delta_k^{(r^{(j)})}(f)|^2\Big)^{\mu_j}.$$

From this we obtain by virtue of Hölder's inequality for the integral of the product of N functions,**

*See inequality 2(3) with $m = N$ and $f_1, \ldots, f_m$ finite-valued functions (cf. derivation of 2(5)).

**For scalar $p^{(j)}$, see 2(3). Generalization of 2(3) to the case of mixed norms is achieved by successive application of inequality 2(3) for each variable.

$$\|f^{(r)}\|_p^* = \Big\|\Big(\sum_k |\delta_k^{(r)}(f)|^2\Big)^{\frac{1}{2}}\Big\|_p^* \leqslant \Big\|\prod_{j=1}^N \Big(\sum_k |\delta_k^{(r^{(j)})}(f)|^2\Big)^{\frac{1}{2}\mu_j}\Big\|_p^*$$
$$\leqslant \prod_{j=1}^N \Big\|\Big(\sum_k |\delta_k^{(r^{(j)})}(f)|^2\Big)^{\frac{1}{2}}\Big\|_{p^{(j)}}^{*\,\mu_j},$$

which is equivalent to the estimate (6).

15.5. Lemma. *Suppose that* $\mathbf{1 < p \leqslant q < \infty}$, $s \geqslant \mathbf{0}$, $s - \frac{1}{q} = r - \frac{1}{p}$. *Then,*

$$\| f^{(s)} \|_q^* \leqslant c_{p,\,q} \| f^{(r)} \|_p^* \tag{7}$$

for all functions $f(x)$ of the form

$$f(x) = \sum_{\text{supp}\, m \supset \text{supp}\, r} c_m e^{imx}, \tag{8}$$

that is, for all functions $f(x)$ with zero mean values with respect to those x_j (almost everywhere on the complement) for which $r_j > 0$.

It will be sufficient to prove this for the case in which p and q differ by only one component (let us say, the jth) with $s = \mathbf{0}$, $r_\nu = \mathbf{0}$ for $\nu \neq j$ and $r_j = \frac{1}{p_j} - \frac{1}{q_j} > 0$.

Consider

$$K^{(j)}(x_j) = K^{(j)}_{\frac{1}{p_j} - \frac{1}{q_j}}(x_j) = \sum_{m_j \neq 0} \frac{1}{(im)^{r_j}} e^{im_j x_j},$$

so that $f^{(s)} = K^{(j)} * f^{(r)}$, where the convolution is only with respect to the variable x_j. Thus, it will be sufficient to show that

$$\| |K^{(j)}| * \varphi \|_q^* \leqslant \tilde{c}_{p,\,q} \| \varphi \|_p^* \,. \tag{9}$$

Since $p_{j+1} = q_{j+1}, \ldots, p_n = q_n$, the estimate (9) follows from the special case of it for which $n = j = 1$. Therefore, we shall concern ourselves with this case.

One can easily obtain an estimate for $K^{(j)}(t)$ by means of a transformation of Abel's series. Specifically (see Zygmund [1], Vol. I, p. 70),

$$|K^{(j)}(t)| \leqslant \frac{\varkappa}{|t|^{1-r_j}},$$

so that all we need to do to obtain (9) with $n = 1$ is to use the Hardy–Littlewood inequality 2(32). This completes the proof of the lemma.

15.6. Theorem. *Let* l *and* $r^{(j)}$, *for* $j = 1, \ldots, N$, *denote* n*-dimensional vectors with nonnegative components. Suppose that*

$$1 < p^{(j)} < \infty, \quad 1 < q < \infty, \quad 0 < \mu_j < 1, \quad \sum_1^N \mu_j = 1,$$

$$\frac{1}{q} \leqslant \sum_1^N \frac{\mu_j}{p^{(j)}}, \quad l - \frac{1}{q} = \sum_1^N \mu_j \left(r^{(j)} - \frac{1}{p^{(j)}}\right).$$

Let f *denote a function that is* 2π*-periodic with respect to all variables and that has zero mean with respect to those* x_ν *for which* $l_\nu > 0$. *Then,*

$$\| f^{(l)} \|_q^* \leqslant C \prod_{j=1}^N \| f^{(r^{(j)})} \|_{p^{(j)}}^{*\,\mu_j}. \tag{10}$$

By virtue of Lemma 15.2, the proof follows from the estimates (6) and (7).

15.7. Theorem. *Let* β *and* $\alpha^{(j)}$, *for* $j = 1, \ldots, N$, *denote* n*-dimensional multi-indices with nonnegative-integer-valued components. Suppose that*

$$1 < p^{(j)} < \infty, \quad 1 < q < \infty, \quad 0 < \mu_j < 1, \quad \sum_1^N \mu_j = 1,$$

$$\frac{1}{q} \leqslant \sum_1^N \frac{\mu_j}{p^{(j)}}, \quad \beta - \frac{1}{q} = \sum_1^N \mu_j \left(\alpha^{(j)} - \frac{1}{p^{(j)}}\right).$$

Then, for $f(x) \in C_0^\infty(E^n)$,

$$\| D^\beta f \|_q \leqslant C \prod_{j=1}^{N} \| D^{\alpha(j)} f \|_{p(j)}^{\mu_j}. \tag{11}$$

PROOF. Suppose first that $f(x)$ is concentrated in the cube

$$\square_1 = \{0 < x_\nu < \pi \ (\nu = 1, \ldots, n)\}.$$

Let us extend it from the cube $\square_1$ to the cube

$$\square = \{|x_\nu| < \pi \ (\nu = 1, \ldots, n)\}$$

in such a way that it is an odd function with respect to each variable x_ν. Then, let us extend $f(x)$ from this cube to all E^n as a 2π-periodic function with respect to each variable. For the function $f_*(x)$ obtained in this way,

$$\| D^\alpha f \|_p^* = 2^{\frac{1}{p_1} + \cdots + \frac{1}{p_n}} \| D^\alpha f \|_p.$$

The estimate (11) now follows from the estimate (10) for f_*.

Suppose now that $f(x) = f(x_1, \ldots, x_n)$ is an arbitrary function in $C_0^\infty(E^n)$. Then, for sufficiently small $\varepsilon > 0$, the function

$$f_\varepsilon(x) = f\left(\frac{x_1 - \frac{\pi}{2}}{\varepsilon}, \ldots, \frac{x_n - \frac{\pi}{2}}{\varepsilon}\right)$$

is concentrated in $\square_1$. If we substitute f_ε into (11), take the norms of the derivatives of f, and then cancel the powers of the parameter ε, we obtain the assertion of the theorem in the general case.

15.8. Since the generalized differentiation operator is closed, the estimate (11) can be carried over to those locally summable functions $f(x)$ that can be approximated to an arbitrary degree of accuracy by sequences of functions in $C_0^\infty(E^n)$ that converge simultaneously in the sense of $L^{\mathrm{loc}}(E^n)$ and in the sense of seminorms, the products of the powers of which appears in the right-hand member of (11). As an example, let us prove the following theorem:*

Theorem. *Suppose that the vector* $l=(l_1, \ldots, l_n)$ *has components that are natural numbers, that* $1 < p^{(j)} < \infty$ *(for* $j = 0, 1, \ldots, n$*),* $1 < q < \infty$, $0 \leqslant \mu_0 < 1$, $0 < \mu_j < 1$ *for* $j = 1, \ldots, n$, $\sum_0^n \mu_j = 1$, $\alpha = (\alpha_1, \ldots, \alpha_n)$, *where the* α_j *are nonnegative integers,*

$$\frac{1}{q} \leqslant \sum_{j=0}^{n} \frac{\mu_j}{p^{(j)}}, \qquad \alpha - \frac{1}{q} = \mu l - \sum_{j=0}^{n} \frac{\mu_j}{p^{(j)}},$$

where $\mu l = (\mu_1 l_1, \ldots, \mu_n l_n)$. *Then, for* $f \in L_{p^{(0)}}$,

$$\| D^\alpha f \|_q \leqslant C \| f \|_{p^{(0)}}^{\mu_0} \prod_{j=1}^{n} \| D_j^{l_j} f \|_{p^{(j)}}^{\mu_j}. \tag{12}$$

PROOF. By the conditions of Theorem 15.7, inequality (12) is satisfied for all functions $f \in C_0^\infty(E^n)$. Since the generalized differentiation operator D^α is closed (see Lemma 6.2), we need only show that an arbitrary function with finite norm

$$\| f \|_{p^{(0)}} + \sum_{j=1}^{n} \| D_j^{l_j} f \|_{p^{(j)}} = \| f \|_{W^l_{p^{(0)}, p^{(1)}, \ldots, p^{(n)}}} \tag{13}$$

*The case of scalar $p^{(j)}$ and q and strict inequality $\frac{1}{q} < \sum_{j=0}^{n} \frac{\mu_j}{p^{(j)}}$ is due to A. D. Džabrailov [1]. The case of scalar $p^{(j)}$ and q with $p^{(1)} = \ldots = p^{(n)} = p$ and $\mu_0 > 0$ is due to V. A. Solonnikov [4]. Here, the values of p, q, and $p^{(0)}$ that are equal to unity under the supplementary conditions $|\alpha : l| + \left(\frac{1}{p} - \frac{1}{q}\right)\left|\frac{1}{l}\right| < 1$, $p \leqslant q$ are not excluded.

can be approximated to an arbitrary degree of accuracy in this norm by functions in $C_0^\infty(E^n)$. And this fact was established in Theorem 14.14.

15.9. Our purpose is to show that the condition $f \in L_{p^{(0)}}$ in Theorem 15.8 with $\mu_0 = 0$ can be removed if in place of the function $f(x)$ we take the difference $f(x) - P(f; x)$, where $P(f; x)$ is a certain polynomial.

Theorem. *Let* $l = (l_1, \ldots, l_n)$ *denote a vector with components that are natural numbers. Suppose that* $\mathbf{1 < p^{(1)} < \infty, 1 < q < \infty}$,

$0 < \mu_j < 1$ *(for* $j = 1, \ldots, n$*),* $\sum_1^n \mu_j = 1$, $\alpha = (\alpha_1, \ldots, \alpha_n)$, *where the* α_j *are integers, and*

$$\frac{1}{q} \leqslant \sum_{j=1}^{n} \frac{\mu_j}{p^{(j)}},$$

$$\alpha - \frac{1}{q} = \mu l - \sum_{j=1}^{n} \frac{\mu_j}{p^{(j)}}, \text{ where } \mu l = (\mu_1 l_1, \ldots, \mu_n l_n).$$

Suppose also that, for some vector $\lambda = (\lambda_1, \ldots, \lambda_n) > \mathbf{0}$,

$$(\alpha, \lambda) > \lambda_j l_j - \left(\lambda, \frac{1}{p^{(j)}}\right) \qquad (j = 1, \ldots, n). \tag{14}$$

Then, there exists a polynomial $P_{l-1}(x; f)$ *of degree not exceeding* $l_j - 1$ *in* x_j *which is linearly independent of* f *and for which*

$$\| D^\alpha (f - P_{l-1}(\cdot\,; f)) \|_q \leqslant C \prod_{j=1}^{n} \| D_j^{l_j} f \|_{p^{(j)}}^{\mu_j}. \tag{15}$$

PROOF. We note first of all that in a number of cases condition (14) follows from the preceding conditions. In such cases it can be omitted from the statement of the theorem. Thus, in the case of scalar $p^{(j)} = (p_j, \ldots, p_j)$, for $j = 1, \ldots, n$, we can take for λ the vector with components

$$\lambda_j = \frac{1}{l_j}\left(1 - \sum_{i=1}^{n} \frac{1}{p_i l_i} + \frac{1}{p_j} \sum_{i=1}^{n} \frac{1}{l_i}\right) > 0 \qquad (j = 1, \ldots, n)$$

provided these are positive. These values of λ_j are easily found from the condition that the right-hand members of (14) be independent of j. Here, λ is the vector normal to the plane passing through the points $a^{(j)} = l_j e_j - \frac{1}{p^{(j)}}$ for $j = 1, \ldots, n$. In this case, the choice was made independently of α.

Another interesting case is the one in which $\sum_{1}^{n} \frac{\alpha_j}{l_j} = 1$. In this case, we can take $\lambda = \left(\frac{1}{l_1}, \ldots, \frac{1}{l_n}\right)$.

Let us set

$$\| f \|_{L^{l}_{p^{(1)}, \ldots, p^{(n)}}} = \sum_{j=1}^{n} \| D_j^{l_j} f \|_{p^{(j)}}.$$

The following relationships (see 7(11), 7(13), and 7(22)) are valid for $f \in L^{l}_{p^{(1)}, \ldots, p^{(n)}}$ with $0 < \varepsilon < \rho < \infty$:

$$D^{\alpha} f_{\rho}(x) = (-1)^{|\alpha|} \int \rho^{-|\lambda| - (\alpha, \lambda)} \Omega^{(\alpha)}(y : \rho^{\lambda}) f(x + y)\, dy, \tag{16}$$

$$- D^{\alpha} f_{(\varepsilon, \rho)}(x) = D^{\alpha} f_{\varepsilon}(x) - D^{\alpha} f_{\rho}(x)$$
$$= (-1)^{|\alpha|} \sum_{1}^{n} \iint_{\varepsilon}^{\rho} h^{-1-|\lambda| + l_j \lambda_j - (\alpha, \lambda)} \mathscr{L}_j^{(\alpha)}(y : h^{\lambda}) D_j^{l_j} f(x + y)\, dy\, dh, \tag{17}$$

where Ω and $\mathscr{L}_j$ belong to $C_0^{\infty}(E^n)$ for $j = 1, 2, \ldots, n$.

For $f \in L^{l}_{p^{(1)}, \ldots, p^{(n)}}$, we obtain from (17) with the aid of Hölder's inequality

$$\sup_x |D^\alpha f_\varepsilon(x) - D^\alpha f_\rho(x)|$$

$$\leqslant C \sum_1^n \int_\varepsilon^\rho h^{-1-(\alpha,\lambda)-|\lambda : p^{(j)}|+l_j\lambda_j} \| D_j^{l_j} f \|_{p^{(j)}} \, dh,$$

from which it follows that, at least for all α for which condition (14) is satisfied, $D^\alpha f_\rho(x)$ converges to a limit uniformly for $x \in E^n$ as $\rho \to \infty$.

We easily obtain from (16) the result that, if $\alpha_j \geqslant l_j$ for at least one j, then $D^\alpha f_\rho \to 0$ uniformly for $x \in E^n$ as $\rho \to \infty$.

Let us set $\varkappa = \max_j \left[\lambda_j l_j - \left(\lambda, \frac{1}{p^{(j)}} \right) \right]$ and

$$c_\alpha = \lim_{\rho \to \infty} \frac{1}{\alpha!} D^\alpha f_\rho(0), \qquad (\alpha, \lambda) > \varkappa,$$

so that $c_\alpha = 0$ if $\alpha_j \geqslant l_j$ for some j.

Consider now the polynomial

$$P_{l-1}(x; f) = \sum_{\substack{(\alpha, \lambda) > \varkappa \\ \min(l_j - \alpha_j) \geqslant 1}} c_\alpha x^\alpha \tag{18}$$

of order not exceeding $l_j - 1$ in the variables x_j for each j.

It follows from Taylor's formula that, for every compact subset K of E^n,

$$D^\alpha f_\rho(x) \to D^\alpha P_{l-1}(x; f) \quad \text{for} \quad (a, \lambda) > \varkappa \tag{19}$$

uniformly for $x \in K$ as $\rho \to \infty$.

We shall show below that inequality (15) is satisfied with the polynomial $P_{l-1}(x; f)$ defined by (19).

To prove the theorem, it will be sufficient to show that

$$\| D^\alpha (f_\varepsilon - f_\rho) \|_{q, K} \leqslant C \| f_\varepsilon - f_\rho \|_{L^l_{p^{(1)}, \ldots, p^{(n)}}} \tag{20}$$

for a constant C independent of f, K, ε, and ρ, where K is any compact set contained in E^n and $0 < \varepsilon < \rho < \infty$.

This is true because, by letting ε and ρ approach 0 in (20), we see that $D^\alpha f_\varepsilon$ converges in $L_q(K)$ as $\varepsilon \to 0$. Then, by Lemma 6.2, $D^\alpha f$ belongs to L_q^{loc} and we can take the limit in (20) as $\varepsilon \to 0$. If we now take the limit in (20) as $\rho \to \infty$, we arrive, by virtue of (19) and Lemma 14.7 at the estimate

$$\| D^\alpha (f - P_{l-1}(\cdot\,;\, f)) \|_{q, K} \leqslant C \| f \|_{L^l_{p^{(1)}, \ldots, p^{(n)}}},$$

in which C is independent of the compact subset K of E^n and which, by virtue of 1(7), is equivalent to the conclusion of the theorem.

Let us prove inequality (20). Shifting if necessary from the function f to its average with sufficiently small parameter, we may assume without loss of generality that it will be sufficient to prove (20) for functions f with the properties

$$f \in C^\infty(E^n), \quad F_j = D_j^{l_j} f \in W^l_{p^{(j)};\, p^{(1)}, \ldots, p^{(n)}}, \quad \| D^l F_j \|_{p^{(i)}} < \infty$$
$$(j = 1, \ldots, n;\ i = 1, \ldots, n).$$

Let us replace the function f with $B_{\delta, H} D^l f$ in 14(30). Since the difference $f_\varepsilon - f_\rho$ is, by virtue of (17), the sum of the convolutions F_j with kernels in C_0^∞ (for $j = 1, \ldots, n$), this is equivalent to replacing each function F_j with $B_{\delta, H} D^l F_j$. It follows from Lemma 14.13 that

$$\max_{1 \leqslant j \leqslant n} \| F_j - B_{\delta, H} D^l F_j \|_{W^l_{p^{(j)};\, p^{(1)}, \ldots, p^{(n)}}}$$

will be sufficiently small for suitably chosen δ and H. It follows that the difference

$$[f_\varepsilon - f_\rho] - [(B_{\delta, H} D^l f)_\varepsilon - (B_{\delta, H} D^l f)_\rho]$$

is sufficiently small in $L^{l}_{p^{(1)}, \ldots, p^{(n)}}$, as is its derivative D^{α} in $C(E^{n})$. (By virtue of Young's inequality 2(18), the norm in $C(E^{n})$ of that derivative is estimated in terms of $\sum_{j=1}^{n} \| F_{j} - B_{\delta, H} D^{l} f \|_{p^{(j)}}$.) Thus, to prove the theorem, it will be sufficient to prove the inequality

$$\| D^{\alpha} [(B_{\delta, H} D^{l} f)_{\varepsilon} - (B_{\delta, H} D^{l} f)_{\rho}] \|_{q, K} \leqslant C \| (B_{\delta, H} D^{l} f)_{\varepsilon} - (B_{\delta, H} D^{l} f)_{\rho} \|_{L^{l}_{p^{(1)}, \ldots, p^{(n)}}}$$

for an arbitrary function $f \in C^{\infty}(E^{n})$ such that $\sum_{j=1}^{n} \| D^{l} f \|_{p^{(j)}} < \infty$ for a constant C independent of f, K, ε, ρ, δ, and H.

Since $D^{l} f$ can be approximated to an arbitrary degree of accuracy in $L_{p^{(j)}}$ (for $j = 1, \ldots, n$) with the function $\chi_{R} D^{l} f$ for sufficiently large R (where χ_{R} is the characteristic function of the ball $|x| < R$), it will be sufficient if we prove the inequality differing from the preceding one only by replacement of $D^{l} f$ with $\chi_{R} D^{l} f$ for a constant C independent of f, K, ε, ρ, δ, H, and R. But such an inequality is an inequality for infinitely differentiable functions of compact support and, by virtue of Theorem 15.7, is valid. This completes the proof of Theorem 15.9.

BIBLIOGRAPHY

Only when a paper has not been translated in any reasonably available publication is a reference to a review of it in Mathematical Reviews listed—in a note of the form (M.R. X, Y), where X is the volume of the review and Y is the page or review number.

Agmon, S.

1. The coerciveness problem for integro-differential forms, *J. Analyse Math.* **6** (1958), 183–223.

Ahmetžanov, A.

1. A particular case in imbedding theorems, *Akad. Nauk Kazah. SSK Trudy Inst. Mat. i Meh.* **1**, 39–50 (1970). (Russian)

Andrienko, V. A.

1. Necessary conditions for imbedding the function classes H_p^ω, *Matem. Sbornik,* **78**, No. 2, 280–300 (1969). Transl.: *Math. of the USSR, Sbornik,* 7 (1969), 273–292.

Aronszajn, N.

1. On coercive integro-differential forms, Conference on Partial Differential Equations, University of Kansas, Summer 1954, Tech. Report No. 14, 94–106 (1955).

Babič, V. M.

1. On theorems of inclusion for a limiting exponent, *Vestnik Leningrad. Univ.,* **11** (1956), No. 19, 186–188. (Russian) (M.R. 188, 877)

Babič, V. M., and L. N. Slobodeckiĭ

1. On the boundedness of the Dirichlet integral, *Dokl. Akad. Nauk SSSR,* **106**, No. 4, 604–607 (1956). (Russian) (M.R. 17, 959)

Bahvalov, N. S.

1. Imbedding theorems for classes of functions with a number of bounded derivatives, *Vestnik Moskov. Univ. Ser. I. Mat. Meh.* 1963, No. 3, 7–16. (Russian) (M.R. 28, 4067)

Barozzi, G. C.

1. Su una generalizzazione degli spazi $L^{(q,\lambda)}$ di Morrey, *Ann. Scuola Norm. Sup. Pisa*, ser III, **19,** Fasc. IV, 609–626 (1965).

Benedek, A., A. P. Calderón and R. Panzone

1. Convolution operators on Banach space valued functions, *Proc. Nat. Acad. Sci. U.S.A.,* **48** (1962), 356–365.

Benedek, A., and R. Panzone

1. The space L_p, with mixed norm, *Duke Math. J.*, **28** (1961), 301–324.

Besov, O. V.

1. Certain conditions for derivatives of periodic functions to belong to L_p, *Nauchn. dokl. vyssei shk., Fiz.-Mat. nauki*, **1**, 13–17 (1959). (Russian)
2. On some properties of spaces $H_p^{(r_1, \ldots, r_m)}$, *Izv. Vyss. Ucebn. Zaved. Mathematika, 1960,* **No. 1** (14), 16–23. (Russian) (M.R. 24, 2233)
3. Investigation of a family of function spaces in connection with theorems of imbedding and extension, *Trudy Mat. Inst. Steklov.*, **60** (1961), 42–81. Transl.: *Amer. Math. Soc. Translations,* ser. 2, **40** (1964), 85–126.
4. An example in the theory of imbedding theorems, *Dokl. Akad. Nauk SSSR,* **143,** No. 5, 1014–1016 (1962). Transl.: *Soviet Math.*, **3** (1962), 507–509.
5. The continuation of functions in L_p^l and W_p^l, *Trudy Mat. Inst. Steklov.*, **89** (1967), 5–17. Transl.: *Proc. Steklov Inst. Math.*, **89** (1967), 1–15.
6. On the density of functions with compact support in $\mathscr{L}_{p,\theta}^l$ and the extension of functions. *Trudy Mat. Inst. Steklov.*, **89** (1967), 18–30. Transl.: *Proc. Steklov Inst. Math.*, **89** (1967), 16–31.
7. On the conditions for existence of a classical solution of the wave equation, *Sibirsk. Mat. Z.*, **8**, No. 2, 243–256 (1967). Transl.: *Siberian Math. J.*, **8**, No. 2, 179–188.
8. On coercivity in nonisotropic Sobolev spaces, *Mat. Sb.*, **73** (115), No. 4, 585–599 (1967). Transl.: *Math. USSR Sb.*, **2** (1967), 521–534.
9. On imbedding and extension theory for classes of differentiable functions, *Mat. Zametki,* **1,** No. 2, 235–250 (1967). Transl.: *Math. Notes,* **1** (1967), 156–161.
10. On the extension of functions beyond the limits of the domain with preservation of the modulus of smoothness in L_p. Proceedings of the conference on the Constructive Theory of Functions (Approximation Theory) Eds. G. Alexits and S. B. Stechkin; Budapest, 1969, pp. 61–67. Akademiai Kiadó, Budapest, 1972. (Russian) (M.R. 53, 6189)
11. Estimates of derivatives in a mixed L_p-norm on a domain and an extension of functions. *Mat. Zametki,* 7 (1970), 147–154. Transl.: *Math. Notes,* 7 (1970), 89–94.
12. Estimates of moduli of smoothness on domains, and imbedding theorems. *Trudy Mat. Inst. Steklov.*, **117** (1972), 22–46. Transl.: *Proc. Steklov Inst. Math.*, **117** (1972), 25–53.
13. The behavior of differentiable functions on a nonsmooth surface. *Trudy Mat. Inst. Steklov.*, **117** (1972), 3–10. Transl.: *Proc. Steklov Inst. Math.*, **117** (1972), 1–9.
14. On traces on a nonsmooth surface of classes of differentiable functions. *Trudy Mat. Inst. Steklov.*, **117** (1972), 11–21. Transl.: *Proc. Steklov Inst. Math.*, **117** (1972), 11–23.

15. Growth of a mixed derivative of a function of $C^{(l_1, l_2)}$, *Mat. Zametki,* **15** (1974), 355–362. Transl. *Math. Notes,* **15** (1974), 201–206.

16. On some families of function spaces. Imbedding and extension theorems. *Dokl. Akad. Nauk SSSR,* **126** (1959), 1163–1165. (Russian) (M.R. 21, 5890)

Besov, O. V., and V. P. Il'in

1. Natural extension of the class of regions in imbedding theorems. *Mat. Sb.,* **75** (117), No. 4, 483–495 (1968). Transl. *Math. USSR Sb.,* **4** (1968), 445–456.

2. An imbedding theorem for the limit exponent, *Mat. Zametki,* **6** (1969), 139–148. Transl.: *Math. Notes,* **6** (1969), 543–548.

Besov, O. V., V. P. Il'in, L. D. Kudryavcev, P. I. Lizorkin, and S. M. Nikol'skiĭ

1. Imbedding theory for classes of differentiable functions of several variables, Partial Differential Equations, Proceedings of a Symposium dedicated to Academician S. L. Sobolev, pp. 38–63, Nauka, Moscow 1970. Transl.: *Amer. Math. Soc. Transl. ser 2.* **105** (1976), 57–94.

Besov, O. V., V. P. Il'in, and P. I. Lizorkin

1. L_p-estimates of certain class of nonisotropic singular integrals, *Dokl. Akad. Nauk SSSR,* **169,** No. 6, 1250–1253 (1966). Transl. *Soviet Math.* **1** (1966), 1065–1069.

Besov, O. V., and P. I. Lizorkin

1. Singular integral operators and sequences of convolutions in L_p spaces, *Mat. Sb.,* **73** (115), No. 1, 65–68. Transl. *Math. USSR Sb.,* **2** (1967), 57–76.

Boman, J.

1. Supremum norm estimates for partial derivatives of functions of several variables, *Illinois J. Math.,* **16,** No. 2 (1972), 203–216.

Brudnyĭ, Yu. A.

1. Spaces defined by means of local approximations, *Trudy Moskov. Mat. Obšč.,* **24,** 69–132 (1971). Transl. *Trans. Moscow Math. Soc.,* **24,** 73–139 (1971).

Brudnyĭ, Yu. A. and V. K. Salašov

1. Lipschitz spaces of functions, *Dokl. Akad. Nauk SSSR,* **197,** No. 1 (1971), 18–20. Transl.: *Soviet Math.* **12** (1971), 383–386.

Bugrov, Ya. S.

1. Function spaces with mixed norm, *Izv. Akad. Nauk SSSR Ser. Mat.,* **35,** No. 5, 1137–1158 (1971). Transl.: *Math. USSR-Izv.,* **5** (1971), 1145–1167.

Burenkov, V. I.

1. Imbedding and extension theorems for classes of differentiable functions of several variables defined on the entire space. Itogi nauki, Matematicheskiy analiz, 1965, Moscow, *Izd. Viniti Akad. Nauk SSSR,* 1966, 71–155 (1966). (Russian) (M.R. 34, 6516).

2. On the additivity of the classes $W_p^{(r)}(\Omega)$, *Trudy Mat. Inst. Steklov.,* **89** (1967), 31–55. Transl.: *Proc. Steklov Inst. Math.,* **89** (1967), 32–62.

3. On the additivity of the spaces W_p^r and B_p^r and on imbedding theorems for regions of a general kind. *Trudy Mat. Inst. Steklov.*, **105** (1969), 30–45. Transl. *Proc. Steklov Inst. Math.*, **105** (1969), 35–53.
4. On approximating functions of the space $C^r(\Omega)$ by functions with compact support, for an arbitrary open set Ω, *Trudy Mat. Inst. Steklov.*, **117** (1972), 62–74. Transl. *Proc. Steklov Inst. Math.*, **117** (1972), 73–87.
5. On approximation of functions in Sobolev spaces by finite functions for an arbitrary open set. *Dokl. Akad. Nauk SSSR*, **202**, No. 2, 259–262 (1972). Transl. *Soviet Math.*, **13** (1972), 60–64.
6. On the density of infinitely differentiable functions in Sobolev spaces for an arbitrary open set. *Trudy Mat. Inst. Steklov.*, **131** (1974), 39–50. Transl. *Proc. Steklov Inst. Math.*, **131** (1974), 39–51.

Calderón, A. P.
1. Lebesgue spaces of differentiable functions and distributions, *Proc. Sympos. Pure Math.*, vol. 4, Amer. Math. Soc., Providence, R.I., 1961, pp. 33–49.

Calderón, A. P., and Zygmund, A.
1. On the existence of certain singular integrals, *Acta Math.*, **88** (1952), 85–139.
2. A note on local properties of solutions of elliptic differential equations, *Proc. Nat. Acad. Sci., U.S.A.* **46** (1960), 1385–1389.
3. Local properties of solutions of elliptic partial differential equations, *Studia Math.*, **20** (1961), 171–225.

Campanato, S.
1. Caratterizzazione delle trace funzioni appartenenti ad una classe di Morrey insieme con le loro derivate prime, *Ann. Scuola Norm. Sup. di Pisa,* ser. 3, **15** (1961), 263–281. (M.R. 24, A3251)
2. Il teorema di immersione di Sobolev per una classe di aperti non dotati della proprietà di cono, *Ricerche Mat.*, **11** (1962), 103–122. (M.R. 25, 3357)
3. Proprietà di inclusione per spazi di Morrey, *Ricerche Mat.*, **12** (1963), 67–86. (M.R. 27, 6157)
4. Proprietà di una famiglia di spazi funzionali, *Ann. Scuola Norm. Sup. Pisa,* ser. 3, **18** (1964), 137–160.

Deny, J., and J. L. Lions
1. Les espaces du type de Beppo Levi, *Ann. Inst. Fourier* (Grenoble) **5**, 305–370 (1953).

Dunford, N. and J. T. Schwartz
1. Linear Operators, part I: General Theory. Interscience (Wiley), New York, 1958.

Dzabrailov, A. D.
1. Investigation of the differential–difference properties of functions defined on an m-dimensional region. Doctoral dissertation, Baku, 1971.

Džafarov, A. S.
1. Some properties of functions of several variables, *Dokl. Akad. Nauk Azerbaĭdžan. SSR,* **14** (1958), 499–503. (Russian) (M.R. 22, 3774)
2. Imbedding theorems for genralized Nikol'skii classes, *Azerbaidzan. Gos. Univ. Ucen. zap. Ser. fiz. -Mat. Nauk,* No. 2, 45–49 (1963).

Ehrling, G.
1. On a type of eigenvalue problems for certain elliptic differential operators, *Math. Scand.,* **2** (1954), 267–285.

Figueiredo, D. G.
1. The coerciveness problem for forms over vector valued functions, *Comm. Pure Appl. Math.,* **16** (1963), 63–94.

Freud, G., and D. Králik
1. Über die Anwendbarkeit des Dirichletschen Prinzips für den Kreis, *Acta Math. Acad. Sci. Hungar.,* **7** (1956), 411–418.

Gagliardo, E.
1. Caratterizzazione delle trace sulla frontiera relative ad alcune classi di funzioni in n variabili, *Rend. Sem. Mat. Univ. Padova,* **27** (1957), 284–305. (M.R. 21, 1525)
2. Proprietà di aclune classi di funzioni in piú variabili, *Ricerche Mat.,* **7** (1958), 102–137. (M.R. 21, 1526)
3. Ulteriori proprietà di alcune classi di funzioni in piú variabili, *Ricerche Mat.,* **8** (1959), 24–51. (M.R. 22, 181)

Galahov, M. A.
1. On summable regions, *Trudy Mat. Inst. Steklov.,* **89** (1967), 69. Transl. *Proc. Steklov Inst. Math.,* **89** (1967), 79.

Gårding, L.
1. Dirichlet's problem for linear elliptic partial differential equations, *Math. Scand.,* **1** (1953), 55–72.

Geronimus, Ya. L.
1. Some properties of functions of class L_p, *Izv. Vysš. Učebn. Zaved. Matematika,* 1958, **No. 1** (2), 24–32. (Russian) (M.R. 27, 1753)

Globenko, I. G.
1. Some questions in the theory of imbedding for domains with singularities on the boundary, *Mat. Sb.,* (N.S.) **57** (99) (1962), 201–224. (Russian) (M.R. 26, 589)

Golovkin, K. K.
1. On imbedding theorems, *Dokl. Akad. Nauk SSSR,* **134,** 1, 19–22 (1960). Transl. *Soviet Math.* **1** (1960), 998–1000.
2. Two classes of inequalities for sufficiently smooth functions of n variables, *Dokl. Akad. Nauk SSSR,* **138,** No. 1, 22–25 (1961). Transl. *Soviet Math.,* **2** (1961), 510–513.

3. On equivalent norms for fractional spaces, *Trudy Mat. Inst. Steklov.*, **66** (1962), 364–383. Transl. *Amer. Math. Soc. Transl.* ser. 2, **81** (1969), 257–280.
4. On the impossibility of certain inequalities between function norms, *Trudy Mat. Inst. Steklov.*, **70** (1964), 5–25. Transl.: *Amer. Math. Soc. Transl.* ser. 2, **67** (1968), 1–24.
5. On the approximation of functions in arbitrary norms, *Trudy Mat. Inst. Steklov.*, **70** (1964), 26–37. Transl. *Amer. Math. Soc. Transl.* ser. 2, **91** (1970), 43–56.
6. Imbedding theorems for fractional spaces, *Trudy Mat. Inst. Steklov.*, **70** (1964), 38–46. Transl. *Amer. Math. Soc. Transl.* ser. 2, **91** (1970), 57–67.
7. Some inequalities for norms of mixed derivatives of functions of several variables, *Dokl. Akad. Nauk SSSR*, **159**, No. 5, 965–967 (1964). Transl. *Soviet Mathematics* **5** (1964), 1628–1630.
8. A generalization of Marcinkiewicz's interpolation theorem, *Trudy Mat. Inst. Steklov.*, **102** (1967), 5–28. Transl. *Proc. Steklov Inst. Math.*, **102** (1967), 1–28.
9. Parametric-normed Spaces and Normed Masses, *Trudy Mat. Inst. Steklov.*, **106** (1969), 1–135. Transl. *Proc. Steklov Inst. Math.*, **106** (1969), 1–121.
10. Approximation of functions of the class $W_p^l(\Omega)$, *Trudy Mat. Inst. Steklov.*, **92** (1966), 50–56. Transl. *Proc. Steklov Inst. Math.*, **92** (1966), 54–62.

Golovkin, K. K., and V. A. Solonnikov

1. Imbedding theorems for fractional spaces, *Dokl. Akad. Nauk SSSR*, **143**, No. 2, 767–770 (1962). Transl. *Soviet Math.*, **3** (1962), 468–471.

Greco, D.

1. Criteri di compattezza per insieme di funzioni in "*n*" variabili independenti, *Ricerche Mat.*, **1**, 124–144 (1952).

Gudiev, A. H.

1. An imbedding theorem for restrictions as abstract functions. *Dokl. Akad. Nauk SSSR*, **147**, No. 4, 764–767 (1962). Transl.: *Soviet Math.*, 3 (1962), 1693–1696.

Hardy, G. H., J. E. Littlewood and G. Polya

1. Inequalities. Cambridge University Press, 1934.

Hörmander, L.

1. Estimates for translation invariant operators in L^p spaces, *Acta Math.*, **104**, No. 1–2, 93–140 (1960).

Il'in, V. P.

1. On a theorem of inclusion for a limiting exponent, *Dokl. Akad. Nauk SSSR*, **96**, No. 5, 905–908 (1954). (Russian) (MR: 16 p. 121)

2. Some inequalities in function spaces and their application to the study of the convergence of variational processes, *Trudy Mat. Inst. Steklov.*, **53** (1959), 64–127. Transl. *Amer. Math. Soc. Transl.* ser. 2, **81** (1969), 1–66.
3. On a theorem of G. H. Hardy and J. E. Littlewood, *Trudy Mat. Inst. Steklov.*, **53** (1969), 128–144. (Russian) (M.R. 22, 11084)
4. On theorems of "imbedding," *Trudy Mat. Inst. Steklov.*, **53** (1969), 359–386. (Russian) (M.R. 22, 3776)
5. On approximation of functions in the spaces $\widetilde{W}_p^{(l)}(D)$ and $W_p^{(l)}(D)$ by continuously differentiable functions, *Trudy Mat. Inst. Steklov.*, **64** (1961), 61–78. Transl. *Amer. Math. Soc. Transl.* ser. 2, **44** (1965), 89–107.
6. The properties of some classes of differentiable functions of several variables defined in an n-dimensional region, *Trudy Mat. Inst. Steklov.*, **66** (1962), 227–363. Transl. *Amer. Math. Soc. Transl.* ser. 2, **81** (1969), 91–256.
7. On some properties of classes of differentiable functions defined in a domain, *Trudy Mat. Inst. Steklov.*, **84** (1965), 93–143. Transl. *Proc. Steklov Inst. Math.*, **84** (1965), 103–160.
8. Integral representations of differentiable functions and their application to questions of extension of functions of the classes $W_p^l(G)$, *Sibirsk. Mat. Ž.*, **8**, No. 3, 573–586 (1967). Transl.: *Siberian Math. J.*, **8** (1967), 421–432.
9. Conditions of validity of inequalities between L_p-norms of partial derivatives of functions of several variables, *Trudy Mat. Inst. Steklov.*, **96** (1968), 205–242. Transl.: *Proc. Steklov Inst. Math.*, **96** (1968), 259–305.
10. Certain integral inequalities and their application to imbedding theorems, *Zap. Naučn. Sem. Leningrad. Otdel. Nat. Inst. Steklov. (LOMI)*, **18** (1970), 191–216. (Russian) (M.R. 44, 3122)
11. Integral representations of functions of the classes $L_p^l(G)$ and imbedding theorems, *Zap. Naučn. Sem. Leningrad. Otdel. Mat. Inst. Steklov. (LOMI)*, **19** (1970), 95–155. (Russian) (M.R. 45, 5740)
12. On some properties of functions from $W_{p,a,\varkappa}^l(G)$ spaces, *Zap. Naučn. Sem. Leningrad. Otdel. Mat. Inst. Steklov. (LOMI)*, **23** (1971), 33–40. Transl. *J. Soviet Math.*, **2**, No. 4, 369–375 (1974).
13. Function spaces $\mathscr{L}_{r,p,\theta}^{\lambda, a; b_s}(G)$, *Zap. Naučn. Sem. Leningrad. Otdel. Mat. Inst. Steklov. (LOMI)*, **30** (1972), 51–75. Transl.: *J. Soviet Math.*, **4**, No. 4, 345–366.
14. On the decomposition of the anisotropic space $W^{(l_1, \ldots, l_n)}(\Omega)$ with the help of a special projection operator, *Sibirsk Mat. Ž.*, **10**, No. 1, 212–216 (1969). Transl. *Siberian Math. J.*, **10** (1969), 148–151.

Il'in, V. P., and V. A. Solonnikov
1. On some properties of differentiable functions of several variables, *Trudy Mat. Inst. Steklov.*, **66** (1962), 205-226. Transl. *Amer. Math. Soc. Transl. ser. 2*, **81** (1969), 67-90.

Jessen, B., J. Marcinkiewicz and A. Zygmund
1. Note on the differentiability of multiple integrals, *Fund. Math.*, **25** (1935), 217-234.

John, F., and L. Nirenberg
1. Functions of bounded mean oscillation, *Comm. Pure Appl. Math.*, **14**, 415-426 (1961).

Jones, F.
1. A class of singular integrals, *Amer. J. Math.*, **86**, No. 2, 441-462 (1964).

Kalugina, Ye. P.
1. Convex functional manifolds. Doctoral dissertation, Leningrad State University, 1952. (Russian)
2. On the classes $H_{\Phi}^{(r_1, \ldots, r_n)}$, *Dokl. Akad. Nauk SSSR*, **96** (1954), 13-15. (Russian) (M.R. 16 p. 834)

Kantorovič, L. V.
1. On integral operators, *Uspehi Mat. Nauk*, **11**, No. 2, 3-29 (1956). (Russian) (M.R. 20, 5432)

Kantorovič, L. V., and G. P. Akilov
1. Functional analysis in normed spaces. Fizmatgiz, Moscow, 1959. Transl.: New York, Macmillan, 1964.

Kazaryan, G. G.
1. The density of smooth functions in $\overset{\circ}{W}{}_p^r$ (Ω), with compact support. *Mat. Zametki*, **2**, No. 1, 45-52 (1967). Transl. *Math. Notes*, **22** (1967), 514-517.

Kocharli, A. F.
1. The compactness of sets of differentiable functions defined in a region. *Akad. Nauk Azerbaĭdžan. SSR Dokl.*, **27** (1971/72), No. 11-12, pp. 10-13. (Russian) (M.R. 46, 4189)

Kolmogorov, A. N.
1. Über Kompaktheit der Funktionenmengen bei der Konvergenz im Mittel, *Nach. Akad. Wiss. Göttingen Math.-Phys. Kl.*, 60-63 (1931).

Kondrašov, V. I.
1. Sur certaines propriétés des fonctions dans l'espace L_p^ν, *Dokl. Akad. Nauk SSSR (C.R.)*, **48**, 415-418 (1945).

Krasnosel'skiĭ, M. A., and Ya. B. Rutickiĭ
1. Convex functions and Orlicz spaces. Fizmatgiz, Moscow, 1958. Transl.: Noordhoff, Groningen, 1961.

Krée, P.

1. Sur les multiplicateurs dans FL_p, *C. R. Acad. Sci. Paris,* 260 (1965), 4400–4403. (M.R. 31, 2634)

Kudryavcev, L. D.

1. On a generalization of S. M. Nikol'skiĭ's theorem on the compactness classes of differentiable functions. *Uspehi Matem. Nauk*, **9**, No. 1, 111-120 (1954). (Russian) (M.R. 16–453)
2. The variation of mappings of regions, in: Metrical Questions of the Theory of Functions and Mappings, No. 1, pp. 34-108, Izd. Naukova Dumka, Kiev, 1969. (Russian) (M.R. 44, 6924)

Lewis, J. E.

1. A class of singular integral operators, *Notices Amer. Math. Soc.,* **12** (1965), 548.

Lizorkin, P. I.

1. Generalized Liouville differentiation and the function spaces $L_p^r(E_n)$. Imbedding theorems, *Mat. Sb.,* **60,** No. 3, 325-353 (1963). (Russian) (M.R. 27, 610)
2. Generalized Hölder spaces $B_{p,\theta}^{(r)}$ and their relation with the Sobolev spaces $L_p^{(r)}$, *Sibirsk. Mat. Ž,* **9,** No. 5, 1127-1152 (1968). Transl. *Siberian Math. J.,* **9** (1968), 837-858.
3. Generalized Liouville differentiation and the method of multipliers in the theory of imbeddings of classes of differentiable functions. *Trudy Mat. Inst. Steklov.,* **105** (1969), 89-167. Transl. *Proc. Steklov Inst. Math.*, **105** (1969), 105-202.
4. Multipliers of Fourier integrals and bounds of convolution in spaces with mixed norms. Applications, *Izv. Akad. Nauk SSSR Ser. Mat.,* **34,** No. 1, 218-247 (1970). Transl. *Math. USSR Izv.,* **4** (1970), 225-255.
5. Boundary properties of functions from "weight" classes, *Dokl. Akad. Nauk SSSR,* **132,** No. 3, 514-517 (1960). Transl.: *Soviet Math.,* **1** (1960), 589-592.

Lizorkin, P. I. and Nikol'skiĭ, S. M.

1. The compactness of sets of differentiable functions, *Trudy Mat. Inst. Steklov.,* **105** (1969), 168-177. Transl. *Proc. Steklov Inst. Math.,* **105** (1969), 203-215.

Marcinkiewicz, J.

1. Sur l'interpolation d'opérateurs, *C. R. Acad. Sci. Paris,* 208, 1272-1273 (1939).

Maz'ya, V. G.

1. Classes of domains and imbedding theorems for function spaces, *Dokl. Akad. Nauk SSSR,* **133,** No. 3, 527-530 (1960). Transl. *Soviet Math.,* **1** (1960), 882 -885.

2. *P*-conductance and theorems for the imbedding of certain function spaces into the space *C*, *Dokl. Akad. Nauk SSSR*, **140**, No. 2, 299–302 (1961). Transl.: *Soviet Math.*, **2** (1961), 1200–1203.
3. Classes of sets and measures that are connected with imbedding theorems, in *Imbedding Theorems and their Applications* (Proc. Sympos., Baku 1966), pp. 142–159. Izd. Nauka, Moscow, 1970. (Russian) (M.R. 47, 2345)

Meyers, C. N.

1. Mean oscillation over cubes and Hölder continuity, *Proc. Amer. Math. Soc.*, **15**, 717–721 (1964).

Mihlin, S. G.

1. Sur la réduction des équations intégrales singuliéres à des équations equivalentes de Fredholm, *Dokl. Akad. Nauk SSSR*, **20** (1938), 93–96.

Mityagin, B. S.

1. Some properties of functions of two variables, *Vestnik Moskov. Univ. Ser. Mat. Meh. Astronom. Fiz. Him.*, 1959, No. 5, 137–152. (Russian) (M.R. 23, A2738)

Morrey, C. B.

1. Multiple integral problems in the calculus of variations and related topics. *Univer. California Publ. Math.*, **1**, 1–130 (1943).
2. Second order elliptic systems of differential equations. in *Contributions to the Theory of Partial Differential Equations*, pp. 101–159. Annals of Mathematics Studies, no. 33, Princeton Univ. Press, Princeton, 1954.
3. Second order elliptic equations in several variables and Hölder continuity, *Math. Z.*, **72** (1959/60), 146–164.

Nečas, J.

1. Les méthodes directes en théorie des équations elliptiques. Prague, Academia, 1967.

Nikol'skiĭ, S. M.

1. Inequalities for entire functions of finite degree and their application to the theory of differentiable functions of several variables, *Trudy Mat. Inst. Steklov.*, **38** (1951), 244–278. Transl. *Amer. Math. Soc. Transl.* ser. 2 **80** (1969), 1–38.
2. Properties of certain classes of functions of several variables on differentiable manifolds, *Mat. Sb.*, **33** (75), No. 2, 261–326 (1953). Transl. *Amer. Math. Soc. Transl.* ser 2, **80** (1969), 38–118.
3. Compactness of classes $H_p^{(r_1, \ldots, r_n)}$ of functions of several variables, *Izv. Akad. Nauk SSSR, ser. Mat.* **20** (1956), 611–622. (Russian) (M.R. 18, 321)
4. Extension of functions of several variables preserving differential properties, *Mat. Sb.*, **40** (82), No. 2, 243–268 (1956). Transl. *Amer. Math. Soc. Transl.* ser 2 **83** (1969), 159–188.

5. Boundary properties of functions defined on a region with angular points. Part I. *Mat. Sb.*, **40** (82), No. 3, 303–318 (1956); Part II. *Mat. Sb.*, **43** (85), No. 1, 127–144 (1957); Part III. *Mat. Sb.*, **45** (87), No. 2, 181–194 (1958). Transl. *Amer. Math. Soc. Transl. ser 2*, pts. I–III, **83** (1969), 101–157.
6. On a problem of S. L. Sobolev, *Sibirsk. Mat. Ž.*, **3** (1962), 845–851. (Russian) (M.R. 27, 5881)
7. Stable boundary-value problems of a differentiable function of several variables, *Mat. Sb.*, **61** (103), No. 2, 224–252 (1963). (Russian) (M.R. 27, 6113)
8. Functions with a dominant mixed derivative satisfying a multiple Hölder condition, *Sibirsk. Mat. Ž.*, **4** (1963), 1342–1364. Transl. *Amer. Math. Soc. Transl. ser. 2*, **102** (1973), 27–51.
9. Approximation of Functions of Several Variables and Imbedding Theorems. Nauka, Moscow 1969. Transl. Grundlehren der math. Wiss. in Einzeldarstell. Band 205, Springer Vlg, New York, 1975.
10. A property of the $H_p^{(r)}$ classes, *Ann. Univ. Sci. Budapest. Eötvös Sect. Math.*, **3–4** (1960/61), 205–216. (Russian) (M.R. 24, A2231)
11. On class preservation under differentiable mappings, *Trudy Mat. Inst. Steklov.*, **112** (1971), 327–336. Transl. *Proc. Steklov Inst. Math.*, **112** (1971), 339–348.
12. Course of Mathematical Analysis, vol. I, Nauka, Moscow, 1973. (Russian)
13. Course of Mathematical Analysis, vol. II, Nauka, Moscow, 1973. (Russian)
14. Nonlinear transformations with conservation of differential properties of functions, in *Linear Operators and Approximation*, Proc. of Conf. Oberwolfach Math. Res. Inst., Aug. 11–22, 1971, (Ed.) P. L. Butzer *et al.* pp. 281–283. ISNM vol. 20, Birkhäuser Vlg. Basel 1972.

Nikol'skii, Yu. S.

1. The behavior at infinity of functions with differential–difference properties given in L_p, *Trudy Mat. Inst. Steklov.*, **131** (1974), 182–198. Transl.: *Proc. Steklov Math. Inst.*, **131** (1974), 189–205.

Nirenberg, L.

1. Estimates and existence of solutions of elliptic equations, *Comm. Pure Appl. Math.*, 9 (1956), 509–530.
2. On elliptic partial differential equations, *Ann. Scuola. Norm. Sup. Pisa*, ser 3, **13**, Fasc. 2 (1959), 115–162.
3. Remarks on strongly elliptic partial differential equations, *Comm. Pure Appl. Math.*, **8**, 648–674 (1955).

Ornstein, D.

1. A non-inequality for differential operators in the L_1-norm, *Arch. Rational Mech. Anal.*, **11**, No. 1 (1962), 40–49.

Otelbaev, M., and Cend, L.

1. Theorems on compactness, *Sibirsk. Mat. Ž.*, **13**, No. 4, 817–822 (1972). Transl.: *Siberian Math. J.*, **13** (1972), 566–570.

Petrovskiĭ, I. G., and K. N. Smirnov

1. Sur une condition suffisante pour qu'une famille de fonctions soit également continue, *Bull. Univ. Moscou Sér. Internat.* A1, No. 10, 1–15 (1937/38).

Picone, M.

1. Sulla derivazione parziale peper serie, *Boll. Un. Mat. Ital.*, ser. 3, **5** (1950), 24–33.

Piccinini, L. C.

1. Su alcune disequaglianze di interpolazione, *Atti Accad. Naz. Lincei Rend. Cl. Sci. Fis. Mat. Natur.*, ser. 8, **42** (1967), 341–346. (M.R. 36, 5685)

Pohozaev, S. I.

1. Eigenfunctions of the equation $Au + \lambda f(u) = 0$, *Dokl. Akad. Nauk SSSR*, **165**, No. 1, 36–39 (1965). Transl.: *Soviet Math.*, **6** (1965), 1408–1411.

Portnov, V. R.

1. Certain integral inequalities, in *Imbedding Theorems and their Applications*, (Proc. Sympos., Baku 1966), pp. 195–203, Izd. Nauka, Moscow, 1970. (Russian) (M.R. 47, 5584)
2. On a Sobolev-type projection operator, *Dokl. Akad. Nauk SSSR*, **189**, No. 2, 258–260 (1969). Transl. *Soviet Math.*, **10** (1969), 1407–1409.

Pucci, C.

1. Compattezza di successioni di funzioni e derivabilità delle funzioni limiti, *Ann. Mat. Pura Appl.*, ser. 4., **36**, 1–25 (1954).

Rellich, F.

1. Ein Satz über mittlere Konvergenz, *Nachr. Akad. Wiss. Göttingen Math.-Phys. Kl.*, 30–35 (1930).

Rešetnyak, Yu. G.

1. Some integral representations of differentiable functions, *Sibirsk. Mat. Ž.*, **12**, No. 2, 420–432 (1971). Transl.: *Siberian Math. J.*, **12** (1971), 299–307.

Rivière, N. M.

1. Singular integrals and multiplier operators, *Ark. Mat.*, **9**, No. 2, 243–278 (1971).

Saks, S.

1. *Theory of the Integral,* 2nd rev. ed. Stechert, New York, 1937 (Dover 1964).

Sapogov, N. A.

1. On an inequality of Čebyšev, *Uspehi Matem. Nauk* (N.S.), **6**, No. 2 (42), 157–159 (1951). (Russian) (M.R. 13–214)

Schechter, M.

1. Integral inequalities for partial differential operators and functions satisfying general boundary conditions, *Comm. Pure Appl. Math.*, **12** (1959), 37–66.

Sedov, V. N.

1. Functions that tend to a polynomial at infinity, in *Imbedding Theorems and their Applications* (Proc. Sympos., Baku 1966) pp. 204–212. Izd. Nauka, Moscow, 1970. (Russian) (M.R. 47, 7416)

Slobodeckiĭ, L. N.

1. Sobolev's spaces of fractional order and their application to boundary problems for partial differential equations, *Dokl. Akad Nauk SSSR,* **118,** No. 2, 243–246 (1958) (Russian) (M.R. 21, 5059)
2. Generalized Sobolev spaces and their applications to boundary problems for partial differential equations, *Leningrad. Gos. Ped. Inst. Učen. Zap.,* **197,** 54–112 (1958). Transl.: *Amer. Math. Soc. Transl.* ser 2, **57** (1966), 207–276.
3. Estimates of the solutions of elliptic and parabolic systems, *Dokl. Akad. Nauk SSSR,* **120,** No. 3, 468–471 (1958). (Russian) (M.R. 21, 5060)
4. Estimates in L_p of solutions of elliptic systems, *Dokl. Akad. Nauk SSSR,* **123,** No. 4, 616–619 (1958). (Russian) (M.R. 21, 5061)

Smirnov, V. I.

1. *A Course of Higher Mathematics,* vol. 5. Fizmatgiz, Moscow, 1959. Transl.: Addison-Wesley, Reading, Mass., 1964.

Smith, K. T.

1. Inequalities for formally positive integro-differential forms, *Bull. Amer. Math. Soc.,* **67,** 368–370 (1961).
2. Formulas to represent functions by their derivatives, *Math. Ann.,* 188, No. 1, 53–77 (1970).

Sobolev, S. L.

1. On a theorem of functional analysis, *Mat. Sb.* (N.S.) **44**(46), (1938), 471–497. Transl.: *Amer. Math. Soc. Transl.* ser. 2, **34** (0000), 39–68.
2. Applications of functional analysis in mathematical physics, Izd. Leningrad Gos. Univ., Leningrad, 1950. Transl.: Transl. Math. Monographs, vol. 7, Amer. Math. Soc., Providence, R.I., 1963.
3. The density of functions with compact support in the space $L_p^m(E_n)$, *Sibirsk. Mat. Ž.,* **4,** No. 3, 673–682 (1963). (Russian) (M.R. 30, 5156)

Solonnikov, V. A.

1. On certain properties of $\mathfrak{M}_p^l$ spaces of fractional order, *Dokl. Akad. Nauk SSSR,* **134,** No. 2, 282–285 (1960). Transl.: *Soviet Math.,* **1** (1960), 1071–1074.
2. A simple proof of an inequality of Hardy and Littlewood for fractional integrals, *Vestnik Leningrad. Univ.,* **13,** No. 3, 190–193 (1962). (Russian) (M.R. 25, 4047)

3. Estimates in L_p of solutions of elliptic and parabolic systems, *Trudy Mat. Inst. Steklov.*, **102** (1967), 137–160. Transl.: *Proc. Steklov Inst. Math.*, **102** (1967), 157–185.

4. Inequalities for functions of the classes $\vec{W}_p(R^n)$, *Zap. Naučn. Sem. Leningrad Otdel. Mat. Inst. Steklov (LOMI)*, **27** (1972), 194–210. Transl.: *J. Soviet Math.*, **3** (1975), 549–564.

Spanne, S.

1. Some function spaces defined using the mean oscillation over cubes, *Ann. Scuola Norm. Sup. Pisa*, ser. 3, **19**, Fasc. 4 (1965), 593–608.

Stampacchia, G.

1. $\mathfrak{L}^{(P,\lambda)}$-spaces and interpolation, *Comm. Pure Appl. Math.*, **17** (1964), 293–306.

2. The spaces $\mathfrak{L}^{(p,\lambda)}$, $\mathfrak{N}^{(p,\lambda)}$ and interpolation, *Ann. Scuola Norm. Sup. Pisa*, ser. 3, **19**, Fasc. 3 (1965) 443–462.

Stein, E. M.

1. *Singular Integrals and Differentiability Properties of Functions*, Princeton University Press, Princeton 1970.

Temirgaliev, N.

1. A connection between imbedding theorems and the uniform convergence of multiple Fourier series, *Mat. Zametki*, **12**, No. 2, 139–148 (1972). Transl.: *Math. Notes*, **12** (1972), 518–523.

Timan, A. F.

1. *Theory of Approximation of Functions of a Real Variable*, Fizmatgiz, Moscow 1960. Transl.: Macmillan, New York 1963.

Trudinger, N. S.

1. On imbeddings into Orlicz spaces and some applications, *J. Math. Mech.*, **17** (1967), 473–483.

Tulaĭkov, A. H.

1. Zur Kompaktheit in Raum L_p für $p = 1$, *Nachr. Akad. Wiss. Göttingen Math. -Phys. Kl.*, 167–170 (1933).

Ul'yanov, P. L.

1. Absolute and uniform convergence of Fourier series, *Mat. Sb.*, **72** (114), No. 2, 193–224 (1967). Transl.: *Math. USSR-Sb.*, **1** (1967), 169–197.

2. Imbedding of certain classes of functions, *Mat. zametki*, **1** (1967), 405–414. Transl.: *Math. Notes* **1** (1967), 270–276.

3. The imbedding of certain classes H_p^ω, *Izv. Akad. Nauk SSSR, Ser. Mat.*, **32**, 649–686 (1968). Transl.: *Math. USSR-Izv.* **2** (1968), 601–637.

Uspenskiĭ, S. V.

1. Properties of W_p^r classes with fractional derivative on differentiable manifolds, *Dokl. Akad. Nauk SSSR,* **132,** No. 1, 60–62 (1960). Transl.: *Soviet Math.,* **1** (1960), 495–497.
2. An imbedding theorem for S. L. Sobolev's classes W_p^r of fractional order, *Dokl. Akad. Nauk SSSR,* **130,** No. 5, 992–993 (1960). Transl.: *Soviet Math.,* **1** (1960), 132–133.
3. Imbedding theorems for weighted classes, *Trudy Mat. Inst. Steklov.,* **60** (1961), 282–283. Transl.: *Amer. Math. Soc. Transl.,* Ser. 2, **87,** (1970), 121–145.
4. Imbedding theorems for the generalized Sobolev class W_p^r, *Sibirsk. Mat. Ž.,* **3,** No. 3, 418–445 (1962). Transl.: *Amer. Math. Soc. Transl.,* ser. 2, **90** (1970), 45–79.
5. Differential properties of solutions of quasielliptic equations in unbounded domains, *Dokl. Akad. Nauk SSSR,* **181,** No. 3, 562–564 (1968). Transl.: *Soviet Math.,* **9** (1968), 900–903.
6. On the summability (boundedness) of solutions of a class of hypoelliptic equations in unbounded domains, *Dokl. Akad. Nauk SSSR,* **187,** No. 5, 998–1001 (1969). Transl.: *Soviet Math.,* **10** (1969), 989–993.
7. Differential properties for solutions of a class of partial differential equations in unbounded regions, *Dokl. Akad. Nauk SSSR,* **196,** No. 1, 61–64 (1971). Transl.: *Soviet Math.,* **12** (1971), 66–70.
8. On the traces, on smooth surfaces, of functions of the Sobolev class $W_p^{l_1, \ldots, l_n}$, *Sibirsk. Mat. Z.,* **13,** No. 2, 429–451 (1972). Transl.: *Siberian Math. J.,* **13** (1972), 298–313.

Van der Waerden, B. L.

1. Moderne Algebra, 2 band, 2 verbess. Aufl., Springer Vlg., Berlin, 1937. Transl. *Modern Algebra,* vol. 2, F. Ungar, New York, 1950.

Yudovič, V. I.

1. Some estimates connected with integral operators and with solutions of elliptic equations, *Dokl. Akad. Nauk SSSR,* **138,** No. 4, 805–808 (1961). Transl.: *Soviet Math.,* **2** (1961), 746–749.

Zygmund, A.

1. *Trigonometric Series,* 2nd ed., vols. I & II. Cambridge University Press, 1959.